TRAINING & REFERENCE

murach's Oracle SQL and PL/SQL

3RD EDITION

Joel Murach

MIKE MURACH & ASSOCIATES, INC.
3730 W Swift Ave. • Fresno, CA 93722
www.murach.com • murachbooks@murach.com

Editorial team

Author:	Joel Murach
Writer/Editor:	Scott McCoy
Editor:	Lisa Cooper
Production:	Juliette Baylon

Murach also has books on these subjects:

Data science

Python for Data Science
R for Data Analysis

Web development

HTML and CSS
Modern JavaScript
JavaScript and jQuery
ASP.NET Core MVC
PHP and MySQL

Programming languages

Python
Java
C++
C#

Databases

MySQL
SQL Server

For more on Murach books, please visit us at www.murach.com

Printed in the United States of America

10 9 8 7 6 5 4 3 2 1
ISBN: 978-1-943873-19-7

Content

Introduction xiii

Section 1 An introduction to SQL

Chapter 1 An introduction to relational databases and SQL 3
Chapter 2 How to use SQL Developer 27

Section 2 The essential SQL skills

Chapter 3 How to retrieve data from a single table 49
Chapter 4 How to retrieve data from two or more tables 89
Chapter 5 How to code summary queries 127
Chapter 6 How to code subqueries 159
Chapter 7 How to insert, update, and delete data 193
Chapter 8 How to work with data types and functions 211

Section 3 Database design and implementation

Chapter 9 How to design a database 265
Chapter 10 How to create a database 293
Chapter 11 How to create views 333
Chapter 12 How to manage database security 353
Chapter 13 How to host a database in the cloud 385

Section 4 The essential PL/SQL skills

Chapter 14 How to write PL/SQL code 407
Chapter 15 How to manage transactions and locking 439
Chapter 16 How to create stored procedures and functions 451
Chapter 17 How to create triggers 491

Appendixes

Appendix A How to set up Windows for this book 515
Appendix B How to set up macOS for this book 533

Expanded contents

Section 1 An introduction to SQL

Chapter 1 An introduction to relational databases and SQL

An introduction to client/server systems....4
The hardware components of a client/server system....4
The software components of a client/server system....6
Other client/server architectures....8

An introduction to the relational database model....10
How a database table is organized....10
How tables are related....12
How columns in a table are defined....14

An introduction to SQL and SQL-based systems....16
A brief history of SQL....16
A comparison of four relational databases....18

SQL statements and comments....20
An introduction to SQL statements....20
Typical SQL statements....22
SQL coding guidelines....24
How to code comments....24

Chapter 2 How to use SQL Developer

How to use SQL Developer to work with a database....28
How to create a database connection....28
How to navigate through the database objects....30
How to view the column definitions and data for a table....32

How to use SQL Developer to run SQL statements....34
How to enter and run a SQL statement....34
How to handle syntax errors....36
How to open and save SQL scripts....38
How to code and run SQL scripts....40

How to use the Oracle Database documentation....42

Section 2 The essential SQL skills

Chapter 3 How to retrieve data from a single table

An introduction to the SELECT statement....50
How to interpret syntax in this book....50
The basic syntax of the SELECT statement....50
SELECT statement examples....52

How to code the SELECT clause....54
How to code column specifications....54
How to name the columns in a result set....56
How to code string expressions....58
How to code arithmetic expressions....60
How to use scalar functions....62
How to use DISTINCT to eliminate duplicate rows....64

How to code the WHERE clause 66
How to use the comparison operators 66
How to use the AND, OR, and NOT logical operators 68
How to use the IN phrase 70
How to use the BETWEEN phrase 72
How to use the LIKE operator 74
How to use the REGEXP_LIKE function 74
How to use the IS NULL condition 76

How to code the ORDER BY clause 78
How to sort by a column name 78
How to sort by an alias, an expression, or a column number 80

Two more skills 82
How to use the row limiting clause 82
How to test expressions 84

Chapter 4 How to retrieve data from two or more tables

How to work with inner joins 90
How to code an inner join 90
When and how to use table aliases 92
How to use compound join conditions 94
How to use a self-join 96
How to join more than two tables 98
How to use the implicit inner join syntax 100

How to work with outer joins 102
How to code an outer join 102
Outer join examples 104
Outer joins that join more than two tables 106
How to use the implicit outer join syntax 108

Other skills for working with joins 110
How to combine inner and outer joins 110
How to join tables with the USING keyword 112
How to join tables with the NATURAL keyword 114
How to use cross joins 116

How to work with unions 118
The syntax of a union 118
Unions that combine data from different tables 118
Unions that combine data from the same table 120
How to use the MINUS and INTERSECT operators 122

Chapter 5 How to code summary queries

How to work with aggregate functions 128
How to code aggregate functions 128
Queries that use aggregate functions 130

How to group and summarize data 132
How to code the GROUP BY and HAVING clauses 132
Queries that use the GROUP BY and HAVING clauses 134
How the HAVING clause compares to the WHERE clause 136
How to code complex search conditions 138

How to summarize data using the ROLLUP CUBE clause 140
How to use ROLLUP 140
How to use CUBE 142

How to code analytic functions 144
How analytic functions work 144
How to code frames 146
Two more examples of frames 148
How to use named windows 150
How to use the ranking functions 152

Chapter 6 How to code subqueries
An introduction to subqueries 160
How to use subqueries 160
How subqueries compare to joins 162
How to code subqueries in search conditions 164
How to use subqueries with the IN operator 164
How to compare a subquery with an expression 166
How to use the ALL keyword 168
How to use the ANY and SOME keywords 170
How to code correlated subqueries 172
How to use the EXISTS operator 174
Other ways to use subqueries 176
How to code subqueries in the FROM clause 176
How to code subqueries in the SELECT clause 178
Guidelines for working with complex queries 180
A complex query that uses subqueries 180
A procedure for building complex queries 182
How to use subquery factoring 184
How to use the WITH clause 184
How to code a recursive query 186
How to use the hierarchical query clause 188

Chapter 7 How to insert, update, and delete data
How to create test tables 194
How to re-create the tables for this book 194
How to create a table from a SELECT statement 194
How to insert new rows 196
How to insert a single row 196
How to insert default values and null values 198
How to use a subquery to insert multiple rows 200
How to update existing rows 202
How to update rows 202
How to use a subquery in an UPDATE statement 204
How to delete existing rows 206
How to delete rows 206
How to use a subquery in a DELETE statement 206
How to commit and roll back changes 208

Chapter 8 How to work with data types and functions
Data type overview 212
How to work with character data 214
The character data types 214
How to use character functions 216
How to parse a string 220

How to work with numeric data 222
The numeric data types 222
Common number format elements 224
How to use numeric functions 226
How to search for floating-point numbers 228
How to work with temporal data 230
The temporal data types 230
Common datetime format elements 232
How to use datetime functions 234
How to perform a date search 236
Common timestamp and interval formats 238
How to use timestamp functions 240
How to use interval functions 242
How to use the EXTRACT function 244
How to convert data from one type to another 246
How to convert characters, numbers, and dates 246
How to sort strings in numerical sequence 248
How to perform a time search 250
How to convert characters to and from their numeric codes 252
Other functions you should know about 254
How to use the CASE expression 254
How to use the COALESCE, NVL, and NVL2 functions 256
How to use the GROUPING function 258

Section 3 Database design and implementation

Chapter 9 How to design a database
How to design a data structure 266
The basic steps for designing a data structure 266
How to identify the data elements 268
How to subdivide the data elements 270
How to identify the tables and assign columns 272
How to identify the primary and foreign keys 274
How to enforce the relationships between tables 276
How normalization works 278
How to identify the columns to be indexed 280
How to normalize a data structure 282
The seven normal forms 282
How to apply the first normal form 284
How to apply the second normal form 286
How to apply the third normal form 288
When and how to denormalize a data structure 290

Chapter 10 How to create a database
How to work with container databases 294
An introduction to CDBs and PDBs 294
How to create and drop a pluggable database 296

How to work with tables 298
How to create a table 298
How to code a primary key constraint 300
How to code a foreign key constraint 302
How to code a check constraint 304
How to alter the columns of a table 306
How to alter the constraints of a table 308
How to rename, truncate, and drop a table 310
How to work with indexes 312
How to create an index 312
How to drop an index 312
How to work with sequences 314
How to create a sequence 314
How to use a sequence 314
How to alter a sequence 316
How to drop a sequence 316
How to automatically generate ID values 318
A script that's used to create a schema 320
An introduction to scripts 320
How the DDL statements work 320
How to use SQL Developer 326
How to work with tables, indexes, and sequences 326
How to display an EER diagram for a table 328

Chapter 11 How to create views
An introduction to views 334
How views work 334
Benefits of using views 336
How to work with views 338
How to create a view 338
How to create an updatable view 342
How to create a read-only view 342
How to use WITH CHECK OPTION 344
How to insert or delete rows through a view 346
How to alter or drop a view 348
How to use SQL Developer with views 350

Chapter 12 How to manage database security
How to work with users and roles 354
How to create a user 354
How to create an admin user 356
How to alter and drop a user 358
How to create and drop a role 360
How to work with privileges and synonyms 362
System privileges and object privileges 362
How to grant privileges 364
How to revoke privileges 366
How to work with synonyms 368
A script that creates roles and users 370
How to view the privileges for users and roles 374

How to use SQL Developer 376
How to create an admin connection for a PDB 376
How to work with users 378
How to grant and revoke roles 380
How to grant and revoke system privileges 382

Chapter 13 How to host a database in the cloud
How to get started with Oracle Cloud 386
The Oracle Cloud portal 386
How to create a database in the cloud 388
How to create a user for a schema 390
How to create the tables for a schema 392
How to view the database objects for a schema 394
How to use SQL Developer with a cloud database 396
How to connect to a cloud database 396
How to run SQL against a cloud database 398
More skills for working with a cloud database 400
How to restore and delete a cloud database 400
How to restart a cloud database 402

Section 4 The essential PL/SQL skills

Chapter 14 How to write PL/SQL code
An introduction to PL/SQL 408
An anonymous PL/SQL block in a script 408
Statements for working with PL/SQL and scripts 410
How to code the basic PL/SQL statements 412
How to print data to an output window 412
How to declare and use variables 414
How to code IF statements 416
How to code CASE statements 418
How to code loops 420
How to use a cursor 422
How to work with composite variables 424
How to use collections 424
How to use records 428
How to work with exceptions and errors 430
How to handle exceptions 430
Predefined exceptions 432
How to drop database objects without displaying errors 434

Chapter 15 How to manage transactions and locking
How to work with transactions 440
How to commit and roll back transactions 440
How to work with save points 442
How to work with concurrency and locking 444
How concurrency and locking are related 444
How to set the transaction isolation level 446
Best practices for concurrency 448

Chapter 16 How to create stored procedures and functions

How to code stored procedures 452
How to create a stored procedure 452
How to call a stored procedure 454
How to code input and output parameters 456
How to code optional parameters 458
How to raise a predefined exception 460
How to raise a user-defined exception 462
A stored procedure that inserts a row 464
How to drop a stored procedure 468

How to code functions 470
How to create and call a function 470
A function that uses multiple RETURN statements 472
How to drop a function 474

How to work with packages 476
How to create a package 476
How to drop a package 478

How to use SQL Developer 480
How to view and drop procedures, functions, and packages 480
How to edit and compile procedures and functions 482
How to grant and revoke privileges 482
How to debug procedures and functions 484

Chapter 17 How to create triggers

How to work with triggers 492
How to create a trigger for a table 492
A trigger that enforces data consistency 494
How to use conditional predicates 496
How to create a trigger for a view 498
How to create a system trigger 500
How to enable, disable, rename, or drop a trigger 502

Other skills for working with triggers 504
How to create a compound trigger 504
A trigger that causes the mutating-table error 506
How to solve the mutating-table problem 508

How to use SQL Developer 510
How to view, enable, disable, rename, or drop a trigger 510
How to edit a trigger 512

Appendix A How to set up Windows for this book

How to install Oracle Database XE 516
How to install Oracle SQL Developer 518
How to download the files for this book 520
How to connect as the sysdba user 522
How to create the schemas for this book 524
How to import connections for the schemas 526
How to make sure your system is set up correctly 528
How to stop and start the database service 530

Appendix B How to set up macOS for this book

How to download the files for this book 534
How to install Oracle SQL Developer 536
How to set up the database for this book 538
How to create the connections for this book 540
How to make sure your system is set up correctly 542

Introduction

Since its release in 1979, Oracle Database has become the leading database management system in the world. Today, Oracle Database has a huge base of customers and continues to dominate the database market, especially for mission-critical enterprise systems. In short, Oracle Database is here to stay.

Who this book is for

This book is designed for anyone who wants to learn how to use SQL and PL/SQL to work with an Oracle database. This includes beginners who have no coding experience as well as experienced programmers who need to develop applications that work with an Oracle database. If you're a beginner, you can read this book carefully and do the exercises at the end of each chapter. If you have more experience, you can skim this book and focus on the parts that present the skills that you need for your applications.

This book is also a good starting point for anyone who wants to become an Oracle database administrator. Although this book doesn't present all of the advanced skills that are needed by a DBA, it presents a solid foundation that will make it easier to learn the advanced skills later.

Why you'll learn faster with this book

- It starts by showing how to query an existing database rather than how to create a new database. Why? Because that's what you're most likely to need to do first on the job.
- It has hundreds of examples that range from the simple to the complex. That way, you can quickly get the idea of how a feature works from the simple examples and also see how the feature is used in the real world from the more complex ones.

- The exercises at the end of each chapter provide a way for you to gain valuable hands-on experience without extra busywork.
- The information is presented in *paired pages*, with the essential syntax, guidelines, and examples on the right page and clear explanations on the left page. This helps you learn faster by reading less.
- The paired-pages format is ideal for reference when you need to refresh your memory about how to do something.
- It shows how to use SQL Developer's intuitive and user-friendly graphical interface instead of showing how to use SQL*Plus, a command-line interface that has a steeper learning curve.

What you'll learn in this book

In section 1, you'll learn the concepts and terms you need for working with any database. Then, you'll learn how to use SQL Developer to work with an Oracle database.

In section 2, you'll learn the essential SQL skills for retrieving data from a database and for adding, updating, and deleting that data. These skills move from the simple to the complex so you won't have any trouble if you're new to SQL. But these skills are also sure to raise your expertise even if you already have SQL experience.

In section 3, you'll learn how to design a database. Then, you'll learn how to implement that design by using SQL statements to create a database. When you're done, you'll be able to design and implement your own database. In addition, you'll learn how to create views and manage database security. Finally, you'll learn how to use Oracle Cloud to host an Oracle database in the cloud. This can make your database more affordable, flexible, and scalable.

In section 4, you'll learn how to use Oracle's procedural language, PL/SQL, to create stored procedures, functions, and triggers. In addition, you'll learn how to manage transactions and locking. Once you master these features, you'll have a powerful set of PL/SQL skills.

What software you need for this book

If you're using Windows, we recommend that you install these two software products:

- Oracle SQL Developer
- The Express Edition of Oracle Database

Both of these products can be downloaded for free from Oracle's web site, and appendix A provides complete instructions for installing them.

If you're using macOS, we also recommend that you install Oracle SQL Developer. However, at the time of this writing, there is no way to install the Express Edition of Oracle Database on macOS natively. Because of that, we recommend that you create a free Oracle database in the cloud. Appendix B provides complete instructions for doing that and for installing SQL Developer.

What you can download from our web site

You can download all the files for this book from our website. These files include:

- A script that creates the database used by this book
- The SQL and PL/SQL for all of examples in this book
- Solutions to the exercises that are at the end of each chapter

Appendixes A (Windows) and B (macOS) provide complete instructions for downloading these items.

Support materials for trainers and instructors

If you're a corporate trainer or a college instructor who would like to use this book for a course, we offer support materials that will help you set up and run your course as effectively as possible. These materials include instructional objectives, test banks, additional exercises, and PowerPoint slides.

To learn more, please go to our website at www.murachforinstructors.com if you're an instructor. If you're a trainer, please go to www.murach.com and click on the *Courseware for Trainers* link, or contact Kelly at 1-800-221-5528 or kelly@murach.com.

Please let us know how this book works for you

When I started writing the 1st edition of book, I had two goals. First, I wanted to show you how to get started with Oracle Database as quickly and easily as possible. Second, I wanted to raise your database development skills to a professional level.

For the 3rd edition of this book, our editorial team worked hard to update and streamline this book to make it easier than ever to get started with Oracle. This included adding a new chapter on using Oracle Cloud to host an Oracle database. Now, I wish you all the best with learning how to use SQL and PL/SQL with Oracle. If you have any comments about this book, I'd love to hear from you.

Joel Murach

Joel Murach, Author
joelmurach@gmail.com

Section 1

An introduction to SQL

Before you begin to learn the fundamentals of programming in SQL, you need to understand some basic concepts and terms related to SQL and relational databases. So, that's what you'll learn in chapter 1. Then, in chapter 2, you'll learn how to use SQL Developer to work with an Oracle database. At that point, you'll have all of the background and skills that you need to work with the rest of this book.

1

An introduction to relational databases and SQL

This chapter presents the basic concepts and terms that you should understand before you begin learning how to use SQL to work with an Oracle database. If you've never worked with relational databases before, this chapter will teach you what they are and how they work.

An introduction to client/server systems **4**
The hardware components of a client/server system 4
The software components of a client/server system 6
Other client/server architectures 8
An introduction to the relational database model **10**
How a database table is organized 10
How tables are related 12
How columns in a table are defined 14
An introduction to SQL and SQL-based systems **16**
A brief history of SQL 16
A comparison of four relational databases 18
SQL statements and comments **20**
An introduction to SQL statements 20
Typical SQL statements 22
SQL coding guidelines 24
How to code comments 24
Perspective **26**

An introduction to client/server systems

If this is your first time working with client/server systems, the topics that follow introduce you to their essential hardware and software components. When you use SQL to access an Oracle database, that system is often a client/server system.

The hardware components of a client/server system

Figure 1-1 presents the three main hardware components of a client/server system: the server, the clients, and the network. The *server* is a computer that has enough processor speed, memory, and disk storage to store the files and databases of the system and provide services to the clients of the system. When its primary purpose is to store databases, the server is often referred to as a *database server*.

The *clients* are usually personal computers or mobile devices like a phone, but they may be any device that sends information to or requests information from the server. The *network* is the cabling, communication lines, network interface cards, hubs, routers, and other components that connect the clients and the server. Because you don't need to know the network's exact configuration to use it, most diagrams use a cloud symbol to represent its unknown nature.

In a simple client/server system, the clients and the server may be part of a *local area network (LAN)* that connects computers that are near each other. However, two or more LANs that reside at separate geographical locations may be connected as part of a larger network such as a *wide area network (WAN)*. In addition, individual systems or networks can be connected over the internet.

When a client/server system consists of private networks and servers, often spread throughout the country or world, it is commonly referred to as an *enterprise system*. Enterprise systems can consist of on-site resources, cloud resources, or a combination of the two.

A *virtual server* provides the same services as a regular server, but its physical location is not fixed to a specific device. Instead, it is located in a *cloud*, which is a group of many computers in a data center. A virtual server's data may be distributed across multiple computers or share a computer with other virtual servers. In the same way a cloud symbol designates an unknown network configuration, the configuration of a server in a cloud is unknown to its users.

Many companies offer virtual servers that you can use to host your server-side data and run services from a cloud almost anywhere in the world. These are known as *cloud computing platforms*. With this approach, you use an online interface to interact with servers maintained by that company. Three of the biggest cloud computing platforms are Amazon Web Services (AWS), Microsoft Azure, and Oracle Cloud.

A simple client/server system

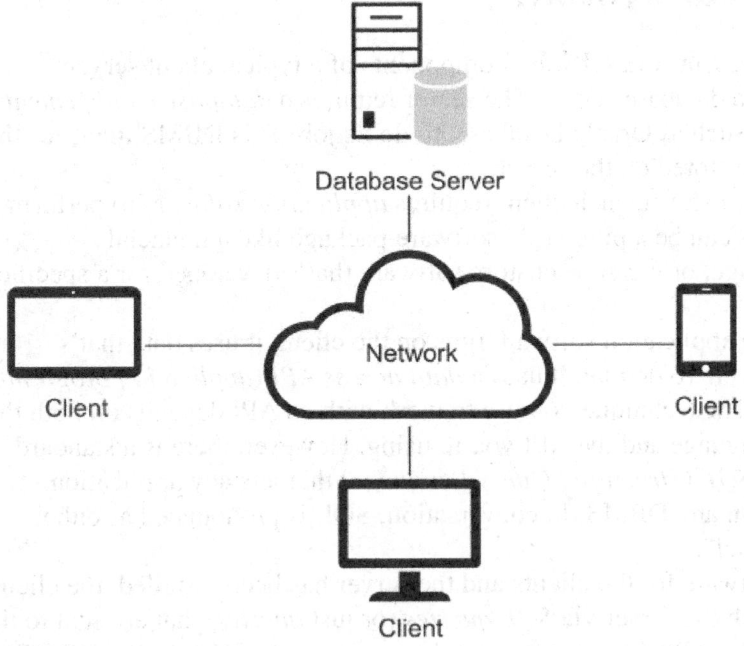

The three hardware components of a client/server system

- The *server* is a specialized computer that stores the files and databases of the system and provides services to the clients. When it is primarily used to store databases, the server is often referred to as a *database server*.
- A *client* refers to any computer, phone, or other device that receives data from the server.
- The *network* consists of the cabling, communication lines, routers, and other components that connect the clients and servers of the system. In diagrams, it is usually represented by a cloud to indicate the unknown nature of the network.

Client/server system implementations

- A simple *client/server system* like the one shown above may use a *local area network (LAN)*, which is a network that directly connects computers that are near each other.
- Individual systems and LANs can be connected and share data over large private networks, such as a *wide area network* (*WAN*), or over a public network like the internet.
- When a system consists of servers and a private network over a wide area, it is commonly called an *enterprise system.*
- A *virtual server* provides the same services as a regular server, but its physical location is not fixed. Instead, it is located in a *cloud*, and its data may be distributed across multiple computers or share a computer with other servers.
- A virtual server is available from a *cloud computing platform.* With this approach, you use an online interface to interact with servers that are maintained by another company.

Figure 1-1 The hardware components of a client/server system

The software components of a client/server system

Figure 1-2 presents the software components of a typical client/server system that uses a database server. The server requires a *database management system (DBMS)*, such as Oracle Database, to do its job. This DBMS manages the databases that are stored on the server.

In contrast to a server, each client requires *application software* to perform useful work. This can be a purchased software package like a financial accounting package, or it can be custom software that's developed for a specific application.

Although the application software runs on the client, it uses data that's stored on the server. To do that, it uses a *data access API* (*application programming interface*). The technique you use to work with an API depends on both the programming language and the API you're using. However, there is a standard language called *SQL (Structured Query Language)* that lets any application communicate with any DBMS. In conversation, SQL is pronounced as either "S-Q-L" or "sequel".

Once the software for the clients and the server has been installed, the clients communicate with the server via *SQL queries* (or just *queries*) that are sent to the DBMS through the API. After a client sends a query to the DBMS, the DBMS processes the query and sends the results back to the client. This processing is typically referred to as *back-end processing*, and the database server is referred to as the *back end*. Conversely, processing done by a client is called *front-end processing*.

The processing done by a client/server system is divided between the clients and the server. In this figure, the DBMS on the server is processing requests made by the application running on the client. This balances the workload between the clients and the server so the system works more efficiently.

Client software, server software, and the SQL interface

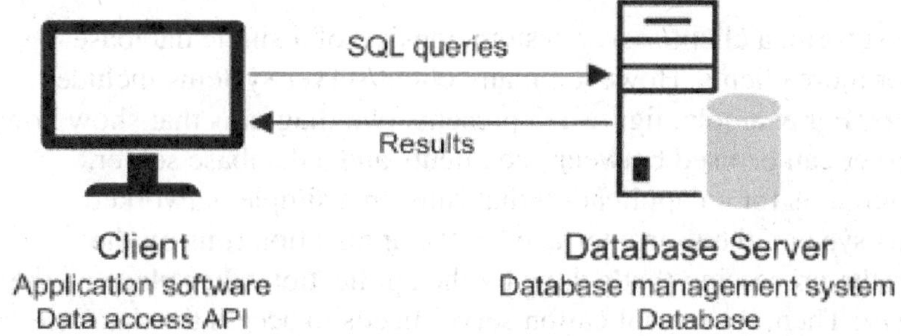

Server software

- To store and manage the databases of the client/server system, each server requires a *database management system* (*DBMS*), such as Oracle Database.
- The database server is referred to as the *back end*, and the processing that's done by the DBMS is typically referred to as *back-end processing*.

Client software

- The *application software* does the work that the user wants to do. This type of software can be purchased or developed.
- The *data access API* (*application programming interface*) provides the interface between the application program and the DBMS.
- A client is typically referred to as the *front end*, and the processing that's done by the client software is typically referred to as *front-end processing*.

The SQL interface

- The application software communicates with the DBMS by sending *SQL queries* through the data access API. When the DBMS receives a query, it provides a service like returning the requested data (the *query results*, or *result set*) to the client.
- *SQL* stands for *Structured Query Language*, which is the standard language for working with a relational database.

Figure 1-2 The software components of a client/server system

Other client/server architectures

In its simplest form, a client/server system consists of a single database server and one or more clients. However, many client/server systems include additional servers. For example, figure 1-3 presents two diagrams that show how an additional server can be used between the clients and a database server.

The first diagram is for an application that runs on a simple networked system. With this system, the user interface for the application runs on the client. However, the processing that's done by the application takes place on the *application server*. Then, if the application server needs to access the database, it uses SQL to query the database server.

Next, the database server sends the results of the query back to the application server. At this point, the application server may further process the results and send an appropriate response back to the client, or it may send another query to the database.

A web-based system like the one shown in the second diagram works similarly. In this diagram, the client, which is typically a web browser such as Chrome, uses a protocol known as HTTP (Hypertext Transfer Protocol) to send a request to a *web server* that's typically running somewhere on the internet. Then, if the web server needs to access data from the database, it uses SQL to send a request for that data to the database server.

Next, the database server sends the results of the query back to the web server. At this point, the web server may do additional processing, send another query to the database server, or send a response back to the web browser.

Although this figure should give you a basic idea of how client/server systems can be configured, these systems can be much more complicated than what's shown here. For example, it's common for a web server to send requests to one or more application servers that in turn send requests to one or more database servers. In most cases, though, it's not necessary for you to know how a system is configured to use SQL.

An application that uses an application server

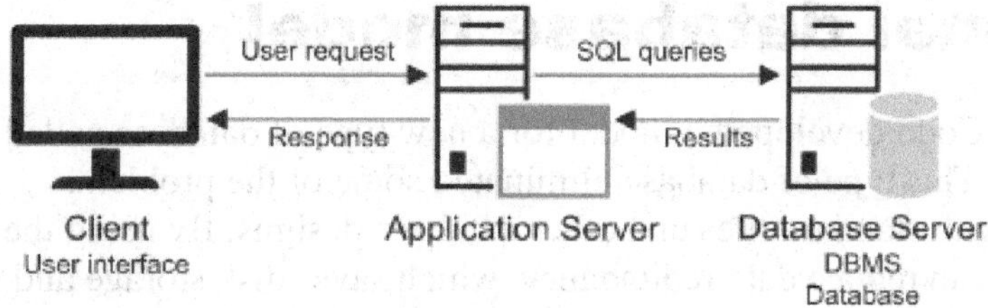

A simple web-based system

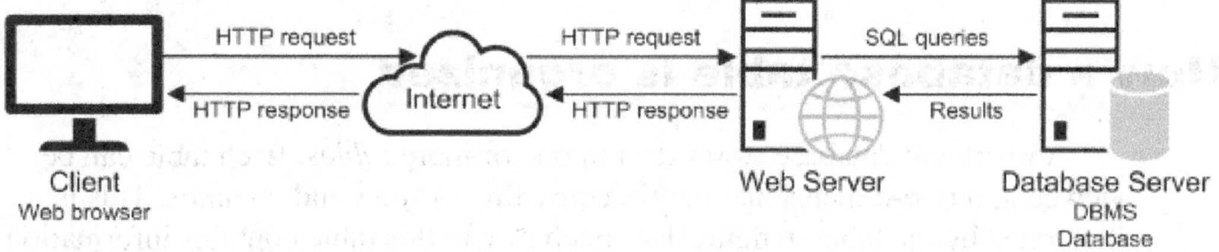

Description

- In addition to a database server and clients, a client-server system can include additional servers, such as application servers and web servers.
- An *application server* handles the processing for an application and uses SQL to communicate with the database server.
- A *web server* uses a protocol known as HTTP (Hypertext Transfer Protocol) to communicate with the client. It can also use SQL to communicate with the database server.
- In a web-based system, it's common for the client to use a web browser to communicate with a web server that's running somewhere on the internet.
- Complex system architectures can include many application servers, web servers, and database servers.

Figure 1-3 Other client/server architectures

An introduction to the relational database model

In 1970, Dr. E. F. Codd developed a model for a new type of database called a *relational database*. This type of database eliminated some of the problems that were associated with standard files and other database designs. By using the relational model, you can reduce data redundancy, which saves disk storage and leads to efficient data retrieval. You can also view and manipulate data in a way that is both intuitive and efficient. Today, relational databases are the de facto standard for database applications.

How a database table is organized

A relational database stores data in one or more *tables*. Each table can be viewed as a two-dimensional matrix consisting of *rows* and *columns*. This is illustrated by the table in figure 1-4. Each row in this table contains information about a single vendor.

In practice, the rows and columns of a relational database table are often referred to by the more traditional terms of *records* and *fields*. Some software packages use one set of terms, some use the other, and some use a combination. This book will use the terms *rows* and *columns* because those are the terms used by Oracle, and because they make the structure of a table easy to visualize.

In general, each table is modeled after a real-world entity such as a vendor or an invoice. Then, the columns of the table represent attributes of the entity such as name, address, and phone number. Each row of the table represents one instance of the entity. A value stored at the intersection of each row and column is sometimes called a *cell*.

If a table contains a column that uniquely identifies each row in the table, you can define this column as the *primary key* of the table. For instance, the primary key of the Vendors table in this figure is the vendor_id column. A primary key can also consist of two or more columns that together uniquely identify a row. In that case, the key is called a *composite primary key*.

In addition to primary keys, some database management systems let you define additional keys that uniquely identify each row in a table. If, for example, the vendor_name column in the Vendors table contains unique data, it can be defined as a *non-primary key*. In Oracle, this is called a *unique key*.

Indexes provide an efficient way of accessing the rows in a table based on the values in one or more columns. Because applications typically access the rows in a table by referring to their key values, an index is automatically created for each key you define. However, you can define indexes for other columns as well. If, for example, you frequently need to sort the Vendor rows by zip code, you can set up an index for that column. Like a key, an index can consist of one or more columns.

The Vendors table in an Accounts Payable database

Primary key

Columns

Rows

	VENDOR_ID	VENDOR_NAME	VENDOR_ADDRESS1	VENDOR_ADDRESS2
1	1	US Postal Service	Attn: Supt. Window Services	PO Box 7005
2	2	National Information Data Ctr	PO Box 96621	NULL
3	3	Register of Copyrights	Library Of Congress	NULL
4	4	Jobtrak	1990 Westwood Blvd Ste 260	NULL
5	5	Newbrige Book Clubs	3000 Cindel Drive	NULL
6	6	California Chamber Of Commerce	3255 Ramos Cir	NULL
7	7	Towne Advertiser's Mailing Svcs	Kevin Minder	3441 W Macarthur Blv
8	8	BFI Industries	PO Box 9369	NULL
9	9	Pacific Gas & Electric	Box 52001	NULL
10	10	Robbins Mobile Lock And Key	4669 N Fresno	NULL
11	11	Bill Marvin Electric Inc	4583 E Home	NULL

Concepts

- A *relational database* consists of *tables*. Tables consist of *rows* and *columns*, which can also be referred to as *records* and *fields*.
- A table is typically modeled after a real-world entity, such as an invoice or a vendor.
- A column represents some attribute of the entity, such as the total of an invoice or a vendor's address.
- A row contains a set of values for a single instance of the entity, such as one invoice or one vendor.
- The intersection of a row and a column is sometimes called a *cell*. A cell stores a single *value*.
- Most tables have a *primary key* that uniquely identifies each row in the table. The primary key is usually a single column, but it can also consist of two or more columns. If a primary key uses two or more columns, it's called a *composite primary key*.
- In addition to primary keys, some database management systems let you define one or more *non-primary keys*. In Oracle, these keys are called *unique keys*. Like a primary key, a non-primary key uniquely identifies each row in the table.
- A table can also be defined with one or more *indexes*. An index provides an efficient way to access data from a table based on the values in specific columns. An index is automatically created for a table's primary and non-primary keys.

Figure 1-4 How a database table is organized

How tables are related

The tables in a relational database can be related to other tables by values in specific columns. The two tables shown in figure 1-5 illustrate this concept. Here, each row in the Vendors table can be related to one or more rows in the Invoices table. This is called a *one-to-many relationship*. That's because one vendor can be related to many invoices, but each invoice can only be related to one vendor.

Typically, relationships exist between the primary key in one table and a *foreign key* in another table. A foreign key is one or more columns in a table that refer to a primary key in another table. In this figure, for example, the vendor_id column in the Invoices column is a foreign key that refers to the primary key of the Vendors table.

Although one-to-many relationships are the most common, two tables can also have a one-to-one or many-to-many relationship. If a table has a *one-to-one relationship* with another table, the data in the two tables could be stored in a single table. However, it's often useful to store large objects such as images, sound, and videos in a separate table. Then, you can join the two tables with the one-to-one relationship only when the large objects are needed.

By contrast, a *many-to-many relationship* is usually implemented by using an intermediate table that has a one-to-many relationship with each of the tables in the many-to-many relationship. In other words, a many-to-many relationship can usually be broken down into two one-to-many relationships.

Oracle helps you manage the relationships between tables by preventing certain kinds of data entry errors. For example, when you define a foreign key for a table in Oracle, you can't add rows to the table with the foreign key unless there's a matching primary key in the related table. So, if you tried to add a row to the Invoices table with a vendor_id value that doesn't exist in the Vendors table, Oracle wouldn't add the row and would instead display an error. This helps to maintain the integrity of the data that's stored in the database and makes it easier for you to focus on coding.

The relationship between the Vendors and Invoices tables

Primary key

	VENDOR_ID	VENDOR_NAME	VENDOR_ADDRESS1
116	117	Suburban Propane	2874 S Cherry Ave
117	118	Unocal	P.O. Box 860070
118	119	Yesmed, Inc	PO Box 2061
119	120	Dataforms/West	1617 W. Shaw Avenue
120	121	Zylka Design	3467 W Shaw Ave #103
121	122	United Parcel Service	P.O. Box 505820
122	123	Federal Express Corporation	P.O. Box 1140

Foreign key

	INVOICE_ID	VENDOR_ID	INVOICE_NUMBER	INVOICE_DATE	INVOICE_TOTAL	PAYMENT_
13	13	122	989319-437	24-APR-24	2765.36	27
14	14	122	989319-427	25-APR-24	2115.81	21
15	15	121	97/553B	26-APR-24	313.55	
16	16	122	989319-417	26-APR-24	2051.59	20
17	17	90	97-1024A	26-APR-24	356.48	3
18	18	121	97/553	27-APR-24	904.14	
19	19	121	97/522	30-APR-24	1962.13	

Concepts

- The tables in a relational database are related to each other through their key columns. For example, the vendor_id column is used to relate the Vendors and Invoices tables.
- The vendor_id column in the Invoices table is called a *foreign key* because it identifies a related row in the Vendors table. A table may contain one or more foreign keys.
- The relationships between the tables in a database correspond to the relationships between the entities they represent.
- The most common type of relationship is a *one-to-many relationship*, as illustrated by the Vendors and Invoices tables.
- A table can also have a *one-to-one relationship* or a *many-to-many relationship* with another table.
- When you define a foreign key for a table in Oracle, you can't add rows to the table with the foreign key unless there's a matching primary key in the related table.

Figure 1-5 How tables are related

How columns in a table are defined

When you add a column to a table, you specify properties that tell Oracle what kind of values the column will store, if a value for the column will be required when adding a new row, and if the column has a default value. To illustrate, figure 1-6 shows these properties for each column in the Invoices table.

The most important property for a column is its *data type*. This determines the type of information that can be stored in the column. This figure lists some of the most basic data types used by Oracle, but there are several others you can use for specialized purposes. In most cases, it's best to assign a data type that minimizes disk storage because that will improve the performance of queries.

In addition to a data type, you must identify whether a column can store *null values* (or just *nulls*). Null represents a value that's unknown, unavailable, or not applicable. If you don't allow null values for a column, all rows must contain a value for that column or else they can't be stored in the table.

You can also assign a *default value* to each column. If you do, the default value is assigned to the column if another value isn't provided when you add a new row.

The columns of the Invoices table

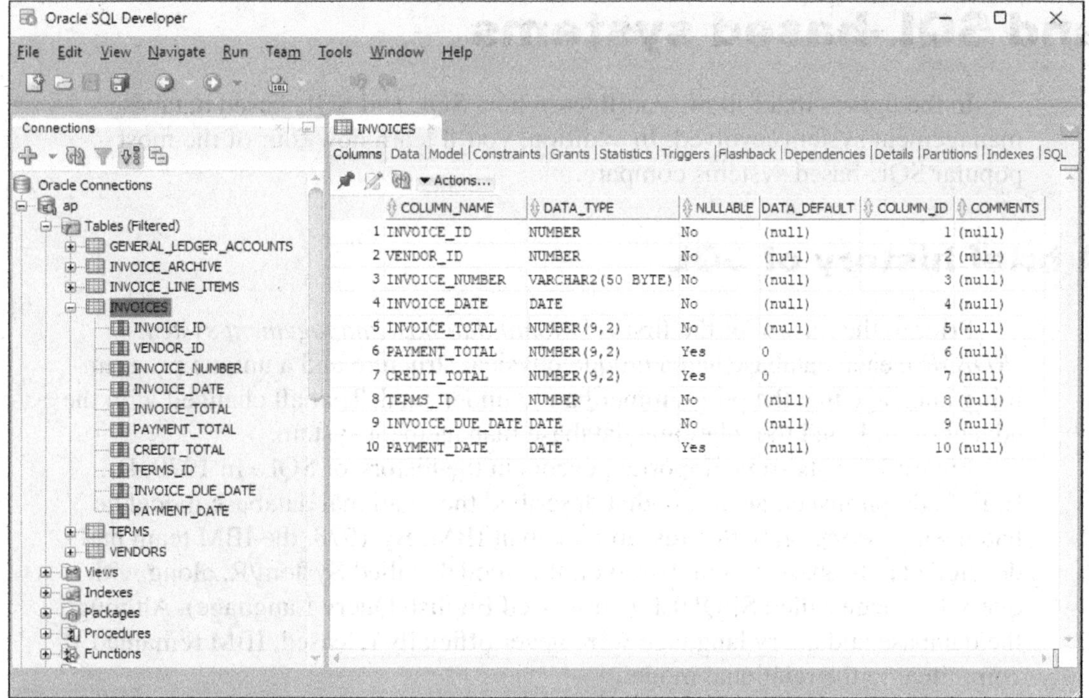

	COLUMN_NAME	DATA_TYPE	NULLABLE	DATA_DEFAULT	COLUMN_ID	COMMENTS
1	INVOICE_ID	NUMBER	No	(null)	1	(null)
2	VENDOR_ID	NUMBER	No	(null)	2	(null)
3	INVOICE_NUMBER	VARCHAR2(50 BYTE)	No	(null)	3	(null)
4	INVOICE_DATE	DATE	No	(null)	4	(null)
5	INVOICE_TOTAL	NUMBER(9,2)	No	(null)	5	(null)
6	PAYMENT_TOTAL	NUMBER(9,2)	Yes	0	6	(null)
7	CREDIT_TOTAL	NUMBER(9,2)	Yes	0	7	(null)
8	TERMS_ID	NUMBER	No	(null)	8	(null)
9	INVOICE_DUE_DATE	DATE	No	(null)	9	(null)
10	PAYMENT_DATE	DATE	Yes	(null)	10	(null)

Common Oracle data types

Type	Description
CHAR, VARCHAR2	A string of letters, symbols, and numbers.
NUMBER	Integer and decimal numbers that contain an exact value.
FLOAT	Floating-point numbers that contain an approximate value.
DATE	Dates and times.

Description

- The *data type* that's assigned to a column determines the type of information that can be stored in the column.
- Each column definition indicates whether or not it can contain *null values*. A null value indicates that a value is unknown, unavailable, or not applicable.
- A column can be defined with a *default value*. Then, that value is used if a value isn't provided when a row is added to the table.

Figure 1-6 How columns in a table are defined

An introduction to SQL and SQL-based systems

In the figures that follow, you'll learn how SQL and SQL-based database management systems evolved. In addition, you'll learn how four of the most popular SQL-based systems compare.

A brief history of SQL

Prior to the release of the first *relational database management system (RDBMS)*, each database had a unique physical structure and a unique programming language that the programmer had to understand. That all changed with the advent of SQL and the relational database management system.

Figure 1-7 lists some important events in the history of SQL. In 1970, Dr. E. F. Codd published an article that described the relational database model he had been working on with a research team at IBM. By 1978, the IBM team had developed a database system based on this model, called System/R, along with a query language called SEQUEL (Structured English Query Language). Although the database and query language were never officially released, IBM remained committed to the relational model.

The following year, Relational Software, Inc. released the first relational database management system, called Oracle. This RDBMS ran on a minicomputer and used SQL as its query language. This product was widely successful, and the company later changed its name to Oracle to reflect that success.

During the 1980s, other SQL-based database systems, including DB2 and SQL Server, were developed. Although each of these systems used SQL as its query language, each implementation was unique. That began to change in 1989, when the *American National Standards Institute* (*ANSI*) published its first set of standards for a database query language. These standards have been revised many times since then. As each database manufacturer has attempted to comply with these standards, their implementations of SQL have become more similar. However, each database still has its own *dialect* of SQL that includes additions, or *extensions*, to the standards.

Although you should be aware of the SQL standards, they will have little effect on your job as a SQL programmer. The main benefit of the standards is that the basic SQL statements are the same in each dialect. As a result, once you've learned one dialect, it's relatively easy to learn another. On the other hand, porting applications that use SQL from one type of database to another often requires substantial changes.

Important events in the history of SQL

Year	Event
1970	Dr. E. F. Codd develops the relational database model.
1978	IBM develops the predecessor to SQL, Structured English Query Language (SEQUEL).
1979	Relational Software, Inc. (later renamed Oracle) releases the first relational database, Oracle.
1985	IBM releases DB2 (Database 2).
1987	Microsoft releases SQL Server.
1989	The American National Standards Institute (ANSI) publishes the first set of standards for a database query language, called ANSI/ISO SQL-89, or SQL 1.
1992	ANSI publishes revised standards (ANSI/ISO SQL-92, or SQL 2) that are more stringent than SQL 1. These standards introduce levels of conformance that indicate the extent to which a dialect meets the standards.
1995	MySQL AB releases MySQL for internal use.
1999	ANSI publishes SQL 3 (ANSI/ISO SQL:1999), which replaces levels of conformance with a core specification along with specifications for nine additional packages.
2000	MySQL is released as an open-source database.
2003	ANSI publishes SQL:2003.
2006	ANSI publishes SQL:2006.
2008	ANSI publishes SQL:2008.
	Sun Microsystems acquires MySQL.
2010	Oracle acquires Sun Microsystems and MySQL.
2011	ANSI publishes SQL:2011.
2016	ANSI publishes SQL:2016.
2019	ANSI publishes SQL:2019.
2023	ANSI publishes SQL:2023.

Description

- Although SQL is a standard language, each vendor has its own *SQL dialect*, or *variant*, that may include *extensions* to the standards.

How knowing "standard SQL" helps you

- The most basic SQL statements are the same for all SQL dialects.
- Once you have learned one SQL dialect, you can easily learn other dialects.

How knowing "standard SQL" does not help you

- Any non-trivial application will require modification when moved from one SQL database to another.

Figure 1-7 A brief history of SQL

A comparison of four relational databases

Although this book is about Oracle Database, you may want to know about some of the other SQL-based relational database management systems. Figure 1-8 compares Oracle with three other popular databases: DB2, Microsoft SQL Server, and MySQL.

Oracle Database has a huge installed base of customers and continues to dominate the marketplace, especially for servers running Unix or Linux operating systems. Oracle Database's software is *proprietary*, which means that the underlying code is not available to the public. To use it, you must license the software from Oracle, usually at a cost. However, Oracle works very well for large systems and has a reputation for being extremely reliable, stable, and secure.

DB2 was originally designed to run on IBM mainframe systems and continues to be the premier database for those systems. It also dominates in hybrid environments where IBM mainframes and newer servers must coexist. Although it is expensive, it also has a reputation for being reliable and easy to use.

SQL Server was designed by Microsoft to run on Windows and is widely used for small- and medium-sized systems. SQL Server 2017 and later also runs on Linux. SQL Server has a reputation for being inexpensive and easy to use.

MySQL runs on all major operating systems and is widely used for web applications. MySQL is an *open-source* database, which means that unlike propriety software, any developer can view and modify its source code. It is also popular for small- and medium-sized systems.

Dozens of other relational database systems are also available. These include proprietary systems like Informix, Sybase, and Teradata and open-source databases like PostgreSQL. There are even non-relational systems, also known as NoSQL databases. Which you choose will largely depend on the needs of your specific business or project.

Oracle Database

- Released in 1979.
- Runs on Unix, z/OS, Windows, Linux, and macOS.
- Typically used for large, mission-critical systems that run on one or more Unix servers.
- Known for high reliability, stability, and security.

DB2

- Released in 1985.
- Runs on Unix, AIX, OS/390, z/OS, Windows, Linux, and macOS.
- Typically used for large, mission-critical systems that run on legacy IBM mainframe systems using the z/OS or OS/390 operating system.

Microsoft SQL Server

- Released in 1987.
- Runs on Windows and Linux.
- Typically used for small- to medium-sized systems that run on one or more Windows or Linux servers.

MySQL

- Released in 2000.
- Runs on Unix, Windows, Linux, and macOS.
- A popular open-source database that runs on all major operating systems and is commonly used for web applications.

Description

- Oracle, DB2, and SQL Server are all *proprietary* database management systems. To use them, you must license the software from the developer.
- MySQL is an *open-source* database management system. Its code can be viewed, modified, and run for free.
- Many other relational databases systems exist, as do non-relational (or "NoSQL") database systems.

Figure 1-8 A comparison of four relational databases

SQL statements and comments

In the figures that follow, you'll learn about some of the basic SQL statements. You can use these statements to manipulate the data in a database and to work with database objects. In addition, you'll learn how to code comments that describe SQL statements.

An introduction to SQL statements

Figure 1-9 summarizes some of the most common *SQL statements*. A statement is an instruction that tells a database to do something.

SQL statements can be divided into two categories. The statements that work with the data in a database are called the *data manipulation language* (*DML*). These statements are presented in the first group in this figure, and these are the statements that application programmers use the most. They are used to get, add, change, and remove data.

The statements that work with the objects in a database are called the *data definition language* (*DDL*). They are used to add, change, and remove objects, like tables and users.

On large systems, DDL statements are used exclusively by *database administrators*, or *DBAs*. It's a DBA's job to maintain existing databases, fine-tune them for faster performance, and create new databases. On smaller systems, though, the SQL programmer may fill the role of DBA.

SQL statements used to work with data (DML)

Statement	Description
`SELECT`	Retrieves data from one or more tables.
`INSERT`	Adds new rows to a table.
`UPDATE`	Changes existing rows in a table.
`DELETE`	Deletes existing rows from a table.

SQL statements used to work with database objects (DDL)

Statement	Description
`CREATE USER`	Creates a new user for a database.
`CREATE TABLE`	Creates a new table in a database.
`CREATE SEQUENCE`	Creates a new sequence that automatically generates numbers.
`CREATE INDEX`	Creates a new index for a table.
`ALTER USER`	Changes the definition of an existing user.
`ALTER TABLE`	Changes the definition of an existing table.
`ALTER SEQUENCE`	Changes the definition of an existing sequence.
`ALTER INDEX`	Changes the structure of an existing index.
`DROP USER`	Deletes an existing user.
`DROP TABLE`	Deletes an existing table.
`DROP SEQUENCE`	Deletes an existing sequence.
`DROP INDEX`	Deletes an existing index.
`GRANT`	Grants privileges to a user.
`REVOKE`	Revokes privileges from a user.

Description

- The *SQL statements* can be divided into two categories: the *data manipulation language (DML)* that lets you work with the data in the database and the *data definition language (DDL)* that lets you work with the objects in the database.
- SQL programmers typically work with the DML statements, while *database administrators* (*DBAs*) use the DDL statements.

Figure 1-9 An introduction to the SQL statements

Typical SQL statements

To give you an idea of how the SQL statements shown in the previous figure work, figure 1-10 presents three typical SQL statements. As you can see, SQL statements are generally self-descriptive. Even without knowing SQL, you can probably guess what the statements in this figure do.

The CREATE TABLE statement creates a table named Invoices with 10 columns. Of these columns, seven are defined as NOT NULL. As a result, you must supply values for these columns when adding a new row. However, the payment_total and credit_total columns are defined with a default value of 0, and the payment_date column allows a null value. As a result, you don't have to supply values for these columns when adding a new row.

The CREATE TABLE statement also defines primary and foreign key columns for the table. Although this statement looks complicated at first glance, closer examination shows that it's only long because it defines multiple columns and keys at once. Each individual column and key definition is short and simple.

The second statement is an INSERT statement that adds a row to the Invoices table. To do that, this statement specifies the name of the table, the names of the seven required columns, and the data for each column. As a result, in the row that's added, the payment_total and credit_total columns store a default value of 0, and the payment_date column stores a null value.

The third statement is a SELECT statement that specifies a list of columns from the Invoices table and limits the results to rows with invoice totals greater than 1000. It also sorts the returned rows by the invoice date. Because a SELECT statement queries a database for information, it can also be called a *query*.

Last, this figure shows the rows returned by the SELECT statement. These rows are in a temporarily created table called a *result set*. If you become a SQL developer, most of the SQL coding you'll do will involve using SELECT statements to create queries that return the desired result sets.

A CREATE TABLE statement that creates a table named Invoices

```
CREATE TABLE invoices
(
  invoice_id          NUMBER          NOT NULL,
  vendor_id           NUMBER          NOT NULL,
  invoice_number      VARCHAR2(50)    NOT NULL,
  invoice_date        DATE            NOT NULL,
  invoice_total       NUMBER(9,2)     NOT NULL,
  payment_total       NUMBER(9,2)                 DEFAULT 0,
  credit_total        NUMBER(9,2)                 DEFAULT 0,
  terms_id            NUMBER          NOT NULL,
  invoice_due_date    DATE            NOT NULL,
  payment_date        DATE,
  CONSTRAINT invoices_pk
    PRIMARY KEY (invoice_id),
  CONSTRAINT invoices_fk_vendors
    FOREIGN KEY (vendor_id)
    REFERENCES vendors (vendor_id),
  CONSTRAINT invoices_fk_terms
    FOREIGN KEY (terms_id)
    REFERENCES terms (terms_id)
)
```

An INSERT statement that adds a row to the Invoices table

```
INSERT INTO invoices
  (invoice_id, vendor_id, invoice_number, invoice_date,
   invoice_total, terms_id, invoice_due_date)
VALUES
  (invoice_id_seq.NEXTVAL, 12, '3289175', '18-JUL-24',
   165, 3, '17-AUG-24')
```

A SELECT statement that gets columns and rows from the Invoices table

```
SELECT invoice_number, invoice_date, invoice_total,
    payment_total, credit_total
FROM invoices
WHERE invoice_total > 1000
ORDER BY invoice_date
```

The result set returned by the statement

	INVOICE_NUMBER	INVOICE_DATE	INVOICE_TOTAL	PAYMENT_TOTAL	CREDIT_TOTAL
1	Q545443	14-MAR-24	1083.58	1083.58	0
2	P-0608	11-APR-24	20551.18	0	1200
3	P-0259	16-APR-24	26881.4	26881.4	0
4	989319-497	17-APR-24	2312.2	0	0
5	989319-487	18-APR-24	1927.54	0	0

Description

- Most SQL statements are self-descriptive, making it easy to figure out what they do.
- A SELECT statement is commonly referred to as a *query*.
- The rows returned by a SELECT statement are called a *result set*.

Figure 1-10 Typical SQL statements

SQL coding guidelines

Although SQL statements are typically easy to understand, they can be difficult to read depending on how you structure your code. This is because SQL is a *freeform language*. That means that you can include line breaks, spaces, and indentation without affecting the way the database interprets the code. In addition, SQL isn't case-sensitive. That means that you can use uppercase or lowercase letters or a combination of the two without changing the way the database interprets the code. All of this means it's possible to write code that works but is extremely hard to read.

For that reason, we suggest that you follow the coding recommendations presented in figure 1-11. The first example presents an unformatted SELECT statement that's difficult to read. By contrast, the second example follows the coding recommendations and is much easier to read.

How to code comments

Most SQL statements are self-descriptive, but you may want to document what a complicated statement does or add other information to your code that you don't want Oracle to process. To do that, you can use *comments*. Comments are skipped by Oracle when you run your code.

You can use two kinds of comments in SQL. The third example in this figure illustrates how to code a *block comment*. This type of comment starts with /* and ends with */. It can span any number of lines and is typically coded at the beginning of a group of statements to document the entire group. Block comments can also be used within a statement to describe blocks of code, but that's not common.

The fourth example in this figure shows a *single-line comment*. This type of comment starts with -- and is typically used to document a single line of code. A single-line comment can be coded on a separate line as shown in this example, or it can be coded at the end of a line of code. In either case, the comment is delimited by the end of the line.

Because comments aren't run by Oracle, it's easy to forget to update them when you change your code. Unfortunately, a comment that incorrectly describes your code is worse than no comment at all. So, be careful to update your comments whenever you update your code.

A SELECT statement that's difficult to read

```
select invoice_number, invoice_date, invoice_total,
payment_total, credit_total, invoice_total - payment_total -
credit_total as balance_due from invoices where invoice_total -
payment_total - credit_total > 0 order by invoice_date
```

A SELECT statement that's coded with a readable style

```
SELECT invoice_number, invoice_date, invoice_total,
    payment_total, credit_total,
    invoice_total - payment_total - credit_total AS balance_due
FROM invoices
WHERE invoice_total - payment_total - credit_total > 0
ORDER BY invoice_date
```

A SELECT statement with a block comment

```
/*
Author: Joel Murach
Date: 5/22/2024
*/
SELECT invoice_number, invoice_date, invoice_total,
    invoice_total - payment_total - credit_total AS balance_due
FROM invoices
```

A SELECT statement with a single-line comment

```
-- The fourth column calculates the balance due
SELECT invoice_number, invoice_date, invoice_total,
    invoice_total - payment_total - credit_total AS balance_due
FROM invoices
```

Coding recommendations

- Capitalize all keywords, and use lowercase for the other code in a SQL statement.
- Separate words in names with underscores, such as invoice_number.
- Start each clause on a new line.
- Break long clauses into multiple lines, and indent continued lines.
- Use *comments* to document code that is difficult to understand.

How to code a comment

- To code a *block comment*, type /* at the start of the block and */ at the end.
- To code a *single-line comment*, type -- followed by the comment.

Description

- Line breaks, white space, indentation, and capitalization have no effect on how the database processes statements.
- Comments can be used to document what a statement does or what specific parts of a statement do. They are not run by the database.

Figure 1-11 SQL coding guidelines

Perspective

To help you understand how SQL is used, this chapter has introduced you to the hardware and software components of a client/server system. It has also described how relational databases are organized and shown you guidelines for coding your SQL statements so they're easy to read. With that as background, you're now ready to start using Oracle.

Terms

server
database server
client
network
client/server system
local area network (LAN)
wide area network (WAN)
enterprise system
virtual server
cloud
cloud computing platform
database management system (DBMS)
back end
back-end processing
application software
data access API (application programming interface)
front end
front-end processing
SQL query
query result
result set
SQL (Structured Query Language)
application server
web server
relational database
table
row
column
record
field
cell
value
primary key
composite primary key
non-primary key
unique key
index
foreign key
one-to-many relationship
one-to-one relationship
many-to-many relationship
data type
null value
default value
SQL dialect
SQL variant
SQL extension
relational database management system (RDBMS)
American National Standards Institute (ANSI)
proprietary software
open-source software
SQL statement
data manipulation language (DML)
data definition language (DDL)
database administrator (DBA)
comment
block comment
single-line comment
freeform language

2

How to use SQL Developer

In the last chapter, you were introduced to relational databases and SQL statements. Now you'll learn how to use Oracle SQL Developer to enter and run SQL statements. In addition, you'll learn how to use the official Oracle Database documentation to get more information about SQL statements.

How to use SQL Developer to work with a database........28
How to create a database connection........28
How to navigate through the database objects........30
How to view the column definitions and data for a table........32
How to use SQL Developer to run SQL statements..........34
How to enter and run a SQL statement........34
How to handle syntax errors........36
How to open and save SQL scripts........38
How to code and run SQL scripts........40
How to use the Oracle Database documentation 42
Perspective 44

How to use SQL Developer to work with a database

This chapter assumes you have already set up your computer for this book according to the instructions in appendix A (Windows) or B (macOS). Those instructions include installing SQL Developer. Now, you'll learn more about how to use SQL Developer. It's a free tool that makes it easy to work with Oracle databases.

How to create a database connection

Before you can work with a database, you need to create a connection to that database. When you start SQL Developer, the Connections window displays all available database connections.

If you followed the directions in appendix A (Windows) or B (macOS), you have already created connections named AP, OM, and EX for users with the same names. As a result, these connections should already be available on your system.

Figure 2-1 shows the dialog box for working with database connections, and it shows the settings for a connection named AP. This connection has a username of AP and a password of "sesame" to connect to a database named XEPDB1 that's running on a local server on port 1521. If you're using Windows and you followed the instructions in appendix A, this is what you should see on your system.

If you're using macOS and you followed the instructions in appendix B, the AP connection should connect to a database that's running in the cloud. As a result, this dialog box should still have a username of AP. However, it should use the password that you specified when you created the AP user in the cloud. In addition, it should use the Cloud Wallet connection type as shown in appendix B.

For the local XEPDB1 database, this book uses a password of "sesame" for the AP, OM, and EX users because that's easy to remember and security is not a concern for this database. In a real-world situation, you should use a longer and more secure password, and you should use a different password for each user.

If you want to create a new local connection or modify an existing one, you can follow the procedure shown in this figure. To start, you display the dialog box for working with database connections. Then, you can set the name of the connection, the username, the password, and the name of the service, which is typically the name of the database.

When you create a database connection, the username is not case-sensitive. As a result, it doesn't matter if you enter it in uppercase or lowercase. However, the password is case-sensitive, so make sure to use the correct case when entering it.

The dialog box for working with database connections

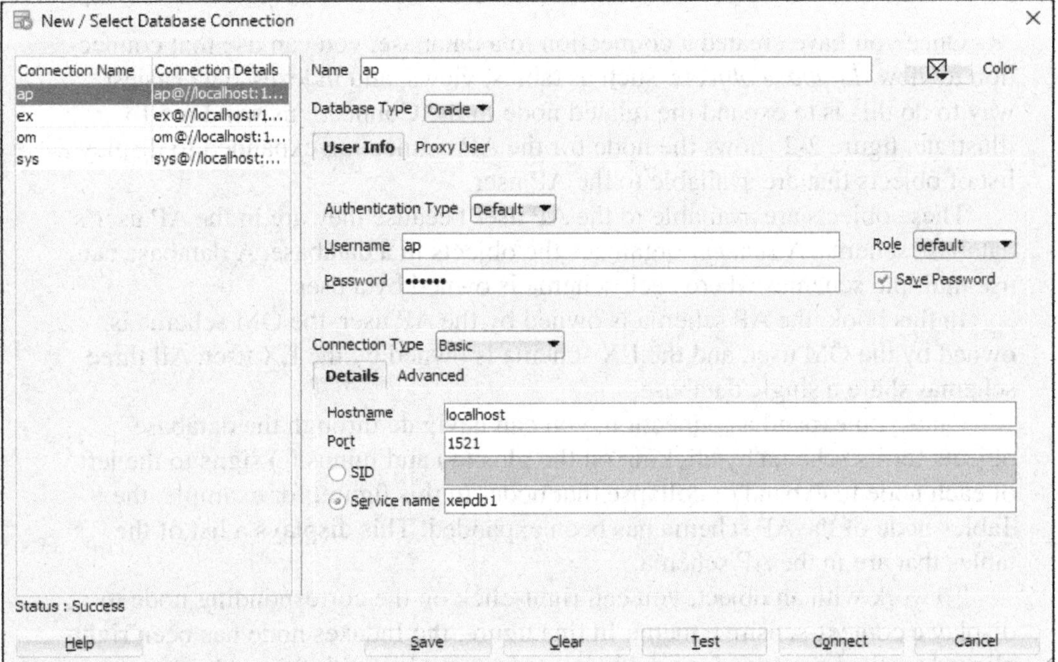

How to create a database connection

1. Start SQL Developer.
2. Right-click on Oracle Connections in the Connections window and select "New Connection" to display the dialog box for working with database connections.
    - Enter a connection name.
    - Enter a username.
    - Enter a password. If you don't want to be prompted for the password every time you connect, check the Save Password box.
    - Select "Service name" and enter the name of the database to connect to.
3. Click the Test button to test the connection. If the connection works, SQL Developer displays a success message below the Connections window.
4. Click the Save button to save the connection. When you do, SQL Developer adds the connection to the Connections window.

Figure 2-1 How to create a database connection

How to navigate through the database objects

Once you have created a connection to a database, you can use that connection to view *database objects*, such as tables, views, and indexes. The easiest way to do this is to expand the related node in the Connections window. To illustrate, figure 2-2 shows the node for the AP connection expanded to display a list of objects that are available to the AP user.

These objects are available to the AP user because they are in the AP user's database schema. A *schema* organizes the objects in a database. A database can use multiple schemas where each schema is owned by a user.

In this book, the AP schema is owned by the AP user, the OM schema is owned by the OM user, and the EX schema is owned by the EX user. All three schemas share a single database.

Once you expand a connection, you can navigate through the database objects for its schema by clicking on the plus (+) and minus (-) signs to the left of each node to expand or collapse that node. In this figure, for example, the Tables node of the AP schema has been expanded. This displays a list of the tables that are in the AP schema.

To work with an object, you can right-click on the corresponding node to display a context-sensitive menu. In this figure, the Indexes node has been right-clicked to display a list of available items for working with this node.

The list of database objects displayed also depends on the permissions of the user that's assigned to the connection you're using. In this book, the AP, EX, and OM users all have all permissions for the objects in their schemas.

The AP connection expanded in the Connections window

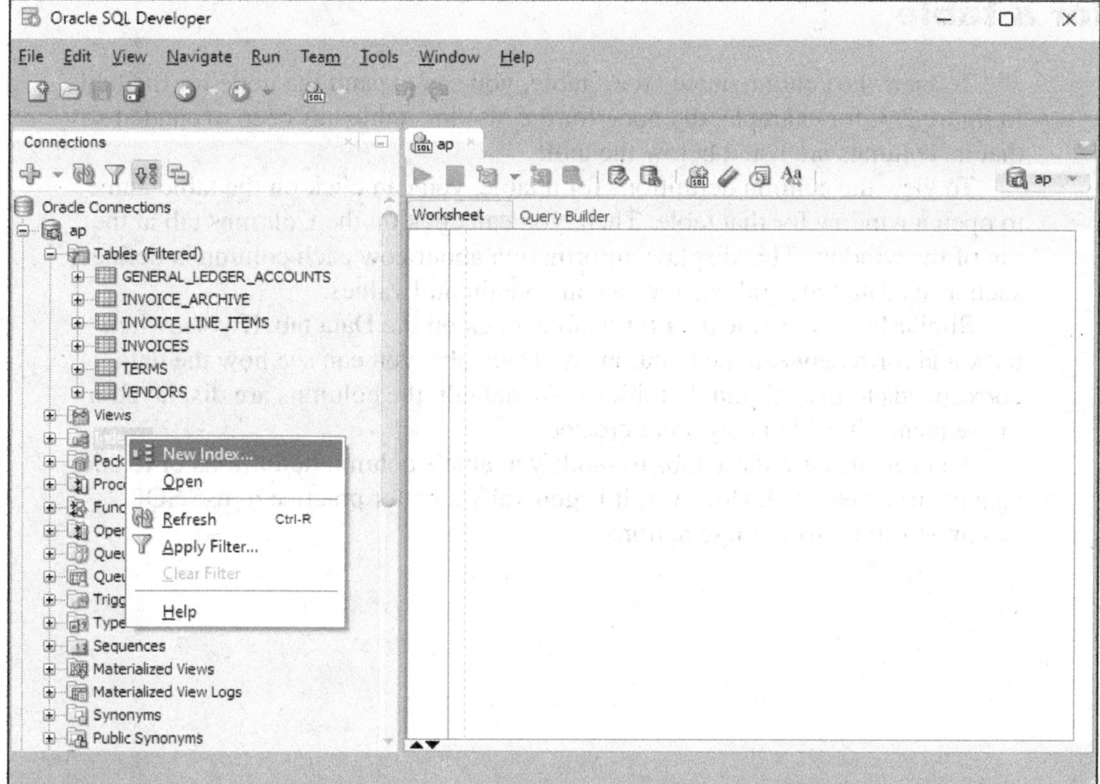

Description

- Each connection provides access to *database objects*, such as tables, views, and indexes.
- Each user is associated with a *database schema* of the same name as the user. A schema organizes database objects.
- To navigate through the database objects for a schema, click on the plus (+) and minus (-) signs to the left of each node to expand or collapse that node.
- To work with an object, right-click on the corresponding node and select an item from the resulting menu.

Figure 2-2 How to navigate through the database objects

How to view the column definitions and data for a table

To view the column names for a table, you can expand the node for the table. In figure 2-3, for example, the node for the Vendors table has been expanded so that its columns are listed below the table.

To view the column definitions for a table, you can click on the table name to open a window for that table. Then, you can click on the Columns tab at the top of the window. This displays information about how each column is defined, such as its data type and whether it can contain null values.

Similarly, to view the data for a table, click on the Data tab. If you switch back and forth between the Column and Data tabs, you can see how the data corresponds to the column definitions. By default, the columns are displayed in the sequence in which they were created.

You can also use these tabs to modify a table's column definitions or to add, change, or delete data. However, it is generally a better practice to use SQL statements to perform these actions.

The Vendors table displayed in the main window

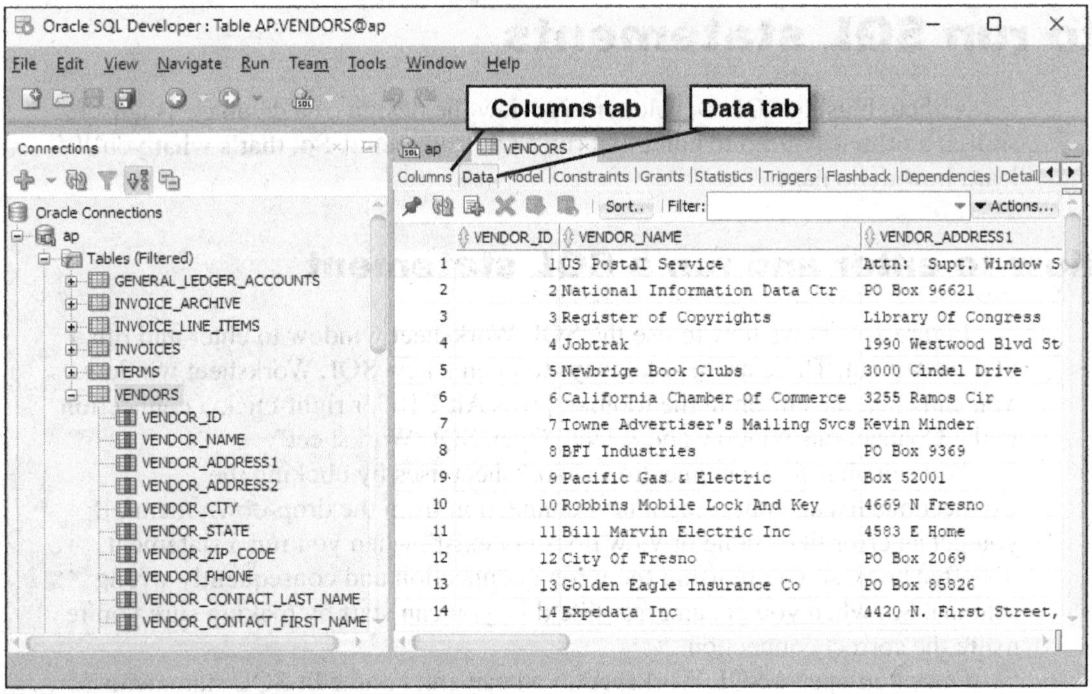

Description

- To view the column names for a specific table, expand the Tables node and then expand the node for the table.
- To view the column definitions for a specific table, click on the name of the table in the Connections window. Then, click the Columns tab in the window for the table.
- To view the data for a specific table, click on the Data tab in the window for the table.
- By default, columns are displayed in the sequence in which they were created.

Figure 2-3 How to view the column definitions and data for a table

How to use SQL Developer to run SQL statements

Although using SQL Developer to review the design of a database is a useful skill, it's primarily used to enter and run SQL statements. So, that's what you'll learn how to do next.

How to enter and run a SQL statement

Figure 2-4 shows how to use the SQL Worksheet window to enter and run a SQL statement. There are several ways to open a new SQL Worksheet window. You can click the button in the toolbar, press Alt-F10, or right-click a connection in the Connections window and select "Open SQL Worksheet".

You can change the connection a worksheet uses by clicking the Connections list and choosing a new connection from the drop-down menu. If you get an error like "table or view does not exist" when you run a statement, it's often because you're using the wrong connection and consequently wrong schema. So, when you get an error like this, you can start by making sure you're using the correct connection.

Once you open a SQL Worksheet, you can enter and edit SQL statements just as you would in any text editor. As you enter statements, SQL Developer automatically applies colors to various elements. For example, keywords like SELECT and FROM are displayed in blue. This makes your statements easier to read and can help you identify coding errors.

In addition, SQL Developer sometimes displays a drop-down list that helps you enter SQL statements. This is the Completion Insight feature, sometimes called "autocomplete". Instead of waiting for it to pop up, you can trigger the suggestion list by pressing Ctrl+Spacebar. Completion Insight is helpful for entering SQL keywords, table names, column names, and so on. Note that entering a table name first enables column name suggestions for that table.

To run a single SQL statement like the one in this figure, press Ctrl+Enter or click the Run Statement button in the toolbar. If the statement returns data, SQL Developer displays that data in a Query Result tab as shown here. If you want, you can adjust the height of the Result tab by dragging the bar that's between the text editor and the query results.

If you're working with a large database and you run a query that returns thousands or even hundreds of thousands of rows, it can take a long time to retrieve every row in the result set. To speed things up, SQL Developer fetches rows in groups. It displays the first group, then the next group as you scroll through the result set until all rows have been fetched or you stop scrolling through the results. This is usually convenient and what you want.

However, in this book, the number of rows returned by a query is often listed to show the differences between running two statements. As a result, you may want to quickly jump to the end of a result set to see if the number of rows matches the number listed in the book. To do that, you can click in the result set and press Ctrl+End to jump to the last row of the result set.

An open SQL Worksheet window

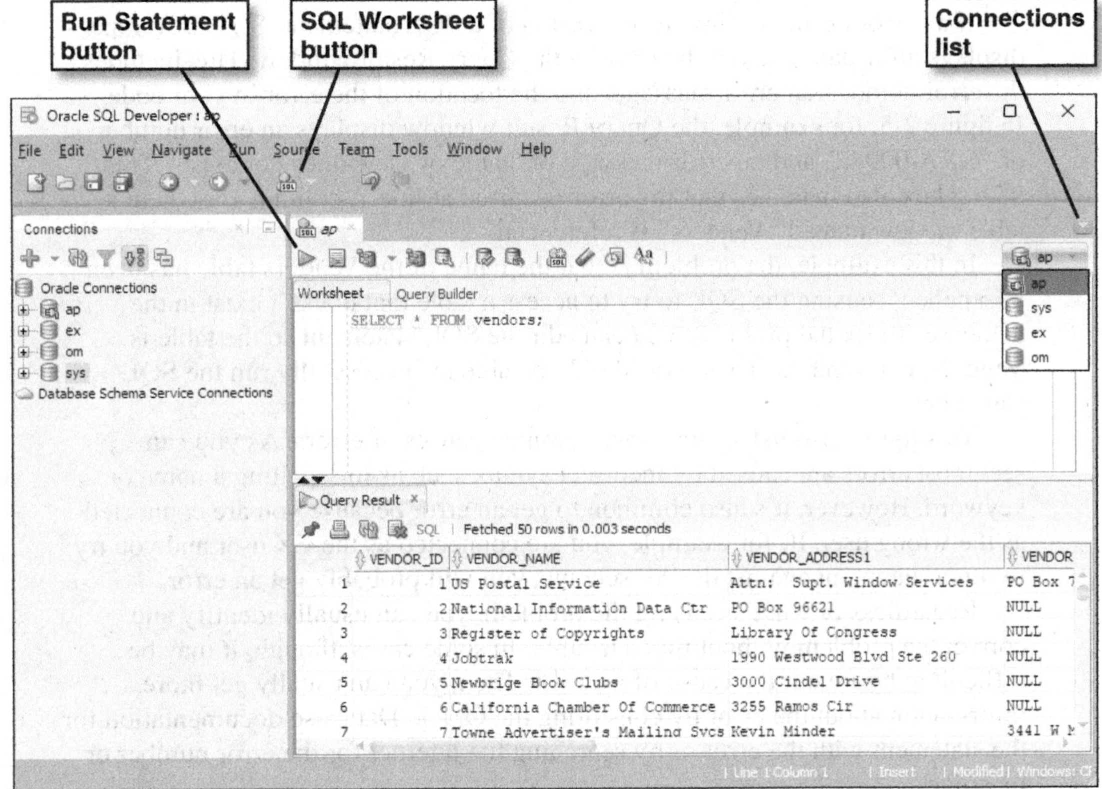

Description

- You can enter and run SQL statements within a SQL Worksheet window.
- To open a new Worksheet window, press Alt-F10 or click the SQL Worksheet button.
- To change the connection for a Worksheet window, use the Connections list drop-down menu.
- To enter a SQL statement, type it into a SQL Worksheet window.
- As you enter the text for a statement, you can use the Completion Insight feature to select SQL keywords, table names, column names, and so on.
- To manually display the code completion list, press Ctrl+spacebar.
- To run a SQL statement, press Ctrl+Enter or click the Run Statement button.
- If the statement retrieves data, the data is displayed in a Query Result tab.
- By default, SQL Developer fetches 50 rows at a time, and displays additional rows after you scroll past the first 50 rows in the Query Result tab.
- To jump to the end of a result set, click on the result set that's displayed in the Query Result tab and press Ctrl+End.

Figure 2-4 How to enter and run a SQL statement

How to handle syntax errors

If an error occurs during the execution of a SQL statement, SQL Developer displays information about the error in the Query Result window. This includes the error number, an error message, and the location of the error in your code. In figure 2-5, for example, the Query Result window displays an error number of "ORA-00942" and an error message of "table or view does not exist." This dialog box also indicates that the error occurred at line 1, column 15, where a table or view named "Venders" is referenced.

In this example, the problem is that the name of the Vendors table has been misspelled, causing the SQL to try to access a table that doesn't exist in the database. To fix the problem, you can edit the SQL statement so the table is Vendors, not Venders. Then, you should be able to successfully run the SQL statement.

This figure also lists some other common causes of errors. As you can see, most errors are caused by incorrect syntax such as misspelling a name or keyword. However, it's also common to get an error because you are connected as the wrong user. If, for example, you are connected as the EX user and you try to run a statement against the AP schema, you will probably get an error.

Regardless of what's causing the problem, you can usually identify and correct the problem without much trouble. In some cases, though, it may be difficult to figure out the cause of an error. Then, you can usually get more information about the error by consulting the Oracle Database documentation for the statement with the error or by searching the internet for the error number or message.

A SQL query with a syntax error

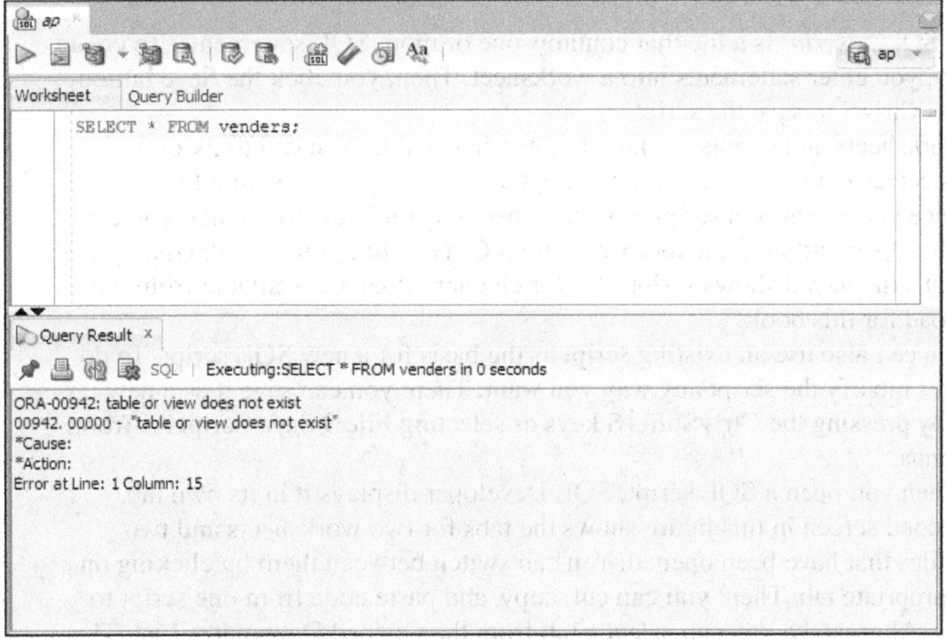

Common causes of errors

- Misspelling the name of a table or column
- Misspelling a keyword
- Omitting the closing quotation mark for a character string
- Having the wrong connection selected

Description

- If an error occurs during the execution of a SQL statement, SQL Developer displays information about the error in the Query Result window.
- Most errors are caused by incorrect syntax and can be easily corrected.
- If you can't immediately solve the problem, try searching the internet or the Oracle Database documentation for more information about the error.

Figure 2-5 How to handle syntax errors

How to open and save SQL scripts

In SQL, a *script* is a file that contains one or more SQL statements. To create a script, you enter statements into a worksheet. Then, you click the Save button or press Ctrl+S to save the script.

Worksheets and scripts are largely interchangeable. You can think of a worksheet as an unsaved script, or a script as a worksheet saved to a file.

Once you've saved a script, you can open it again later. To do that, you can click the Open button in the toolbar or press Ctrl+O. In figure 2-6, the dialog box that's displayed shows script files for chapter 2 that are available from the download for this book.

You can also use an existing script as the basis for a new SQL script. To do that, just modify the script any way you want. Then, you can save it as a new script by pressing the Ctrl+Shift+S keys or selecting File→Save Script As from the menus.

When you open a SQL script, SQL Developer displays it in its own tab. The second screen in this figure shows the tabs for two worksheets and two script files that have been opened. You can switch between them by clicking on the appropriate tab. Then, you can cut, copy, and paste code from one script to another. Alternately, you can select a tab from the Opened Documents List. This is helpful if you have so many scripts open that the tabs can't all fit across the top of the window.

When you open a SQL script, you must specify a connection before you can run any SQL statements stored within the script. To do that, you can use the Connections list to select a connection. Or, SQL Developer will prompt you to select a connection when you try to run a statement.

The dialog box for opening scripts

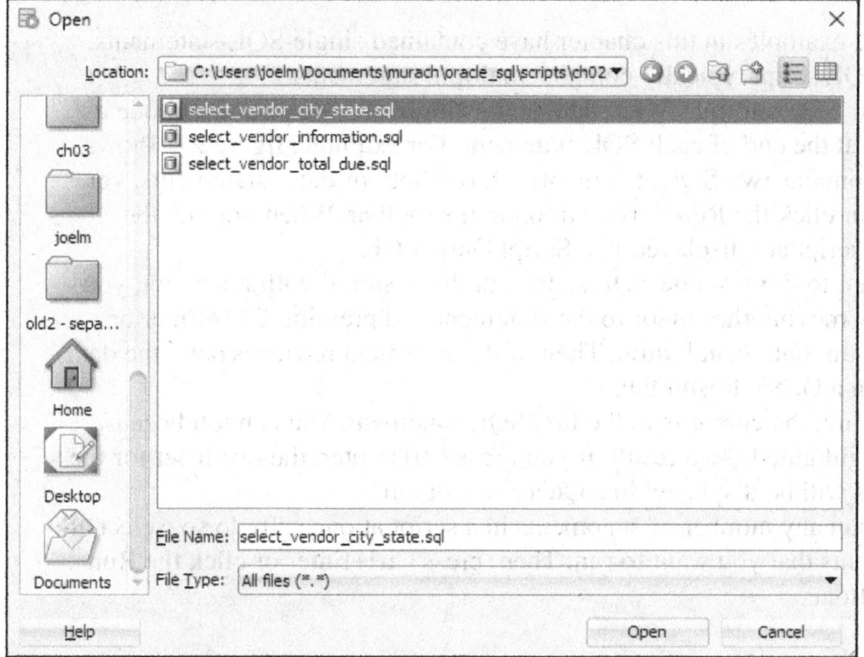

Multiple open files and worksheets

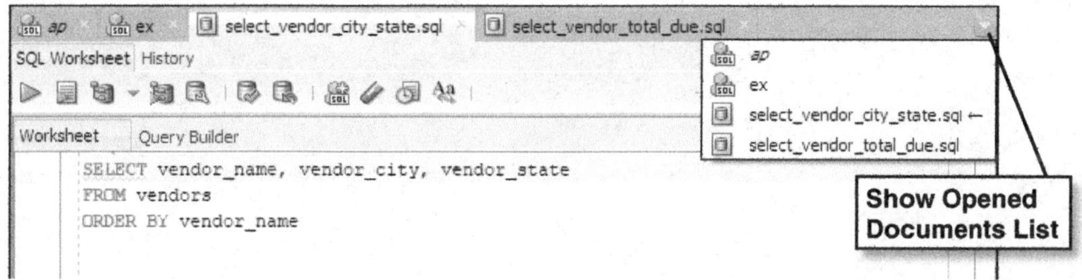

Description

- A *SQL script* is a file that contains one or more SQL statements.
- To open a SQL script, click the Open button in the toolbar or press Ctrl+O. Then, use the Open dialog box to locate and open the script.
- You can have multiple scripts and worksheets open at once. If the name of a script or worksheet is displayed in italics, it contains unsaved changes.
- You can cut, copy, and paste code from one script or worksheet to another.
- To save a worksheet to a file, click the Save button in the toolbar or press Ctrl+S.
- To save a script to a new file, press Ctrl+Shift+S or select File→Save Script As.

Figure 2-6 How to open and save SQL scripts

How to code and run SQL scripts

So far, the examples in this chapter have contained single SQL statements. However, a SQL script typically contains multiple statements.

When you code multiple SQL statements within a script, you must code a semicolon (;) at the end of each SQL statement. For example, figure 2-7 shows a script that contains two SQL statements. To run both of these statements, you can press F5 or click the Run Script button in the toolbar. When you do, the results of the script are displayed in a Script Output tab.

If you want to run just one SQL statement that's stored within a script, you can do that by moving the cursor to the statement and pressing Ctrl+Enter or clicking the Run Statement button. Then, if the statement retrieves data, the data is displayed in a Query Result tab.

In this figure, the cursor is in the first SQL statement. You can tell because it is lightly highlighted. As a result, if you press Ctrl+Enter, the result set for the first statement will be displayed in a Query Result tab.

You can run any number of statements in a script at once. To do so, select all of the statements that you want to run. Then, press Ctrl+Enter or click the Run Statement button.

A SQL script and its results

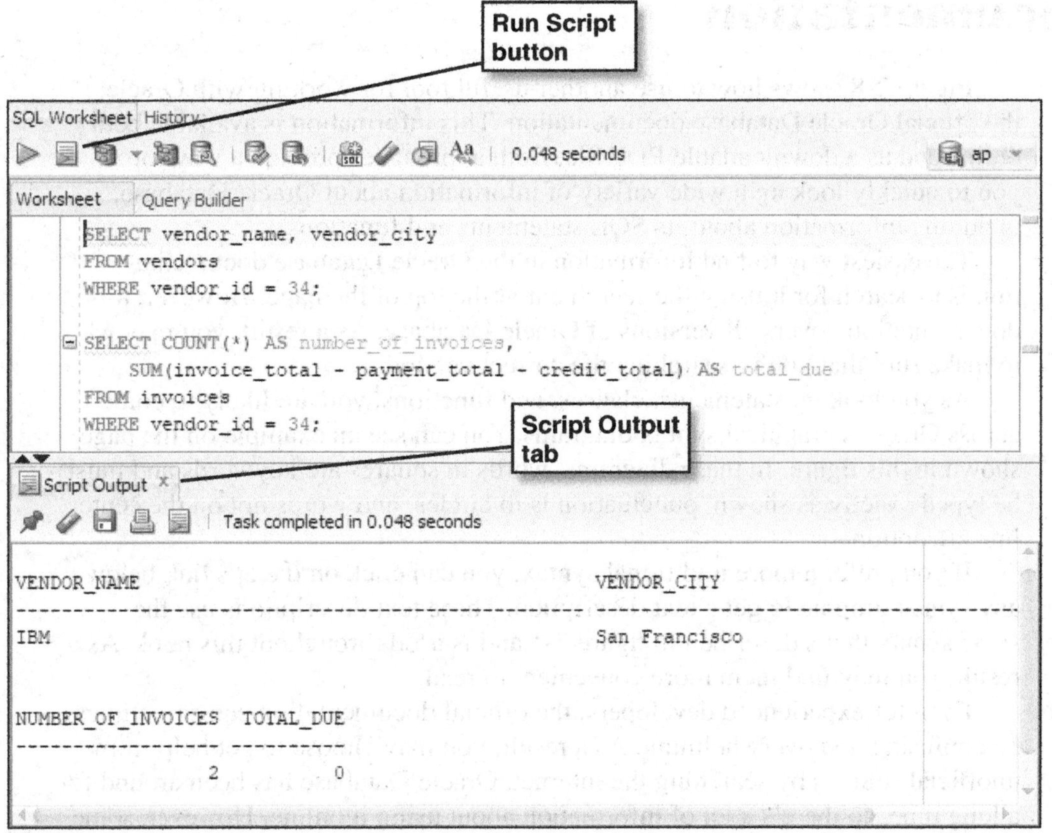

Description

- When you code more than one SQL statement in the same script, you must code a semicolon at the end of each statement to separate them.
- To run an entire SQL script, press F5 or click the Run Script button in the toolbar. The results are displayed in a Script Output tab.
- To run just one SQL statement within a script, move the cursor into the statement. Then, press Ctrl+Enter or click the Run Statement button.
- To run two or more statements without running the whole script, select them in the editor and then press Ctrl+Enter or click the Run Statement button.

Figure 2-7 How to code and run SQL scripts

How to use the Oracle Database documentation

Figure 2-8 shows how to use another useful tool for working with Oracle: the official Oracle Database documentation. This information is available both online and as a downloadable PDF file. Both approaches provide a way for you to quickly look up a wide variety of information about Oracle Database, including information about its SQL statements and functions.

The easiest way to find information in the Oracle Database documentation is to search for it using the search bar at the top of the page. However, this documentation covers all versions of Oracle Database. As a result, you may want to make sure that you're searching the correct version.

As you look up statements, clauses, and functions, you are likely to come across Oracle's graphical syntax diagrams. You can see an example on the page shown in this figure. In these diagrams, words in squares are keywords and must be typed exactly as shown, punctuation is in circles, and words not on the center line are optional.

If you prefer a more traditional syntax, you can click on the eps link below any syntax graphic to get a text description. These text descriptions use the same syntax that's described in figure 3-1 and is used throughout this book. As a result, you may find them more convenient to read.

Even for experienced developers, the official documentation can sometimes be confusing and overwhelming. As a result, you may choose to get help from unofficial sources by searching the internet. Oracle Database has been around for a long time, so there's a lot of information about using it online. However, some of this information is old and out of date, so keep that in mind when searching for solutions. Whenever necessary, consult the official documentation to make sure you're getting up-to-date information.

The web address for the Oracle Database documentation

```
https://docs.oracle.com/en/database/oracle/oracle-database/index.html
```

A web page from the Oracle Database documentation

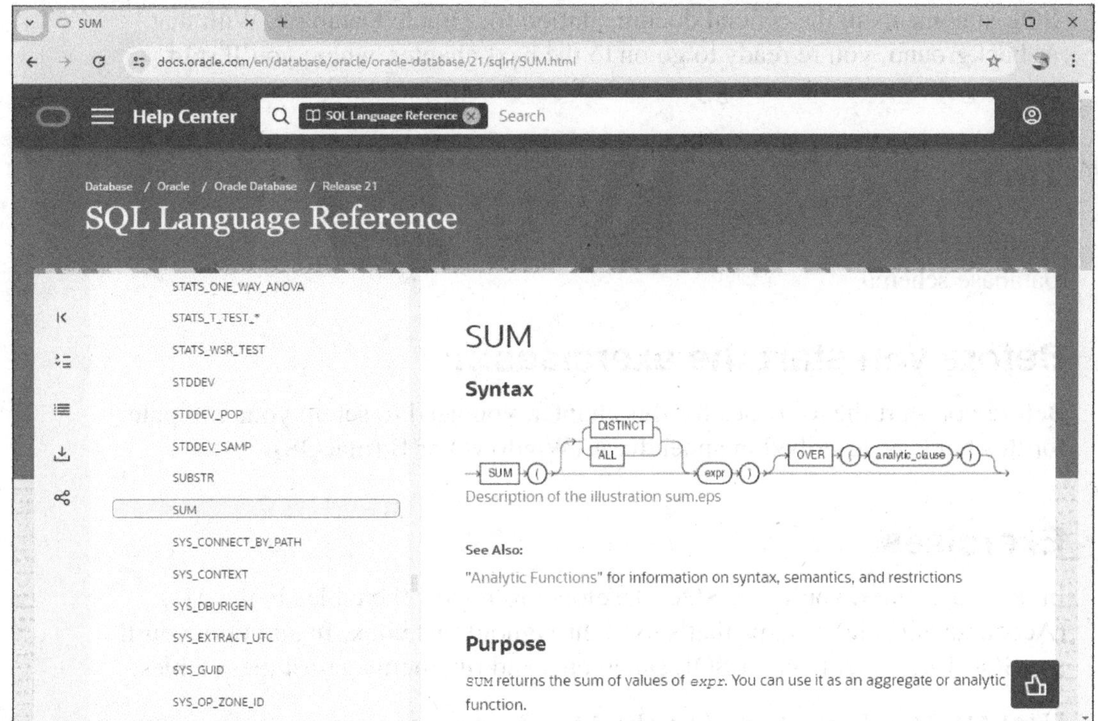

The text for the syntax

Description of the illustration sum.eps

```
SUM([ DISTINCT | ALL ] expr)
  [ OVER (analytic_clause) ]
```

Description

- To view the Oracle Database documentation, go to URL shown above. Or, go to Oracle's website and click Resources in the top bar, then Documentation, then Oracle Database.
- To select the version of Oracle Database for the documentation, use the drop-down list that's shown on the home page for the documentation.
- To find information on a specific topic, use the search bar at the top of the page.
- Oracle uses graphics to illustrate the syntax of statements, clauses, and functions. To view text for the syntax, you can click the eps link below the graphic.

Figure 2-8 How to use the Oracle Database documentation

Perspective

In this chapter, you learned how to use SQL Developer to enter and run SQL statements. In addition, you learned how to look up information about SQL statements in the official documentation for Oracle Database. With that as background, you're ready to go on to the next chapter, where you'll start learning the details of coding your own SQL statements.

Terms

database object
database schema
SQL script

Before you start the exercises...

Before you start the exercises for this chapter, you need to set up your computer for this book as described in appendix A (Windows) or B (macOS).

Exercises

In these exercises, you'll use SQL Developer to review the tables in the AP (Accounts Payable) schema that's used throughout this book. In addition, you'll use SQL Developer to enter SQL statements and run them against these tables.

Use SQL Developer to review the AP schema

1. Start SQL Developer.
2. In the Connections window, click on the AP connection to expand it. That should expand the connection so you can see the folders for all of the database objects in the AP schema.
3. Navigate through the database objects and view the column definitions and data for at least the Vendors and Invoices tables. To do that, you can start by expanding the Tables folder.

Use SQL Developer to enter and run SQL statements

4. Open a SQL Worksheet window for the AP connection. Then, enter and run this SQL statement:

```
SELECT vendor_name FROM vendors
```

5. Add some column names and an ORDER BY clause to the statement. As you do that, try using Completion Insight to select and enter the keywords and column names.

```
SELECT vendor_name, vendor_address1, vendor_city, vendor_state,
    vendor_zip_code
FROM vendors
ORDER BY vendor_name
```

6. Run the modified statement.

7. Modify the statement by commenting out the ORDER BY clause by putting two dashes in front of it, like this:

```
SELECT vendor_name, vendor_address1, vendor_city, vendor_state,
    vendor_zip_code
FROM vendors
-- ORDER BY vendor_name
```

8. Run the statement again, comparing the new results to the previous, ordered results.
9. Modify the statement by deleting the "e" at the end of vendor_zip_code and run the statement again. This should display a syntax error.
10. Read the syntax error closely and note that it explains that the column isn't a valid identifier.
11. Click the Save button to save this statement. Then, click the Cancel button in the Save dialog box to cancel the operation.

Use SQL Developer to open and run scripts

12. Open the select_vendor_city_state script that's in the oracle_sql\scripts\ch02 directory. Note that this script contains just one SQL statement.
13. Run the statement. If you didn't specify a connection for this statement, SQL Developer should ask you to select one before it runs the statement.
14. Open the select_vendor_total_due script in the same folder. Note that this opens another tab.
15. Select the AP connection from the connection list and run this script.
16. Open the select_vendor_information script that's in the ch02 directory. Notice that this script contains two SQL statements that end with semicolons.
17. Move the insertion point to the first statement and press Ctrl+Enter to run that statement.
18. Move the insertion point to the second statement and press Ctrl+Enter to run that statement.
19. Press F5 to run both of the statements that are stored in this script. If you scroll through the Script Output window, this should show the query results for both statements.

Close and restart SQL Developer

20. Continue to experiment on your own.
21. Make sure to leave at least one saved script open. Then, exit SQL Developer.
22. Restart SQL Developer. When it starts, note that any saved scripts that you left open are automatically opened.
23. Run one of the open scripts. Note that you have to select a connection before SQL Developer can run the script.
24. Exit SQL Developer.

Section 2

The essential SQL skills

This section teaches you the essential SQL coding skills for working with the data in an Oracle database. The first four chapters show you how to retrieve data from a database using the SELECT statement. In chapter 3, you'll learn how to code the basic clauses of the SELECT statement to retrieve data from a single table. In chapter 4, you'll learn how to retrieve data from two or more tables. In chapter 5, you'll learn how to summarize the data that you retrieve. And in chapter 6, you'll learn how to code subqueries, which are SELECT statements that are coded within other statements.

Next, chapter 7 shows you how to use the INSERT, UPDATE, and DELETE statements to add new rows to a table, to modify rows in a table, and to remove rows from a table. Then, chapter 8 shows you how to work with the various types of data that Oracle supports and how to use some of the built-in functions that Oracle provides for working with these data types. When you complete the six chapters in this section, you'll have all the skills you need to code SELECT, INSERT, UPDATE, and DELETE statements.

3

How to retrieve data from a single table

In this chapter, you'll learn how to code SELECT statements that retrieve data from a single table. The skills covered here are the essential ones that apply to any SELECT statement you code, no matter how many tables it operates on or how complex the retrieval. So, you'll want to be sure you have a good understanding of the material in this chapter before you go on to the chapters that follow.

An introduction to the SELECT statement 50
How to interpret syntax in this book 50
The basic syntax of the SELECT statement 50
SELECT statement examples 52
How to code the SELECT clause 54
How to code column specifications 54
How to name the columns in a result set 56
How to code string expressions 58
How to code arithmetic expressions 60
How to use scalar functions 62
How to use DISTINCT to eliminate duplicate rows 64
How to code the WHERE clause 66
How to use the comparison operators 66
How to use the AND, OR, and NOT logical operators 68
How to use the IN phrase 70
How to use the BETWEEN phrase 72
How to use the LIKE operator 74
How to use the REGEXP_LIKE function 74
How to use the IS NULL condition 76
How to code the ORDER BY clause 78
How to sort by a column name 78
How to sort by an alias, an expression, or a column number 80
Two more skills 82
How to use the row limiting clause 82
How to test expressions 84
Perspective 86

An introduction to the SELECT statement

This chapter starts by presenting the basic syntax of the SELECT statement. Then, it presents several examples that will give you an idea of what you can do with this useful statement.

How to interpret syntax in this book

Over the course of reading this book, you'll be introduced to many new SQL statements. In most cases, when the book introduces a statement, it will show the statement's *syntax*. In programming, syntax refers to how to structure a statement or piece of code so that it runs the way you want.

The table at the top of this figure lists the syntax conventions used in this book. Capitalized words are *keywords* that you must type exactly as shown. In contrast, lowercase words are used as placeholders, and should be replaced as appropriate.

Beyond that, you can omit clauses enclosed in brackets ([ ]), and you can choose between items enclosed in braces ({ }) and separated by pipes (|). If you have a choice between two or more items, the default item is underlined. And if an element can be coded multiple times in a statement, it's followed by an ellipsis (...).

The basic syntax of the SELECT statement

This syntax summary has been simplified so you can focus on four main clauses of the SELECT statement: SELECT, FROM, WHERE, and ORDER BY. Most SELECT statements contain all of these clauses. However, only the SELECT and FROM clauses are required, and as of Oracle 23ai, the FROM clause can also be omitted.

The SELECT clause is always the first clause in a SELECT statement. It identifies the columns in the result set. These columns are retrieved from one or more *base tables* named in the FROM clause.

The WHERE and ORDER BY clauses are optional. The WHERE clause determines which rows in the base tables are included in the result set by specifying a *search condition* that's used to *filter* the rows in the base tables. When the search condition evaluates to True, the row is included in the result set. If the WHERE clause is omitted, all rows are included in the result set.

The ORDER BY clause determines how the rows in the result set are sorted. If omitted, the result set will use the default order for rows.

The syntax used in this book

Syntax	Meaning
`KEYWORD`	Keywords have each letter capitalized.
`element_name`	Words in lowercase should be replaced with the appropriate element, such as a list of columns or the name of a table.
`{A|B|C}`	Choose only one of the elements in the brackets separated by the pipes.
`[A]`	The element in brackets is optional.
<u>`DEFAULT`</u>	Default elements are underlined.
`...`	More elements can be added to the clause.

The basic syntax of the SELECT statement

```
SELECT select_list
[FROM table_source]
[WHERE search_condition]
[ORDER BY order_by_list]
```

Four clauses of a basic SELECT statement

Clause	Description
`SELECT`	Selects the columns in the result set.
`FROM`	Names the base table from which the query retrieves the data.
`WHERE`	Specifies the condition for a row to be included in the result set.
`ORDER BY`	Specifies how to sort the rows in the result set.

Description

- The SELECT statement retrieves the columns specified in the SELECT clause from the *base tables* specified in the FROM clause and stores them in a result set.
- The WHERE clause is used to *filter* the rows in the base tables so that only those rows that match the search condition are included in the result set.
- The *search condition* of a WHERE clause consists of one or more Boolean expressions that result in a True, False, or null value.
- If you omit the WHERE clause, all of the rows in the base tables are included.
- If you include the ORDER BY clause, the rows in the result set are sorted in the specified sequence.
- You must code the clauses in the order shown or you'll get a syntax error.

Notes

- As of Oracle 23ai, the FROM clause is no longer required in SELECT statements.
- The syntax shown above only includes the most common clauses of the SELECT statement. You'll learn about other clauses later in this book.

Figure 3-1 The basic syntax of the SELECT statement

SELECT statement examples

Figure 3-2 presents five SELECT statement examples. All of these statements retrieve data from the Invoices table in the AP schema. After each statement, you can see its result set as displayed by SQL Developer. In these examples, a horizontal or vertical scroll bar indicates that the result set contains more columns or rows than can be displayed at one time.

The first statement in this figure retrieves all of the columns and rows from the Invoices table. Here, an asterisk (*) indicates that all of the columns should be retrieved, and the WHERE clause is omitted so all of the rows in the table are retrieved. In addition, this statement doesn't include an ORDER BY clause, so the rows are in their default sequence.

The second statement retrieves selected columns from the Invoices table. These columns are listed in the SELECT clause. Like the first statement, this statement doesn't include a WHERE clause, so all the rows are retrieved. Then, the ORDER BY clause causes the rows to be sorted by the invoice_total column in descending order, from largest to smallest.

The third statement also lists the columns to be retrieved. In this case, though, the last column is calculated from two columns in the base table, credit_total and payment_total, and the resulting column is given the name total_credits. In addition, the WHERE clause specifies that only the invoice whose invoice_id column has a value of 17 should be retrieved.

The fourth SELECT statement includes a WHERE clause whose condition specifies a range of values. In this case, only invoices with invoice dates between 05/01/2024 and 05/31/2024 are retrieved. In addition, the rows in the result set are sorted by invoice date.

The last statement in this figure shows another example of the WHERE clause. In this case, only those rows with invoice totals greater than 50,000 are retrieved. Since none of the rows in the Invoices table satisfy this condition, the result set is empty.

Retrieve all the data from the Invoices table

```
SELECT * FROM invoices
```

	INVOICE_ID	VENDOR_ID	INVOICE_NUMBER	INVOICE_DATE	INVOICE_TOTAL	PAYMENT_TOTAL	CREDIT_
1	1	34	QP58872	25-FEB-24	116.54	116.54	
2	2	34	Q545443	14-MAR-24	1083.58	1083.58	
3	3	110	P-0608	11-APR-24	20551.18	0	
4	4	110	P-0259	16-APR-24	26881.4	26881.4	

```
(114 rows)
```

Retrieve three columns and sort the rows by invoice total

```
SELECT invoice_number, invoice_date, invoice_total
FROM invoices
ORDER BY invoice_total DESC
```

	INVOICE_NUMBER	INVOICE_DATE	INVOICE_TOTAL
1	0-2058	08-MAY-24	37966.19
2	P-0259	16-APR-24	26881.4
3	0-2060	08-MAY-24	23517.58
4	40318	18-JUL-24	21842

```
(114 rows)
```

Retrieve two columns and a calculated value for one row

```
SELECT invoice_id, invoice_total,
       (credit_total + payment_total) AS total_credits
FROM invoices
WHERE invoice_id = 17
```

	INVOICE_ID	INVOICE_TOTAL	TOTAL_CREDITS
1	17	356.48	356.48

```
(1 row)
```

Retrieve three columns for rows between given dates

```
SELECT invoice_number, invoice_date, invoice_total
FROM invoices
WHERE invoice_date BETWEEN '01-MAY-2024' AND '31-MAY-2024'
ORDER BY invoice_date
```

	INVOICE_NUMBER	INVOICE_DATE	INVOICE_TOTAL
1	7548906-20	01-MAY-24	27
2	4-321-2596	01-MAY-24	10
3	4-327-7357	01-MAY-24	162.75
4	4-342-8069	01-MAY-24	10

```
(70 rows)
```

Retrieve three columns with invoice totals greater than 50,000

```
SELECT invoice_number, invoice_date, invoice_total
FROM invoices
WHERE invoice_total > 50000
```

INVOICE_...	INVOICE_...	INVOICE_...

```
(0 rows)
```

Figure 3-2 SELECT statement examples

How to code the SELECT clause

Now that you have a general idea of how the main clauses of a SELECT statement work, you're ready to learn the details for coding the first clause, the SELECT clause. You can use this clause to specify the columns for a result set.

How to code column specifications

Figure 3-3 begins by presenting a more detailed syntax for the SELECT clause. Then, it continues by summarizing five techniques you can use to specify the columns for a result set. First, you can code an asterisk in the SELECT clause to retrieve all of the columns in the base table. When you use this technique, Oracle returns the columns in the order that they are defined in the base table.

Second, you can code a list of column names from the base table separated by commas. In this figure, for instance, the second example specifies three columns that are in the Vendors table. When you use this technique, Oracle returns the columns in the order you list them in the SELECT clause.

Third, you can code an *expression* that concatenates strings. The result of an expression is a single value. The third example demonstrates this by combining first and last name columns with a space between to create a new column containing full names.

Fourth, you can code an expression that uses arithmetic operators. In this figure, for instance, the fourth example uses an expression to subtract the payment_total and credit_total columns from the invoice_total column and return the balance due.

Finally, you can code an expression that uses scalar functions. Like an expression, a scalar function returns a single value. For example, the last example in this figure returns the average invoice total.

When you code the SELECT clause, you should include only the columns you need. For example, you shouldn't code an asterisk to retrieve all the columns unless you need all the columns. That's because the amount of data that's retrieved can affect system performance. This is particularly important if you're developing SQL statements that will be used by applications.

The expanded syntax of the SELECT clause

```
SELECT [ALL|DISTINCT]
    column_specification [[AS] result_column]
    [, column_specification [[AS] result_column]] ...
```

Five ways to code column specifications

Source	Option	Syntax
Base table value	All columns	*
	Column name	column_name
Calculated value	Result of a concatenation	String expressions
	Result of a calculation	Arithmetic expressions
	Result of a scalar function	Scalar functions

Column specifications that use base table values

Retrieve all columns

```
SELECT *
```

Retrieve specific columns

```
SELECT vendor_name, vendor_city, vendor_state
```

Column specifications that use calculated values

A string expression that returns the full name

```
SELECT vendor_contact_first_name || ' ' || vendor_contact_last_name
    AS full_name
```

An arithmetic expression that calculates the balance due

```
SELECT invoice_total - payment_total - credit_total AS balance_due
```

A scalar function that calculates the average invoice total

```
SELECT AVG(invoice_total) AS average_total
```

Description

- Use SELECT * only when you need to retrieve all columns from a table. Otherwise, list the names of the columns you need.
- An *expression* is a combination of column names and operators that evaluate to a single value. In the SELECT clause, you can code string expressions, arithmetic expressions, and expressions that include one or more functions.
- After each column specification, you can code an AS clause to specify the name for the column in the result set.

Note

- The ALL and DISTINCT keywords specify whether or not duplicate rows are returned.

Figure 3-3 How to code column specifications

How to name the columns in a result set

By default, a column in a result set is given the same name as the column in the base table. However, you can specify a different name if you want to. You can also name a column that contains a calculated value. When you do that, the new column name is called a *column alias*. Figure 3-4 presents two techniques for creating column aliases.

The first technique is to code the column specification followed by the AS keyword and the column alias. This is illustrated by the first example in this figure.

The second technique is to code the column specification followed by a space and the column alias. This coding technique is illustrated by the second example. Whenever possible, though, you should use the first technique since the AS keyword makes it easier to identify the alias for the column, which makes your SQL statement overall easier to read and maintain.

When you code an alias, you must enclose the alias in double quotes if the alias contains a space or is a keyword that's reserved by Oracle. In this figure, the first two examples specify an alias for the invoice_number column that uses two words with a space between them.

In addition, these two examples specify an alias for the invoice_date column that uses a keyword that's reserved by Oracle: the DATE keyword. If you don't enclose this keyword in double quotes, you will get an error when you attempt to run either of these SQL statements. When you enter a statement into SQL Developer, keywords reserved by Oracle are colored blue. This makes it easy to identify Oracle keywords when you're writing SQL statements.

When you enclose an alias in double quotes, the result set uses the capitalization specified by the alias. Otherwise, the result set capitalizes all letters in the column name. In this figure, for instance, the first two columns in the first result set use the capitalization specified by the aliases. However, since no alias is specified for the third column, all letters in the name of this column are capitalized.

When you code a column that contains a calculated value, it's a good practice to specify an alias for the calculated column. If you don't, Oracle will assign the entire calculation as the name, which can be unwieldy, as shown in the third example. As a result, you usually assign a name to any column that's calculated from other columns in the base table.

Name the columns in the result set

Using the AS keyword

```
-- DATE is a reserved keyword.
-- As a result, it must be enclosed in quotations.
SELECT invoice_number AS "Invoice Number", invoice_date AS "Date",
    invoice_total AS total
FROM invoices
```

Omitting the AS keyword

```
SELECT invoice_number "Invoice Number", invoice_date "Date",
    invoice_total total
FROM invoices
```

The result set for both

	Invoice Number	Date	TOTAL
1	QP58872	25-FEB-24	116.54
2	Q545443	14-MAR-24	1083.58
3	P-0608	11-APR-24	20551.18
4	P-0259	16-APR-24	26881.4

A calculated column that doesn't have an alias

```
SELECT invoice_number, invoice_date, invoice_total,
    invoice_total - payment_total - credit_total
FROM invoices
```

	INVOICE_NUMBER	INVOICE_DATE	INVOICE_TOTAL	INVOICE_TOTAL-PAYMENT_TOTAL-CREDIT_TOTAL
1	QP58872	25-FEB-24	116.54	0
2	Q545443	14-MAR-24	1083.58	0
3	P-0608	11-APR-24	20551.18	19351.18
4	P-0259	16-APR-24	26881.4	0

Description

- By default, a column in the result set is given the same name as the column in the base table. If that's not what you want, you can specify a substitute name, or *column alias*, for the column.
- To specify an alias for a column, use the AS phrase. Although the AS keyword is optional, it enhances readability.
- If an alias includes spaces or an Oracle reserved keyword, you must enclose it in double quotes.
- When you enclose an alias in quotes, the result set uses the capitalization specified by the alias. Otherwise, the result set capitalizes all letters in the column name.
- If you don't specify an alias for a column that's based on a calculated value, Oracle uses the expression for the calculated value as the column name.

Figure 3-4 How to name the columns in a result set

How to code string expressions

A *string expression* consists of a combination of one or more character columns and *literal values*. A literal value is a constant value that isn't calculated or stored in a column. A literal can be any data type, including strings and numbers. For example, 'Hello' is a literal value for a string, and 58 is a literal value for a number.

To combine, or *concatenate*, columns and values, you use the *concatenation operator* (||). This is illustrated by the examples in figure 3-5.

The first example shows how to concatenate the vendor_city and vendor_state columns in the Vendors table. Because no alias is assigned to this column, Oracle assigns a name, which is the entire expression.

The second example shows how to format a string expression by adding spaces and punctuation. Here, the vendor_city column is concatenated with a *string literal*, or *string constant*, that contains a comma and a space. Then, the vendor_state column is concatenated with that result, followed by a string literal that contains a single space and the vendor_zip_code column.

Occasionally you may need to include a single quotation mark or an apostrophe within a literal string. If you type a single quote, however, the system will misinterpret it as the end of the literal string. As a result, you must code two single quotation marks in a row. This is illustrated by the third example in this figure.

Concatenate string data

```
SELECT vendor_city, vendor_state, vendor_city || vendor_state
FROM vendors
```

	VENDOR_CITY	VENDOR_STATE	VENDOR_CITY||VENDOR_STATE
1	Madison	WI	MadisonWI
2	Washington	DC	WashingtonDC
3	Washington	DC	WashingtonDC
4	Los Angeles	CA	Los AngelesCA

Format string data using literal values

```
SELECT vendor_name,
    vendor_city || ', ' || vendor_state || ' ' || vendor_zip_code AS address
FROM vendors
```

	VENDOR_NAME	ADDRESS
1	US Postal Service	Madison, WI 53707
2	National Information Data Ctr	Washington, DC 20090
3	Register of Copyrights	Washington, DC 20559
4	Jobtrak	Los Angeles, CA 90025

Include apostrophes in literal values

```
SELECT vendor_name || '''s address: ',
    vendor_city || ', ' || vendor_state || ' ' || vendor_zip_code
FROM vendors
```

	VENDOR_NAME||'''SADDRESS:'	VENDOR_CITY||','||VENDOR_STATE||''||VENDOR_ZIP_CODE
1	US Postal Service's address:	Madison, WI 53707
2	National Information Data Ctr's address:	Washington, DC 20090
3	Register of Copyrights's address:	Washington, DC 20559
4	Jobtrak's address:	Los Angeles, CA 90025

Description

- A *string expression* can consist of one or more character columns, one or more *literal values*, or a combination of character columns and literal values.
- The literal values in a string expression can be called *string literals* or *string constants*. To create a literal value, enclose one or more characters within single quotation marks (').
- To combine, or *concatenate*, columns and literals in a string expression, you can use the *concatenation operator* (||).
- You can include a single quotation mark within a literal value by coding two single quotation marks in a row.

Figure 3-5 How to code string expressions

How to code arithmetic expressions

Figure 3-6 shows how to code *arithmetic expressions*. To start, it summarizes the four *arithmetic operators* you can use in this type of expression. Then, it presents two examples that show how to use these operators.

The SELECT statement in the first example includes an arithmetic expression that calculates the balance due for an invoice. This expression subtracts the payment_total and credit_total columns from the invoice_total column. The resulting column is given the name balance_due.

When Oracle evaluates an arithmetic expression, it performs the operations from left to right based on the *order of precedence*. This order says that multiplication and division are done first, followed by addition and subtraction. If that's not what you want, you can use parentheses to specify how you want an expression evaluated. Then, the expressions in the innermost sets of parentheses are evaluated first, followed by the expressions in outer sets of parentheses. Within each set of parentheses, the expression is evaluated from left to right in the order of precedence.

To illustrate how parentheses and the order of precedence affect the evaluation of an expression, consider the second example in this figure. Here, the expressions in the second and third columns both use the same operators. However, when Oracle evaluates the expression in the second column it performs the multiplication operation before the addition operation because multiplication comes before addition in the order of precedence.

In contrast, when Oracle evaluates the expression in the third column, it performs the addition operation first because it's enclosed in parentheses. As you can see in the result set, these two expressions result in different values.

Unlike some other databases, Oracle doesn't provide a modulo operator that can be used to return the remainder of a division operation. Instead, you must use the MOD function.

The arithmetic operators in order of precedence

*	Multiplication
/	Division
+	Addition
-	Subtraction

Calculate the balance due

```
SELECT invoice_total, payment_total, credit_total,
    invoice_total - payment_total - credit_total AS balance_due
FROM invoices
```

	INVOICE_TOTAL	PAYMENT_TOTAL	CREDIT_TOTAL	BALANCE_DUE
1	116.54	116.54	0	0
2	1083.58	1083.58	0	0
3	20551.18	0	1200	19351.18
4	26881.4	26881.4	0	0

Use parentheses to control the sequence of operations

```
SELECT invoice_id,
    invoice_id + 7 * 3 AS order_of_precedence,
    (invoice_id + 7) * 3 AS add_first
FROM invoices
ORDER BY invoice_id
```

	INVOICE_ID	ORDER_OF_PRECEDENCE	ADD_FIRST
1	1	22	24
2	2	23	27
3	3	24	30
4	4	25	33
5	5	26	36
6	6	27	39

Description

- The operations in an expression take place from left to right in the *order of precedence*. For *arithmetic expressions*, multiplication and division are done first, followed by addition and subtraction.
- Whenever necessary, you can use parentheses to clarify or override the sequence of operations. Then, the operations in the innermost sets of parentheses are done first, followed by the operations in the next sets, and so on.

Figure 3-6 How to code arithmetic expressions

How to use scalar functions

Figure 3-7 introduces *scalar functions*, which operate on a single value and return a single value. These functions work differently than aggregate functions, which are used to summarize data. For now, don't worry about the details of how these functions work. You'll learn more about them throughout this book. Instead, just focus on how they're used in column specifications.

To code a function, you begin by entering its name followed by a set of parentheses. Within the parentheses, you code any column names or other information required by the function, separated by commas. The values given to the function this way are called *parameters* (or *arguments*).

The first example in this figure shows how to use the SUBSTR function (short for "substring") to extract the first character of two different columns. The first parameter of this function specifies the column name, the second parameter specifies the starting position, and the third parameter specifies the number of characters to return. The results of the two functions are then concatenated to form initials as shown in the result set for this statement.

The second example shows how to use the TO_CHAR function. This function converts a column with a DATE or NUMBER data type to a character string. A common use for it is in concatenation operations in which you want to specify a certain date or number format. This function has two parameters. The first parameter, which specifies the column name, is required. The second parameter, which specifies a format mask for the column, is optional. In this example, a format mask of 'MM/DD/YYYY' is used to convert the payment_date column from a DATE type to a CHAR type with the specified format.

The third example uses the SYSDATE function to return the current date. Since this function doesn't accept any parameters, you don't need to code parentheses after the name of the function. In fact, if you do code parentheses after the name of the function, you will get an error when you run the statement.

In this example, the second column uses the SYSDATE function to return the current date, and the third column uses the SYSDATE function to calculate the number of days between the two dates. Here, the third column also uses the ROUND function to round the decimal value that's returned by the calculation to an integer.

The fourth example shows how to use the MOD function to return the remainder of a division of two integers. Here, the second column contains an expression that returns the remainder of the division operation when the invoice_id column is divided by 10. In the result set, you can see the results for IDs 8 through 11 (the remainders are 8, 9, 0, and 1).

A function that returns a substring

```
SELECT vendor_contact_first_name, vendor_contact_last_name,
    SUBSTR(vendor_contact_first_name, 1, 1) ||
    SUBSTR(vendor_contact_last_name, 1, 1) AS initials
FROM vendors
```

	VENDOR_CONTACT_FIRST_NAME	VENDOR_CONTACT_LAST_NAME	INITIALS
1	Francesco	Alberto	FA
2	Ania	Irvin	AI
3	Lukas	Liana	LL
4	Kenzie	Quinn	KQ

A function that converts a number to text

```
SELECT 'Invoice: # ' || invoice_number || ', dated ' ||
    TO_CHAR(payment_date, 'MM/DD/YYYY') ||
    ' for $' || TO_CHAR(payment_total) AS "Invoice Text"
FROM invoices
```

	Invoice Text
1	Invoice: # QP58872, dated 04/11/2024 for $116.54
2	Invoice: # Q545443, dated 05/14/2024 for $1083.58
3	Invoice: # P-0608, dated for $0
4	Invoice: # P-0259, dated 05/12/2024 for $26881.4

Functions that generate a timestamp and round a number

```
SELECT invoice_date,
    SYSDATE AS today,
    ROUND(SYSDATE - invoice_date)  AS invoice_age_in_days
FROM invoices
```

	INVOICE_DATE	TODAY	INVOICE_AGE_IN_DAYS
1	18-JUL-24	13-MAY-24	-66
2	20-JUN-24	13-MAY-24	-38
3	14-JUN-24	13-MAY-24	-32
4	11-JUN-24	13-MAY-24	-29

A function that gets a remainder (modulus)

```
SELECT invoice_id,
    MOD(invoice_id, 10) AS Remainder
FROM invoices
ORDER BY invoice_id
```

	INVOICE_ID	REMAINDER
8	8	8
9	9	9
10	10	0
11	11	1

Description

- SQL statements can include *functions*, which perform an operation and return a value.
- You'll learn more about functions throughout this book.

Figure 3-7 How to use scalar functions

How to use DISTINCT to eliminate duplicate rows

By default, all of the rows in the base table that satisfy the search condition in the WHERE clause are included in the result set. In some cases, though, that means that the result set will contain duplicate rows, or rows with identical column values. If that's not what you want, you can include the DISTINCT keyword in the SELECT clause to eliminate the duplicate rows.

Figure 3-8 illustrates how this works. Here, both SELECT statements retrieve the vendor_city and vendor_state columns from the Vendors table. The first statement, however, doesn't include the DISTINCT keyword. Because multiple vendors can share a city and state, the same city and state pair can appear in the result set multiple times. In the results shown in this figure, for example, you can see that "Anaheim CA" occurs twice. In contrast, the second statement includes the DISTINCT keyword, so each city/state combination is included only once.

Return all rows from a table

```
SELECT vendor_city, vendor_state
FROM vendors
ORDER BY vendor_city
```

	VENDOR_CITY	VENDOR_STATE
1	Anaheim	CA
2	Anaheim	CA
3	Ann Arbor	MI
4	Auburn Hills	MI

(122 rows)

Return only unique rows

```
SELECT DISTINCT vendor_city, vendor_state
FROM vendors
ORDER BY vendor_city
```

	VENDOR_CITY	VENDOR_STATE
1	Anaheim	CA
2	Ann Arbor	MI
3	Auburn Hills	MI
4	Boston	MA

(53 rows)

Description

- The DISTINCT keyword prevents duplicate (identical) rows from being included in the result set.
- The ALL keyword causes all rows matching the search condition to be included in the result set, regardless of whether rows are duplicated.
- To use the DISTINCT or ALL keywords, code them immediately after the SELECT keyword.
- Since ALL is the default, it's common to omit the ALL keyword.

Figure 3-8 How to use DISTINCT to eliminate duplicate rows

How to code the WHERE clause

The WHERE clause in a SELECT statement filters the rows in the base table so that only the rows you need are retrieved. In the topics that follow, you'll learn a variety of ways to code this clause.

How to use the comparison operators

Figure 3-9 shows how to use the *comparison operators* in the search condition of a WHERE clause to compare two expressions. If the result of the comparison is True, the row being tested is included in the query results.

The examples in this figure show how to use the comparison operators. The first WHERE clause uses the equal operator (=) to retrieve only those rows whose vendor_state column has a value of 'IA'. Here, the state code is a string literal so it must be enclosed in single quotes.

In contrast, the second WHERE clause uses the greater than (>) operator to retrieve only those rows that have a balance greater than zero. In this case, 0 is a numeric literal so it isn't enclosed in quotes.

The third WHERE clause illustrates another way to retrieve all the invoices with a balance due. Like the second clause, it uses the greater than operator. Instead of comparing the balance due to a value of zero, however, it compares the invoice total to the total of the payments and credits that have been applied to the invoice.

The fourth WHERE clause illustrates how you can use comparison operators other than the equal operator with string data. In this example, the less than operator (<) is used to compare the value of the vendor_name column to a literal string that has the letter M in the first position. That will cause the query to return all vendors with names that begin with the letters A through L.

You can also use the comparison operators with *date literals*, as illustrated by the fifth and sixth WHERE clauses. The fifth clause retrieves rows with invoice dates on or before May 31, 2024, and the sixth clause retrieves rows with invoice dates on or after May 1, 2024. Like string literals, date literals must be enclosed in single quotes. In addition, you can use different formats to specify dates, as shown by the two date literals shown in this figure.

The last WHERE clause shows how you can test for a not equal condition. To do that, you code a less than sign followed by a greater than sign. In this case, only rows with a credit total that's not equal to zero will be retrieved.

Whenever possible, you should compare expressions that have similar data types. If you attempt to compare expressions that have different data types, Oracle may convert the data types for you. Although these conversions are often acceptable, they will occasionally yield unexpected results. In chapter 8, you'll learn how to explicitly convert data types so your comparisons will always yield the results that you want.

The syntax of the WHERE clause with comparison operators

```
WHERE expression_1 operator expression_2
```

The comparison operators

=	Equal
>	Greater than
<	Less than
<=	Less than or equal to
>=	Greater than or equal to
<>	Not equal

WHERE clauses that retrieve...

Vendors located in Iowa

```
WHERE vendor_state = 'IA'
```

Invoices with a balance due (two variations)

```
WHERE invoice_total - payment_total - credit_total > 0
WHERE invoice_total > payment_total + credit_total
```

Vendors with names from A to L

```
WHERE vendor_name < 'M'
```

Invoices on or before a specified date

```
WHERE invoice_date <= '31-MAY-24'
```

Invoices on or after a specified date

```
WHERE invoice_date >= '1/MAY/2024'
```

Invoices with credits that don't equal zero

```
WHERE credit_total <> 0
```

Description

- You can use a *comparison operator* to compare any two expressions.
- Since Oracle automatically converts the data for comparison, the expressions don't need to have the same data type. However, mismatched data types may cause unexpected results.
- If the result of a comparison results in a True value, the row being tested is included in the result set. If it's False or Unknown, the row isn't included.
- To use a string literal or a *date literal* in a comparison, enclose it in quotes. To use a numeric literal, enter the number without quotes.
- Character comparisons are case-sensitive. 'CA' and 'Ca', for example, are not equivalent.
- If you compare a null value using one of these comparison operators, the result is always a null value.

Figure 3-9 How to use the comparison operators

How to use the AND, OR, and NOT logical operators

Figure 3-10 shows how to use *logical operators* in a WHERE clause. You can use the AND and OR operators to combine two or more search conditions into a *compound condition*. You can use the NOT operator to negate a search condition. The examples in this figure illustrate how these operators work.

The first two examples show the AND and OR operators. When you use the AND operator, both conditions must be true. So, in the first example, only those vendors in the state of New Jersey and the city of Springfield are retrieved from the Vendors table. When you use the OR operator, though, only one of the conditions must be true. So, in the second example, all the vendors in the state of New Jersey and all the vendors in a city named Pittsburgh (no matter what state) are retrieved.

The third example shows a compound condition that uses two NOT operators. As you can see, this expression is difficult to understand. Because of that, you should generally avoid using this operator. The fourth example in this figure, for instance, shows how the search condition in the third example can be rephrased to eliminate the NOT operator. As a result, the condition in the fourth example is much easier to understand.

The last two examples in this figure show how the order of precedence for the logical operators and the use of parentheses affect the result of a search condition. By default, the NOT operator is evaluated first, followed by AND, and then by OR. However, you can use parentheses to override the order of precedence or to clarify a logical expression just as you can with arithmetic expressions.

In the next to last example, for instance, no parentheses are used, so the two conditions connected by the AND operator are evaluated first. In the last example, though, parentheses are used so the two conditions connected by the OR operator are evaluated first.

The syntax of the WHERE clause with logical operators

```
WHERE [NOT] search_condition_1 {AND|OR} [NOT] search_condition_2 ...
```

Logical operators

The AND operator

```
WHERE vendor_state = 'NJ' AND vendor_city = 'Springfield'
```

The OR operator

```
WHERE vendor_state = 'NJ' OR vendor_city = 'Pittsburgh'
```

The NOT operator

```
WHERE NOT (invoice_total >= 5000 OR NOT invoice_date <= '01-JUL-2024')
```

The same condition rephrased to eliminate the NOT operator

```
WHERE invoice_total < 5000 AND invoice_date <= '01-JUL-2024'
```

A compound condition without parentheses

```
SELECT invoice_number, invoice_date, invoice_total
FROM invoices
WHERE invoice_date > '01-MAY-2024' OR invoice_total > 500
AND invoice_total - payment_total - credit_total > 0
ORDER BY invoice_number
```

	INVOICE_NUMBER	INVOICE_DATE	INVOICE_TOTAL
1	0-2058	08-MAY-24	37966.19
2	0-2060	08-MAY-24	23517.58
3	0-2436	07-MAY-24	10976.06

(91 rows)

The same compound condition with parentheses

```
WHERE (invoice_date > '01-MAY-2024' OR invoice_total > 500)
AND invoice_total - payment_total - credit_total > 0
ORDER BY invoice_number
```

	INVOICE_NUMBER	INVOICE_DATE	INVOICE_TOTAL
1	0-2436	07-MAY-24	10976.06
2	109596	14-JUN-24	41.8
3	111-92R-10092	04-JUN-24	46.21

(39 rows)

Description

- The AND and OR *logical operators* create *compound conditions* that consist of two or more conditions. The AND operator specifies that the search must satisfy both conditions. The OR operator specifies that the search must satisfy at least one condition.
- You can use the NOT operator to negate a condition. Because this can make the search condition unclear, you should rephrase the condition if possible so it doesn't use NOT.
- When MySQL evaluates a compound condition, it evaluates the operators in this order: (1) NOT, (2) AND, (3) OR. You can use parentheses to override this order.

Figure 3-10 How to use the AND, OR, and NOT logical operators

How to use the IN phrase

Figure 3-11 shows how to code a WHERE clause that uses the IN operator to create an IN phrase. When you use this phrase, the value of the test expression is compared with the list of expressions in the IN phrase. If the test expression is equal to one of the expressions in the list, the row is included in the query results. This is illustrated by the first example in this figure, which returns all rows whose terms_id column is equal to 1, 3, or 4.

You can also use the NOT operator with the IN phrase to test for a value that's not in a list of expressions. This is illustrated by the second example. In this case, only those vendors who are not in California, Nevada, or Oregon are retrieved.

If you look at the syntax of the IN phrase shown at the top of this figure, you'll see that you can code a *subquery* in place of a list of expressions. A subquery is a query that's coded within another statement. For instance, the third example in this figure uses a subquery to return a list of vendor_id values for vendors who have invoices dated May 1, 2024. Then, the WHERE clause retrieves a vendor row only if the vendor is in that list. For this to work, the subquery must return values from a single column. In this case, it returns values from the vendor_id column.

The syntax of the WHERE clause with an IN phrase

```
WHERE test_expression [NOT] IN ({subquery|expression_1 [, expression_2]...})
```

The IN phrase

With a list of numeric literals

```
WHERE terms_id IN (1, 3, 4)
```

Preceded by NOT

```
WHERE vendor_state NOT IN ('CA', 'NV', 'OR')
```

With a subquery

```
WHERE vendor_id IN
    (SELECT vendor_id
     FROM invoices
     WHERE invoice_date = '01-MAY-2024')
```

Description

- You can use the IN phrase to test whether an expression is equal to a value in a list of expressions. Each expression in the list is automatically converted to the same type of data as the test expression.
- The list of expressions can be coded in any order without affecting the order of the rows in the result set.
- You can use the NOT operator to test for an expression that's not in the list of expressions.
- You can compare the test expression to the items in a list returned by a *subquery*, which is when one query is coded within another query.

Figure 3-11 How to use the IN phrase

How to use the BETWEEN phrase

Figure 3-12 shows how to code a WHERE clause that uses the BETWEEN operator to create a BETWEEN phrase. When you use this phrase, the value of a test expression is compared to the range of values specified in the BETWEEN phrase. If the value falls within this range, the row is included in the query results.

The first example in this figure shows a simple WHERE clause that uses the BETWEEN operator. It retrieves invoices with invoice dates between May 1, 2024 and May 31, 2024. Note that the range is inclusive, so invoices with invoice dates of May 1st and May 31st are included in the results.

The second example shows how to use the NOT operator to select rows that are not within a given range. In this case, vendors with zip codes that aren't between 93600 and 93799 are included in the results.

The third example shows how you can use a calculated value in the test expression. Here, the payment_total and credit_total columns are subtracted from the invoice_total column to give the balance due. Then, this value is compared to the range specified in the BETWEEN phrase.

The last example shows how you can use calculated values in the BETWEEN phrase. Here, the first value selects the function SYSDATE (which retrieves the current date), and the second value is SYSDATE plus 30 days. So the query results will include all those invoices that are due between the current date and 30 days from the current date.

The syntax of the WHERE clause with a BETWEEN phrase

```
WHERE test_expression [NOT] BETWEEN begin_expression AND end_expression
```

The BETWEEN phrase

With literal values

```
WHERE invoice_date BETWEEN '01-MAY-2024' AND '31-MAY-2024'
```

Preceded by NOT

```
WHERE vendor_zip_code NOT BETWEEN 93600 AND 93799
```

With a test expression coded as a calculated value

```
WHERE invoice_total - payment_total - credit_total BETWEEN 200 AND 500
```

With the upper and lower limits coded as calculated values

```
WHERE invoice_due_date BETWEEN SYSDATE AND (SYSDATE + 30)
```

Description

- You can use the BETWEEN phrase to test whether an expression falls within a range of values.
- The lower limit must be coded as the first expression and the upper limit must be coded as the second expression. Otherwise, the result set will be empty.
- The two expressions used in the BETWEEN phrase for the range of values are inclusive. That is, the result set will include values that are equal to the lower or upper limit.
- You can use the NOT operator to test for an expression that's not within the given range.

Figure 3-12 How to use the BETWEEN phrase

How to use the LIKE operator

To retrieve rows that match a specific *string pattern*, or *mask*, you can use the LIKE operator or REGEXP_LIKE function as shown in figure 3-13. The LIKE operator lets you search for simple string patterns. When you use this operator, the mask can contain the *wildcard* symbols shown in the first table in this figure.

In the first example, the LIKE phrase specifies that all vendors in cities that start with the letters "San" should be included in the query results. Here, the percent sign (%) indicates that any characters can follow these three letters. So San Diego and Santa Ana are both included in the results.

The second example selects all vendors whose vendor name starts with the letters "Compu", followed by any one character, the letters "er", and any characters after that. Two vendor names that match that pattern are Compuserve and Computerworld.

How to use the REGEXP_LIKE function

In contrast to the LIKE operator, the REGEXP_LIKE function allows you to create complex string patterns known as *regular expressions*. To do that, you can use the special characters and constructs shown in the second table in this figure. Although creating regular expressions can be tricky at first, they allow you to search for virtually any string pattern.

In the third example, the REGEXP_LIKE phrase searches for the letters "sa" within the vendor_city column. Since the letters can be in any position within the string but must both be lowercase, Pasadena is included in the results but Santa Ana is not.

The next two examples demonstrate how to use REGEXP_LIKE to match a pattern to the beginning or end of the string being tested. In the fourth example, the mask ^Sa matches the letters "Sa" at the beginning of vendor_city, as in Santa Ana and Sacramento. In contrast, the mask na$ matches the letters "na" at the end of vendor_city, as shown in the fifth example.

The sixth example uses the pipe (|) character to search for either of two string patterns: "rs" or "sn". In this case, the first pattern would match Traverse City and the second would match Fresno, so both are included in the result set.

The last three examples use brackets to specify multiple values. In the seventh example, the vendor_state column is searched for values that contain the letter N followed by either C or V. That excludes NJ and NY. In contrast, the eighth example searches for states that contain the letter N followed by any letter from A to J. This excludes NV and NY.

In the final example, the REGEXP_LIKE function searches for a vendor_city that ends with any lowercase letter, a vowel, and then the letter "n".

Both the LIKE and REGEXP operators provide powerful functionality for finding information in a database. However, searches that use these operators sometimes run slowly since they can't use a table's indexes. As a result, you should only use these operators when necessary.

The syntax of the WHERE clause with the LIKE operator

```
WHERE match_expression [NOT] LIKE pattern
```

The syntax of the WHERE clause with a REGEXP_LIKE function

```
WHERE [NOT] REGEXP_LIKE(match_expression, pattern)
```

LIKE wildcards

Symbol	Description
%	Matches any string of zero or more characters.
_	Matches any single character.

REGEXP_LIKE special characters and constructs

<table>
<tr><th>Character/Construct</th><th>Description</th></tr>
<tr><td>^</td><td>Matches the pattern to the beginning of the value being tested.</td></tr>
<tr><td>$</td><td>Matches the pattern to the end of the value being tested.</td></tr>
<tr><td>.</td><td>Matches any single character.</td></tr>
<tr><td>[charlist]</td><td>Matches any single character listed within the brackets.</td></tr>
<tr><td>[char1-char2]</td><td>Matches any single character within the given range.</td></tr>
<tr><td>|</td><td>Separates two string patterns and matches either one.</td></tr>
</table>

WHERE clauses that use the LIKE operator and REGEXP_LIKE function

<table>
<tr><th>Example</th><th>Results that match the mask</th></tr>
<tr><td>WHERE vendor_city LIKE 'San%'</td><td>“San Diego”, “Santa Ana”</td></tr>
<tr><td>WHERE vendor_name LIKE 'Compu_er%'</td><td>“Compuserve”, “Computerworld”</td></tr>
<tr><td>WHERE REGEXP_LIKE(vendor_city, 'sa')</td><td>“Pasadena”</td></tr>
<tr><td>WHERE REGEXP_LIKE(vendor_city, '^Sa')</td><td>“Santa Ana”, “Sacramento”</td></tr>
<tr><td>WHERE REGEXP_LIKE(vendor_city, 'na$')</td><td>“Gardena”, “Pasadena”, “Santa Ana”</td></tr>
<tr><td>WHERE REGEXP_LIKE(vendor_city, 'rs|sn')</td><td>“Traverse City”, “Fresno”</td></tr>
<tr><td>WHERE REGEXP_LIKE(vendor_state, 'N[CV]')</td><td>“NC” and “NV” but not “NJ” or “NY”</td></tr>
<tr><td>WHERE REGEXP_LIKE(vendor_state, 'N[A-J]')</td><td>“NC” and “NJ” but not “NV” or “NY”</td></tr>
<tr><td>WHERE REGEXP_LIKE(vendor_city, '[a-z][aeiou]n$')</td><td>“Boston”, “Mclean”, “Oberlin”</td></tr>
</table>

Description

- You use the LIKE operator and REGEXP_LIKE function to retrieve rows that match a *string pattern*, called a *mask*. The mask determines which values satisfy the condition.
- The mask for a LIKE phrase can contain special symbols, called *wildcards*. The mask for a REGEXP_LIKE phrase can contain special characters and constructs.
- In Oracle Database, masks are case-sensitive.
- If you use the NOT keyword, only those rows with values that don't match the string pattern are included in the result set.

Figure 3-13 How to use the LIKE operator and REGEXP_LIKE function

For the sake of brevity, this chapter only presents the most common symbols that are used in regular expressions. However, Oracle supports most of the symbols that are standard for creating regular expressions.

How to use the IS NULL condition

In chapter 1, you learned that a column can contain a *null value*. A null value is typically used to indicate that a value is not known. A null value is not the same as an empty string (''). An empty string is typically used to indicate that the value is known, and it doesn't exist.

If you're working with a table that allows null values, you need to know how to test for them in search conditions. To do that, you use the IS NULL clause as shown in figure 3-14.

This figure uses a table named Null_Sample from the EX schema to show how to search for null values. This table contains two columns: invoice_id and invoice_total. The values in this table are displayed in the first example.

The second example shows what happens when you retrieve all the rows where invoice_total is equal to zero. In this case, the row that has a null value isn't included in the result set. And as the third example shows, this row isn't included in the result set where invoice_total isn't equal to zero either. Instead, you have to use the IS NULL clause to retrieve rows with null values as shown by the fourth example.

You can also use the NOT operator with the IS NULL clause as shown by the last example. When you use this operator, all of the rows that don't contain null values are included in the query results.

The syntax of the WHERE clause with the IS NULL condition

```
WHERE expression IS [NOT] NULL
```

The Null_Sample table

```
SELECT * FROM null_sample
```

	INVOICE_ID	INVOICE_TOTAL
1	1	125
2	2	0
3	3	(null)
4	4	2199.99
5	5	0

Retrieve rows with zero values

```
SELECT * FROM null_sample
WHERE invoice_total = 0
```

	INVOICE_ID	INVOICE_TOTAL
1	2	0
2	5	0

Retrieve rows with non-zero values

```
SELECT * FROM null_sample
WHERE invoice_total <> 0
```

	INVOICE_ID	INVOICE_TOTAL
1	1	125
2	4	2199.99

Retrieve rows with null values

```
SELECT * FROM null_sample
WHERE invoice_total IS NULL
```

	INVOICE_ID	INVOICE_TOTAL
1	3	(null)

Retrieve rows without null values

```
SELECT * FROM null_sample
WHERE invoice_total IS NOT NULL
```

	INVOI...	INVOICE_TOTAL
1	1	125
2	2	0
3	4	2199.99
4	5	0

Description

- A *null value* represents a value that's unknown, unavailable, or not applicable. It isn't the same as a zero or an empty string ('').

Figure 3-14 How to use the IS NULL condition

How to code the ORDER BY clause

The ORDER BY clause specifies the sort order for the rows in a result set. In most cases, you can use column names from the base table to specify the sort order as shown by some of the examples earlier in this chapter. However, you can also use other techniques to sort the rows in a result set as shown in the figures that follow.

How to sort by a column name

Figure 3-15 presents the expanded syntax of the ORDER BY clause. This syntax shows that you can sort by one or more expressions in either ascending or descending sequence. The three examples in this figure show how to code this clause for expressions that involve column names.

The first two examples show how to sort the rows in a result set by a single column. In the first example, the rows in the Vendors table are sorted in ascending sequence by the vendor_name column. Since ascending is the default sequence, the ASC keyword can be omitted. In the second example, the rows are sorted by the vendor_name column in descending sequence.

To sort by more than one column, you list the names in the ORDER BY clause separated by commas as shown in the third example. Here, the rows in the Vendors table are first sorted by the vendor_state column in ascending sequence. Then, within each state, the rows are sorted by the vendor_city column in ascending sequence. Finally, within each city, the rows are sorted by the vendor_name column in ascending sequence.

Although all of the columns in this example are sorted in ascending sequence, that doesn't have to be the case. For example, this example could have sorted by the vendor_name column in descending sequence like this:

```
ORDER BY vendor_state, vendor_city, vendor_name DESC
```

In this example, the DESC keyword applies only to the vendor_name column. The vendor_state and vendor_city columns are still sorted in ascending sequence.

If you study the first example in this figure, you can see that capital letters come before lowercase letters in an ascending sort. As a result, "ASC Signs" comes before "Abbey Office Furnishings" in the result set. For some business applications, this is acceptable. But if it isn't, you can use the LOWER function to convert the column to lowercase letters in the ORDER BY clause like this:

```
ORDER BY LOWER(vendor_name)
```

Then, the rows will be sorted in the correct alphabetical sequence.

The expanded syntax of the ORDER BY clause

```
ORDER BY expression [ASC|DESC] [, expression [ASC|DESC]] ...
```

Sort by one column in ascending sequence

```
SELECT vendor_name,
    vendor_city || ', ' || vendor_state || ' ' || vendor_zip_code AS address
FROM vendors
ORDER BY vendor_name
```

	VENDOR_NAME	ADDRESS
1	ASC Signs	Fresno, CA 93703
2	AT&T	Phoenix, AZ 85062
3	Abbey Office Furnishings	Fresno, CA 93722

Sort by one column in descending sequence

```
SELECT vendor_name,
    vendor_city || ', ' || vendor_state || ' ' || vendor_zip_code AS address
FROM vendors
ORDER BY vendor_name DESC
```

	VENDOR_NAME	ADDRESS
1	Zylka Design	Fresno, CA 93711
2	Zip Print & Copy Center	Fresno, CA 93777
3	Zee Medical Service Co	Washington, IA 52353

Sort by three columns

```
SELECT vendor_name,
    vendor_city || ', ' || vendor_state || ' ' || vendor_zip_code AS address
FROM vendors
ORDER BY vendor_state, vendor_city, vendor_name
```

	VENDOR_NAME	ADDRESS
1	AT&T	Phoenix, AZ 85062
2	Computer Library	Phoenix, AZ 85023
3	Wells Fargo Bank	Phoenix, AZ 85038
4	Aztek Label	Anaheim, CA 92807
5	Blue Shield of California	Anaheim, CA 92850
6	Diversified Printing & Pub	Brea, CA 92621

Description

- The ORDER BY clause specifies how you want the rows in the result set sorted. You can sort by one or more columns, and you can sort each column in either ascending (ASC) or descending (DESC) sequence. ASC is the default.
- By default, in an ascending sort special characters appear first in the sort sequence, followed by numbers, then capital letters, then lowercase letters, and finally null values. In a descending sort, this sequence is reversed.
- With one exception, you can sort by any column in the base table, regardless of whether it's included in the SELECT clause. The exception is if the query includes the DISTINCT keyword. Then, you can only sort by columns included in the SELECT clause.

Figure 3-15 How to sort by a column name

How to sort by an alias, an expression, or a column number

Figure 3-16 presents three more techniques that you can use to sort rows. First, you can use a column alias that's defined in the SELECT clause. The first SELECT statement in this figure, for example, sorts by a column named address, which is an alias for the concatenation of the vendor_city, vendor_state, and vendor_zip_code columns. Notice that besides the address column, the result set is also sorted by the vendor_name column.

You can also use an arithmetic or string expression in the ORDER BY clause, as shown by the second example in this figure. Here, the expression consists of the vendor_contact_last_name column concatenated with the vendor_contact_first_name column. Notice that neither of these columns is included in the SELECT clause.

The last example in this figure shows how you can use column numbers to specify a sort order. To use this technique, you code the number that corresponds to the column of the result set, where 1 is the first column, 2 is the second column, and so on. In this example, the ORDER BY clause sorts the result set by the second column, which contains the concatenated address, then by the first column, which contains the vendor name. As a result, this statement returns the same result set as the first statement.

However, the statement that uses column numbers is more difficult to read because you have to look at the SELECT clause to see what columns the numbers refer to. In addition, if you add or remove columns from the SELECT clause, you may also have to change the ORDER BY clause to reflect the new column positions. As a result, you should avoid using this technique in most situations.

Sort by an alias

```
SELECT vendor_name,
    vendor_city || ', ' || vendor_state || ' ' || vendor_zip_code AS address
FROM vendors
ORDER BY address, vendor_name
```

	VENDOR_NAME	ADDRESS
1	Aztek Label	Anaheim, CA 92807
2	Blue Shield of California	Anaheim, CA 92850
3	Malloy Lithographing Inc	Ann Arbor, MI 48106
4	Data Reproductions Corp	Auburn Hills, MI 48326

Sort by an expression

```
SELECT vendor_name,
    vendor_city || ', ' || vendor_state || ' ' || vendor_zip_code AS address
FROM vendors
ORDER BY vendor_contact_last_name || vendor_contact_first_name
```

	VENDOR_NAME	ADDRESS
1	Dristas Groom & McCormick	Fresno, CA 93720
2	Internal Revenue Service	Fresno, CA 93888
3	US Postal Service	Madison, WI 53707
4	Yale Industrial Trucks-Fresno	Fresno, CA 93706

Sort by column positions

```
SELECT vendor_name,
    vendor_city || ', ' || vendor_state || ' ' || vendor_zip_code AS address
FROM vendors
ORDER BY 2, 1
```

	VENDOR_NAME	ADDRESS
1	Aztek Label	Anaheim, CA 92807
2	Blue Shield of California	Anaheim, CA 92850
3	Malloy Lithographing Inc	Ann Arbor, MI 48106
4	Data Reproductions Corp	Auburn Hills, MI 48326

Description

- The ORDER BY clause can include a column alias that's specified in the SELECT clause. If the alias contains a space, it must be enclosed in double quotes.
- The ORDER BY clause can include any valid expression. The expression can refer to any column in the base table, even if it isn't included in the result set.
- The ORDER BY clause can use numbers to specify the columns to use for sorting. In that case, 1 represents the first column in the result set, 2 represents the second column, and so on.

Figure 3-16 How to sort by an alias, an expression, or a column number

Two more skills

This chapter ends by presenting two more skills that are related to coding the clauses of a SELECT statement. First, it shows how to limit the number of rows that are returned by a query. Second, it shows how to test expressions in the SELECT clause.

How to use the row limiting clause

When you don't need all the rows in a result set, limiting the number of rows speeds up your query and makes the results easier to read. One way to limit the number of rows retrieved by a SELECT statement is to use the *row limiting clause* as shown in figure 3-17. The row limiting clause refers to the combination of the OFFSET and FETCH clauses, but using both clauses isn't required.

The first example shows how to use FETCH to limit the number of rows in a result set to the first three rows. Without the FETCH clause, this statement would return all 114 rows in the table.

The second example shows how to use the OFFSET clause to skip a number of rows before returning a result set. In this example, the OFFSET clause skips the first 15 rows in the table. Then, the FETCH clause gets the next 3 rows. The result set shows that this query gets rows 16 to 18, as you would expect.

To sort a table before returning a limited number of rows, add an ORDER BY clause. In the third example, the statement sorts the rows by invoice total in descending order. Then, it gets the top 10 percent of the rows. Since there are 114 rows in the Invoices table, this returns 12 rows. As a result, these rows represent the top 10% of the largest invoices.

If necessary, you can specify whether the FETCH clause should consider ties when retrieving rows by using WITH TIES instead of the ONLY keyword. WITH TIES only affects the result set if an ORDER BY clause has been included and some rows have the same sort order.

Since the examples in this figure work with multiple rows, they use the ROWS keyword, not the ROW keyword. However, the ROW and ROWS keywords are interchangeable. As a result, you can use the ROW keyword instead if it's more readable. For example, if you wanted to skip a single row and get a single row, you could use the ROW keyword like this:

```
SELECT invoice_id, invoice_total
FROM invoices
OFFSET 1 ROW
FETCH NEXT 1 ROW ONLY;
```

Similarly, the FIRST and NEXT keywords are also interchangeable. However, using the FIRST keyword makes sense if there's no OFFSET clause as shown by the first example. Conversely, using the NEXT keyword makes sense if there is an OFFSET clause, as shown by the second example.

The syntax of the row limiting clause

```
[OFFSET offset {ROW|ROWS}]
[FETCH {FIRST|NEXT} [{rowcount|percent PERCENT}]
    {ROW|ROWS} {ONLY|WITH TIES}]
```

Get the first 3 rows from the Invoices table

```
SELECT invoice_id, invoice_total
FROM invoices
FETCH FIRST 3 ROWS ONLY
```

	INVOICE_ID	INVOICE_TOTAL
1	1	116.54
2	2	1083.58
3	3	20551.18

Skip 15 rows then get the next 3 rows

```
SELECT invoice_id, invoice_total
FROM invoices
OFFSET 15 ROWS
FETCH NEXT 3 ROWS ONLY
```

	INVOICE_ID	INVOICE_TOTAL
1	16	2051.59
2	17	356.48
3	18	904.14

Sort the table before limiting the rows returned to 10% of total

```
SELECT invoice_id, invoice_total
FROM invoices
ORDER BY invoice_total DESC
FETCH FIRST 10 PERCENT ROWS ONLY
```

	INVOICE_ID	INVOICE_TOTAL
1	37	37966.19
2	4	26881.4
3	36	23517.58

(12 rows)

Description

- Together, the OFFSET and FETCH clauses are called the *row limiting clause*. You use it to limit the number of rows returned by a query.
- You can choose to specify a specific number of rows or a percentage of the total number of rows.
- WITH TIES can be used instead of ONLY if you have sorted the table with an ORDER BY clause.

Figure 3-17 How to use the row limiting clause

How to test expressions

Figure 3-18 shows how to test expressions by entering them in the SELECT clause. To do that, you can specify the Dual table in the FROM clause.

The *Dual table* contains just one column named dummy and one row with a value of "X" as shown by the first example. This table is automatically available to all users and is useful for testing expressions that use strings, arithmetic calculations, and functions.

In the second example, the first column displays the result of a string expression, the second column shows the result of an arithmetic calculation (10 minus 7), and the third column shows the date that's returned by the SYSDATE function.

All of the columns in the second example use aliases. However, that's not necessary. For instance, the third example doesn't use aliases. This can be preferable because the result set shows the expression in the column header and its result in the row that's returned.

As of Oracle Database 23ai, the FROM clause is no longer required in SELECT statements. That means you can test expressions by coding queries that only have a SELECT clause as shown in the fourth example. In other words, with 23ai and later, you no longer need to code a FROM clause that specifies the Dual table.

The Dual table

```
SELECT * FROM dual
```

	DUMMY
1	X

Test three expressions using the Dual table

```
SELECT 'test'  AS test_string,
       10-7    AS test_calculation,
       SYSDATE AS test_date
FROM dual
```

	TEST_STRING	TEST_CALCULATION	TEST_DATE
1	test	3	13-MAY-24

The same expressions without using column aliases

```
SELECT 'test', 10-7, SYSDATE
FROM dual
```

	'TEST'	10-7	SYSDATE
1	test	3	13-MAY-24

The same expressions without a FROM clause (23ai and later)

```
SELECT 'test', 10-7, SYSDATE
```

	'TEST'	10-7	SYSDATE
1	test	3	13-MAY-24

Description

- The *Dual table* is automatically created by Oracle Database and contains a single row with a value of "X".
- The Dual table is useful for testing expressions that use strings, arithmetic expressions, and functions.
- As of Oracle Database 23ai, the FROM clause is no longer required in SELECT statements.

Figure 3-18 How to test expressions

Perspective

The goal of this chapter has been to teach you the basic skills for coding SELECT statements. You'll use these skills in almost every SELECT statement you code. However, there's a lot more to coding SELECT statements than what you've learned so far. In the next three chapters, then, you'll learn more skills for coding SELECT statements.

Terms

base table
filter a table
search condition
syntax
keyword
expression
column alias
string expression
literal value
string literal
string constant
concatenation
concatenation operator
order of precedence
arithmetic expression
arithmetic operator
function
scalar function
parameter
argument
comparison operator
date literal
logical operator
compound condition
subquery
string pattern
mask
wildcard
regular expression
null value
row limiting clause
Dual table

Exercises

In these exercises, you'll enter and run your own SELECT statements. Save all five statements in one SQL script.

1. Write a SELECT statement that returns the following columns from the Vendors table: vendor_name, vendor_contact_last_name, and vendor_contact_first_name.

   Add code to this statement so it sorts the result set by last name and then first name. Then, run this statement again. This is a good way to build and test a statement; one clause at a time.

2. Write a SELECT statement that returns one column from the Vendors table named full_name that concatenates the vendor_contact_first_name and vendor_contact_last_name columns. Format this column with the last name followed by a comma, a space, and the first name like this:

   ```
   Doe, John
   ```

   Sort the result set by last name and then first name.

Modify the statement to include a WHERE clause that limits the result set to contacts whose last name begins with the letter A, B, C, or E. To do this, write a compound condition that uses the comparison operators and the LIKE operator.

3. Write a SELECT statement that returns four columns from the Invoices table named Due Date, Invoice Total, 10%, and Plus 10%. These columns should contain this data:

Due Date	The invoice_due_date column
Invoice Total	The invoice_total column
10%	10% of the value of invoice_total
Plus 10%	The value of invoice_total plus 10%

Use the arithmetic operators to calculate the values for the 10% and Plus 10% columns. For example, if invoice_total is 100, the value of 10% is 10, and Plus 10% is 110.

Filter the result set so it returns only those rows with an invoice total that's greater than or equal to $500 and less than or equal to $1,000. Then, sort the result set in descending sequence by invoice_due_date.

4. Write and run a SELECT statement that returns four columns from the Invoices table named Number, Total, Credits, and Balance Due. These columns should include this data:

Number	The invoice_number column
Total	The invoice_total column
Credits	Sum of the payment_total and credit_total columns
Balance Due	Invoice_total minus payment_total and credit_total

Use the arithmetic operators to calculate the values for the Credits and Balance Due columns.

Sort the result set by the balance due in descending sequence.

Use the row limiting clause so the result set contains only the rows with the 10 largest balance dues (no ties).

5. Write a SELECT statement that returns the invoice due date, balance due, and payment date from the Invoices table, but only where the payment_date column contains a null value. Run this statement to verify it works as expected.

4

How to retrieve data from two or more tables

In the last chapter, you learned how to create result sets that contain data from a single table. Now, this chapter shows you how to create result sets that contain data from two or more tables. To do that, you can use either a join or a union.

How to work with inner joins **90**
How to code an inner join 90
When and how to use table aliases 92
How to use compound join conditions 94
How to use a self-join 96
How to join more than two tables 98
How to use the implicit inner join syntax 100
How to work with outer joins **102**
How to code an outer join 102
Outer join examples 104
Outer joins that join more than two tables 106
How to use the implicit outer join syntax 108
Other skills for working with joins **110**
How to combine inner and outer joins 110
How to join tables with the USING keyword 112
How to join tables with the NATURAL keyword 114
How to use cross joins 116
How to work with unions **118**
The syntax of a union 118
Unions that combine data from different tables 118
Unions that combine data from the same table 120
How to use the MINUS and INTERSECT operators 122
Perspective **124**

How to work with inner joins

A *join* lets you combine columns from two or more tables into a single result set. To start, this chapter shows how to code the most common type of join used with relational databases, an inner join.

How to code an inner join

To join data from two tables, you name two tables in the FROM clause of a SELECT statement and specify a *join condition*. A join condition compares two columns to determine how the rows should be joined. In most cases, a join condition is based on the relationship between the primary key of the first table and a foreign key of the second table. When you code an *inner join*, only rows that satisfy the join condition are included in the result set.

For example, the Vendors table in the AP schema contains information about vendors, including each vendor's name. The Invoices table, meanwhile, contains information about individual invoices. It contains a column for the ID of the vendor associated with each invoice, but not any other vendor information.

To get the name of the vendor associated with an invoice, an inner join can be used to generate a result set that contains the invoice_number column from the Invoices table and the corresponding vendor_name column from the Vendors table. The example in figure 4-1 shows how this works.

The join condition in this example compares the vendor_id columns in the Vendors and Invoices tables. Because the equal operator is used in this condition, the value of the vendor_id column in a row in the Vendors table must match the vendor_id in a row in the Invoices table for that row to be included in the result set. In other words, only vendors with one or more invoices will be included. The same is true of the invoices – only invoices with a vendor ID will be included in the results – but because the Invoices table does not allow the vendor_id column to be null, all rows from the Invoices table will be included anyway.

Note that in this example, the tables are joined using columns that have the same name: vendor_id. Columns that have the same name in both tables must be qualified to indicate which table they come from. You code a *qualified column name* by entering the table name and a period in front of the column name. If you don't, Oracle will return an error indicating that the column name is ambiguous. You can also qualify column names that aren't ambiguous, but it's not necessary.

Finally, although you'll code most inner joins using the equal operator in the join condition, you should know that you can compare two tables based on other conditions too. However, using operators like less than and greater than in a join condition are likely to return more rows than you were expecting. So, carefully check your results to make sure they're correct and what you wanted when using a less common join condition.

The explicit syntax for an inner join

```
SELECT select_list
FROM table_1
    [INNER] JOIN table_2
        ON join_condition_1
   [[INNER] JOIN table_3
        ON join_condition_2]...
```

The syntax of a column name that's qualified with a table name

```
table_name.column_name
```

An inner join of the Vendors and Invoices tables

```
-- invoice_number is a column in the Invoices table
-- vendor_id is a column in both the Invoices and Vendors tables
-- vendor_name is a column in the Vendors table

SELECT invoice_number, invoices.vendor_id, vendor_name
FROM vendors INNER JOIN invoices
    ON vendors.vendor_id = invoices.vendor_id
```

The result set

	INVOICE_NUMBER	VENDOR_ID	VENDOR_NAME
1	QP58872	34	IBM
2	Q545443	34	IBM
3	547479217	37	Blue Cross
4	547480102	37	Blue Cross

(114 rows)

Description

- A *join* combines rows and columns from two or more tables into a result set.
- A *join condition* compares a column from each table to determine how the rows should be joined. Any comparison operator can be used, but equal is the most common.
- Join conditions are most often based on the relationship between the primary key in one table and a foreign key in the other table.
- For an *inner join*, only rows that satisfy the join condition are included in the result set.
- If the two columns in a join condition have the same name, you must *qualify* them with the table name so Oracle can distinguish between them.

Notes

- The INNER keyword is optional and is seldom used.
- This syntax for coding an inner join can be referred to as the *explicit syntax*.

Figure 4-1 How to code an inner join

When and how to use table aliases

When you name a table to be joined in the FROM clause, you can specify an alias for that table. A *table alias* is similar to a column alias, except that you do not use the word AS, like you do when you assign a column alias. After you assign a table alias, you must use the alias in place of the original table name throughout the query. This is illustrated in figure 4-2.

The first SELECT statement in this figure joins data from the Vendors and Invoices tables. Here, both tables have been assigned aliases that consist of a single letter. Table aliases are usually used to reduce typing and to make a query more understandable, particularly when table names are lengthy.

The alias used in the second SELECT statement in this figure, for example, simplifies the name of the Invoice_Line_Items table to just Line_Items. That way, the shorter name can be used to refer to the invoice_id column of the table in the join condition. Although this doesn't improve the query in this example much, it can have a dramatic effect on a query that refers to the Invoice_Line_Items table several times.

The syntax for an inner join that uses table aliases

```
SELECT select_list
FROM table_1 n1
    [INNER] JOIN table_2 n2
        ON n1.column_name operator n2.column_name
   [[INNER] JOIN table_3 n3
        ON n2.column_name operator n3.column_name]...
```

An inner join with aliases for all tables

```
SELECT invoice_number, vendor_name, invoice_due_date,
    (invoice_total - payment_total - credit_total) AS balance_due
FROM vendors v JOIN invoices i
    ON v.vendor_id = i.vendor_id
WHERE (invoice_total - payment_total - credit_total) > 0
ORDER BY invoice_due_date DESC
```

The result set

	INVOICE_NUMBER	VENDOR_NAME	INVOICE_DUE_DATE	BALANCE_DUE
1	39104	Data Reproductions Corp	20-JUL-24	85.31
2	40318	Data Reproductions Corp	20-JUL-24	21842
3	0-2436	Malloy Lithographing Inc	17-JUL-24	10976.06
4	109596	Coffee Break Service	11-JUL-24	41.8

(40 rows)

An inner join with an alias for only one table

```
SELECT invoice_number, line_item_amt, line_item_description
FROM invoices JOIN invoice_line_items line_items
    ON invoices.invoice_id = line_items.invoice_id
WHERE account_number = 540
ORDER BY invoice_date
```

The result set

	INVOICE_NUMBER	LINE_ITEM_AMT	LINE_ITEM_DESCRIPTION
1	97/553B	313.55	Card revision
2	97/553	904.14	DB2 Card decks
3	97/522	765.13	SCMD Flyer
4	97/222	1000.46	Crash Course Cover

(8 rows)

Description

- A *table alias* is an alternative table name assigned in the FROM clause.
- Table aliases are used to make a SQL statement easier to code and read.
- If you assign an alias to a table, you must use that alias to refer to the table throughout your query. You can't use the original table name.
- You can use an alias for a table in a join without using an alias for any other table.

Figure 4-2 When and how to use table aliases

How to use compound join conditions

Although a join condition typically consists of a single comparison, you can include two or more comparisons in a join condition using the AND and OR operators. Figure 4-3 shows how this works using two tables from the EX schema.

The query in this figure uses the AND operator to return the first and last names of all customers in the Customers table whose first and last names also exist in the Employees table. Since Thomas Hardy is the only name that exists in both tables, this is the only row that's returned in the result set for this query.

Another way to code compound join conditions is to code one join condition in the FROM clause and any other conditions in the WHERE clause. This is illustrated by the second SELECT statement in this figure.

When you code compound join conditions like this, you may wonder which technique is most efficient. Actually, Oracle typically does a good job of optimizing queries so they run efficiently regardless of how you write the query. As a result, you can usually focus on writing a query so the code is easy to read and maintain. Then, if you encounter performance problems with your queries, you can work on optimizing the query later.

The Customers table

	CUSTOMER_ID	CUSTOMER_LAST_NAME	CUSTOMER_FIRST_NAME	CUSTOMER_ADDRESS	CL
1	1	Anders	Maria	345 Winchell Pl	Ande
2	2	Trujillo	Ana	1298 E Smathers St	Bent
3	3	Moreno	Antonio	6925 N Parkland Ave	Puya
4	4	Hardy	Thomas	83 d'Urberville Ln	Cast

The Employees table

	EMPLOYEE_ID	LAST_NAME	FIRST_NAME	DEPARTMENT_NUMBER	MANAGER_ID
1	1	Smith	Cindy	2	(null)
2	2	Jones	Elmer	4	1
3	3	Simonian	Ralph	2	2
4	4	Hernandez	Olivia	1	9

An inner join with two conditions

```
SELECT customer_first_name, customer_last_name
FROM customers c JOIN employees e
    ON c.customer_first_name = e.first_name
   AND c.customer_last_name = e.last_name
```

The same join coded with the second condition in the WHERE clause

```
SELECT customer_first_name, customer_last_name
FROM customers c JOIN employees e
    ON c.customer_first_name = e.first_name
WHERE c.customer_last_name = e.last_name
```

The result set

	CUSTOMER_FIRST_NAME	CUSTOMER_LAST_NAME
1	Thomas	Hardy

Description

- A join condition can include two or more conditions connected by AND or OR operators.
- A join condition for an inner join can alternately be coded in the WHERE clause.
- In most cases, your code will be easier to read if you code join conditions in the ON expression and search conditions in the WHERE clause.

Figure 4-3 How to use compound join conditions

How to use a self-join

A *self-join* joins a table to itself. Although self-joins are rare, they are sometimes useful for retrieving data that can't be retrieved any other way. For example, figure 4-4 presents a self-join that returns rows from the Vendors table where the vendor is in a city and state that has at least one other vendor. In other words, it does not return a vendor if that vendor is the only vendor in that city and state.

Since this example uses the same table twice, it must use aliases to distinguish each occurrence of the table from the other. In addition, this query must qualify each column name with a table alias since every column naturally occurs in both tables.

Then, the join condition uses three comparisons. The first two match the vendor_city and vendor_state columns in the two tables. As a result, the query returns rows for vendors that are in the same city and state as another vendor. However, since a vendor resides in the same city and state as itself, a third comparison is included to exclude rows that match a vendor with itself. To do that, this condition uses the not-equal operator to compare the vendor_name columns in the two tables.

In addition, this statement includes the DISTINCT keyword. That way, each vendor appears only once in the result set. Otherwise, a vendor would appear once for every other row with a matching city and state. For example, if a vendor is in a city and state that has nine other vendors in that city and state, this query would return nine rows for that vendor.

This example also shows how you can use columns other than key columns in a join condition. Keep in mind, however, that this is an unusual situation and you're not likely to code joins like this often.

A self-join that returns vendors from cities in common with other vendors

```
SELECT DISTINCT v1.vendor_name, v1.vendor_city,
    v1.vendor_state
FROM vendors v1 JOIN vendors v2
    ON (v1.vendor_city = v2.vendor_city) AND
       (v1.vendor_state = v2.vendor_state) AND
       (v1.vendor_id <> v2.vendor_id)
ORDER BY v1.vendor_state, v1.vendor_city
```

The result set

	VENDOR_NAME	VENDOR_CITY	VENDOR_STATE
1	AT&T	Phoenix	AZ
2	Computer Library	Phoenix	AZ
3	Wells Fargo Bank	Phoenix	AZ
4	Aztek Label	Anaheim	CA
5	Blue Shield of California	Anaheim	CA
6	ASC Signs	Fresno	CA
7	Abbey Office Furnishings	Fresno	CA
8	BFI Industries	Fresno	CA

(84 rows)

Description

- A *self-join* is a join that joins a table with itself.
- When you code a self-join, you must use aliases for the tables, and you must qualify each column name with the alias.

Figure 4-4 How to use a self-join

How to join more than two tables

So far, this chapter has shown how to join data from two tables. Depending on the requirement, however, you may have to join many more tables. For example, you sometimes need to join 10 or more tables. Fortunately, once you code the join condition correctly, you can often reuse it.

The SELECT statement in figure 4-5 joins data from four tables: Vendors, Invoices, Invoice_Line_Items, and General_Ledger_Accounts. Each of the joins is based on the relationship between the primary key of one table and a foreign key of the other table. For example, the account_number column is the primary key of the General_Ledger_Accounts table and a foreign key of the Invoice_Line_Items table.

This SELECT statement also shows how table aliases make a statement easier to code and read. Here, the one-letter and two-letter aliases that are used by the tables allow the ON clause to be coded more concisely.

A SELECT statement that joins four tables

```
SELECT vendor_name, invoice_number, invoice_date,
    line_item_amt, account_description
FROM vendors v
    JOIN invoices i
        ON v.vendor_id = i.vendor_id
    JOIN invoice_line_items li
        ON i.invoice_id = li.invoice_id
    JOIN general_ledger_accounts gl
        ON li.account_number = gl.account_number
WHERE (invoice_total - payment_total - credit_total) > 0
ORDER BY vendor_name, line_item_amt DESC
```

The result set

	VENDOR_NAME	INVOICE_NUMBER	INVOICE_DATE	LINE_ITEM_AMT	ACCOUNT_DESCRI
1	Abbey Office Furnishings	203339-13	02-MAY-24	17.5	Office Supplies
2	Blue Cross	547480102	19-MAY-24	224	Group Insurance
3	Blue Cross	547481328	20-MAY-24	224	Group Insurance
4	Blue Cross	547479217	17-MAY-24	116	Group Insurance
5	Cardinal Business Media, Inc.	134116	01-JUN-24	90.36	Card Deck Adver
6	Coffee Break Service	109596	14-JUN-24	41.8	Meals
7	Compuserve	21-4748363	09-MAY-24	9.95	Books, Dues, an

(44 rows)

Description

- You can think of a multi-table join as a series of two-table joins proceeding from left to right.

How to use the implicit inner join syntax

Although it's generally considered a best practice to use the explicit inner join syntax used so far in this chapter, Oracle also provides the *implicit inner join syntax* shown in figure 4-6. This syntax was widely used prior to the introduction of the explicit syntax. You should be familiar with the older implicit syntax mainly because you may need to maintain existing SQL statements that use it.

When you use the implicit syntax for an inner join, you code the tables in the FROM clause separated by commas. Then, you code the join conditions in the WHERE clause.

The first SELECT statement in this figure joins data from the Vendors and Invoices tables. Like the SELECT statement shown in figure 4-1, this statement joins these tables on an equal comparison between the vendor_id columns in the two tables. In this case, however, the comparison is coded in the search condition of the WHERE clause. However, both of these statements return the same result set.

The second SELECT statement uses the implicit syntax to join data from four tables. This is the same join you saw in figure 4-5. In this example, the three join conditions are combined in the WHERE clause using the AND operator. In addition, an AND operator is used to combine the join conditions with the search condition.

Because the explicit syntax for joins lets you separate join conditions from search conditions, statements that use the explicit syntax are typically easier to read than those that use the implicit syntax. In addition, the explicit syntax helps you avoid a common coding mistake with the implicit syntax: omitting the join condition. An implicit join without a join condition results in a cross join, which can return a very large number of rows. For these reasons, we recommend that you always use the explicit syntax.

The implicit syntax for an inner join

```
SELECT select_list
FROM table_1, table_2 [, table_3]...
WHERE table_1.column_name operator table_2.column_name
    [AND table_2.column_name operator table_3.column_name]...
```

Join the Vendors and Invoices tables

```
SELECT invoice_number, vendor_name
FROM vendors v, invoices i
WHERE v.vendor_id = i.vendor_id
```

The result set

	INVOICE_NUMBER	VENDOR_NAME
1	QP58872	IBM
2	Q545443	IBM
3	547479217	Blue Cross
4	547480102	Blue Cross
5	547481328	Blue Cross
6	P02-88D77S7	Fresno County Tax Collector

(114 rows)

Join four tables

```
SELECT vendor_name, invoice_number, invoice_date,
    line_item_amt, account_description
FROM  vendors v, invoices i, invoice_line_items  li,
    general_ledger_accounts gl
WHERE v.vendor_id = i.vendor_id
  AND i.invoice_id = li.invoice_id
  AND li.account_number = gl.account_number
  AND (invoice_total - payment_total - credit_total) > 0
ORDER BY vendor_name, line_item_amt DESC
```

The result set

	VENDOR_NAME	INVOICE_NUMBER	INVOICE_DATE	LINE_ITEM_AMT	ACCOUNT_DESCRI
1	Abbey Office Furnishings	203339-13	02-MAY-24	17.5	Office Supplies
2	Blue Cross	547480102	19-MAY-24	224	Group Insurance
3	Blue Cross	547481328	20-MAY-24	224	Group Insurance
4	Blue Cross	547479217	17-MAY-24	116	Group Insurance
5	Cardinal Business Media, Inc.	134116	01-JUN-24	90.36	Card Deck Adver

(44 rows)

Description

- Instead of coding a join condition for an inner join in the FROM clause, you can code it in the WHERE clause along with any search conditions. Then, you list the tables you want to join in the FROM clause separated by commas.
- This syntax for coding joins is referred to as the *implicit syntax*. It was used prior to the SQL-92 standards, which introduced the explicit syntax.

Figure 4-6 How to use the implicit inner join syntax

How to work with outer joins

Although inner joins are the type of join you'll use most often, Oracle also supports *outer joins*. Unlike an inner join, an outer join returns all of the rows from one or both tables involved in the join, regardless of whether the join condition is true.

How to code an outer join

Figure 4-7 presents the explicit syntax for coding an outer join. As you can see, this syntax is very similar to the explicit syntax for inner joins. The main difference is that you include the LEFT, RIGHT, or FULL keyword to specify the type of outer join you want to perform. You can also include the OUTER keyword, but it's optional and is usually omitted.

The table in this figure summarizes the differences between left, right, and full outer joins. When you use a *left outer join*, the result set includes all the rows from the first, or left, table. Similarly, when you use a *right outer join*, the result set includes all the rows from the second, or right, table. And when you use a *full outer join*, the result set includes all the rows from both tables.

The example in this figure illustrates a left outer join. Here, the Vendors table is joined with the Invoices table. In addition, the result set includes vendor rows even if no matching invoices are found. In that case, null values are returned for the columns in the Invoices table.

Note that when coding outer joins, it's a common practice to avoid using right joins. To do that, you can substitute a left outer join for a right outer join by reversing the order of the tables in the FROM clause and using the LEFT keyword instead of RIGHT. This often makes it easier to read statements that join more than two tables.

The explicit syntax for an outer join

```
SELECT select_list
FROM table_1
    {LEFT|RIGHT|FULL} [OUTER] JOIN table_2
        ON join_condition_1
   [{LEFT|RIGHT|FULL} [OUTER] JOIN table_3
        ON join_condition_2]...
```

What outer joins do

Joins of this type	Keep unmatched rows from
Left outer join	The first (left) table
Right outer join	The second (right) table
Full outer join	Both tables

A SELECT statement that uses a left outer join

```
SELECT vendor_name, invoice_number, invoice_total
FROM vendors LEFT JOIN invoices
    ON vendors.vendor_id = invoices.vendor_id
ORDER BY vendor_name
```

The result set

	VENDOR_NAME	INVOICE_NUMBER	INVOICE_TOTAL
1	ASC Signs	(null)	(null)
2	AT&T	(null)	(null)
3	Abbey Office Furnishings	203339-13	17.5
4	American Booksellers Assoc	(null)	(null)
5	American Express	(null)	(null)

(202 rows)

Description

- An *outer join* retrieves all rows that satisfy the join condition, plus unmatched rows in the left, the right, or both tables depending on the join type.
- In most cases, you use the equal operator to retrieve rows with matching columns. However, you can also use any of the other comparison operators.
- When a row with unmatched columns is retrieved, any columns from the other table that are included in the result set are given null values.

Notes

- The OUTER keyword is optional and typically omitted.

Figure 4-7 How to code an outer join

Outer join examples

To give you a better understanding of how outer joins work, figure 4-8 presents three more examples. These examples use the Departments and Employees tables shown at the top of this figure. In each case, the join condition joins the tables based on the values in their respective department_number columns.

The first SELECT statement performs a left outer join on these two tables. In the result set produced by this statement, you can see that department number 3 (Operations) is included in the result set even though none of the employees in the Employees table work in that department. Because of that, a null value is assigned to the last_name column from that table.

The second SELECT statement uses a right outer join. In this case, all of the rows from the Employees table are included in the result set. However, two of the employees, Locario and Watson, are assigned to a department that doesn't exist in the Departments table and so a null value is assigned to the department_name column. Note that if the department_number column in the Employees table had been defined as a foreign key to the Departments table, Oracle would have prevented this by throwing an error when these rows were inserted.

The third SELECT statement in this figure illustrates a full outer join. If you compare the results of this query with the results of the queries that use a left and right outer join, you'll see that this is a combination of the two joins. In other words, each row in the Departments table is included in the result set along with each row in the Employees table. Because the department_number column from both tables is included in this example, you can clearly identify the row in the Departments table that doesn't have a matching row in the Employees table and the two rows in the Employees table that don't have matching rows in the Departments table.

The Departments table

	DEPARTMENT_NUMBER	DEPARTMENT_NAME
1	1	Accounting
2	2	Payroll
3	3	Operations
4	4	Personnel
5	5	Maintenance

The Employees table

	EMPLOYEE_ID	LAST_NAME	FIRST_NAME	DEPARTMENT_NUMBER	MANAGER_ID
1	1	Smith	Cindy	2	(null)
2	2	Jones	Elmer	4	1
3	3	Simonian	Ralph	2	2
4	4	Hernandez	Olivia	1	9
5	5	Aaronsen	Robert	2	4
6	6	Watson	Denise	6	8
7	7	Hardy	Thomas	5	2
8	8	O'Leary	Rhea	4	9
9	9	Locario	Paulo	6	1

A left outer join

```
SELECT department_name AS dept_name,
    d.department_number AS dept_no, last_name
FROM departments d LEFT JOIN employees e
    ON d.department_number = e.department_number
ORDER BY department_name
```

	DEPT_NAME	DEPT_NO	LAST_NAME
1	Accounting	1	Hernandez
2	Maintenance	5	Hardy
3	Operations	3	(null)
4	Payroll	2	Smith
5	Payroll	2	Simonian
6	Payroll	2	Aaronsen
7	Personnel	4	Jones
8	Personnel	4	O'Leary

A right outer join

```
SELECT department_name AS dept_name,
    e.department_number AS dept_no, last_name
FROM departments d
    RIGHT JOIN employees e
    ON d.department_number = e.department_number
ORDER BY department_name
```

	DEPT_NAME	DEPT_NO	LAST_NAME
1	Accounting	1	Hernandez
2	Maintenance	5	Hardy
3	Payroll	2	Smith
4	Payroll	2	Simonian
5	Payroll	2	Aaronsen
6	Personnel	4	O'Leary
7	Personnel	4	Jones
8	(null)	6	Watson
9	(null)	6	Locario

A full outer join

```
SELECT department_name AS dept_name,
    d.department_number AS d_dept_no,
    e.department_number AS e_dept_no,
    last_name
FROM departments d
    FULL JOIN employees e
    ON d.department_number = e.department_number
ORDER BY department_name
```

	DEPT_NAME	D_DEPT_NO	E_DEPT_NO	LAST_NAME
1	Accounting	1	1	Hernandez
2	Maintenance	5	5	Hardy
3	Operations	3	(null)	(null)
4	Payroll	2	2	Simonian
5	Payroll	2	2	Smith
6	Payroll	2	2	Aaronsen
7	Personnel	4	4	Jones
8	Personnel	4	4	O'Leary
9	(null)	(null)	6	Locario
10	(null)	(null)	6	Watson

Description

- The examples in this figure and the next use the Departments, Employees, and Projects tables from the EX schema.

Figure 4-8 Outer join examples

Outer joins that join more than two tables

Like inner joins, you can use outer joins to join data from more than two tables. The two examples in figure 4-9 illustrate how this works. These examples use the Departments and Employees tables you saw in the previous figure along with a Projects table. All three tables are in the EX schema and are shown at the top of this figure.

The first example in this figure uses left outer joins to join the data in the three tables. Here, you can see once again that none of the employees in the Employees table are assigned to the Operations department. Because of that, null values are returned for the columns in both the Employees and Projects tables. In addition, you can see that two employees, Hardy and Jones, aren't assigned to a project.

The second example in this figure uses full outer joins to join the three tables. This result set includes unmatched rows from the Departments and Employees table, just like the result set you saw in figure 4-8 that was created using a full outer join. In addition, the result set in this example includes an unmatched row from the Projects table: the one for project number P1014. In other words, no employee is assigned to this project.

The Departments table

	DEPARTMENT_NUMBER	DEPARTMENT_NAME
1	1	Accounting
2	2	Payroll
3	3	Operations
4	4	Personnel
5	5	Maintenance

The Employees table

	EMPLOYEE_ID	LAST_NAME	FIRST_NAME	DEPARTMENT_NUMBER	MANAGER_ID
1	1	Smith	Cindy	2	(null)
2	2	Jones	Elmer	4	1
3	3	Simonian	Ralph	2	2
4	4	Hernandez	Olivia	1	9
5	5	Aaronsen	Robert	2	4
6	6	Watson	Denise	6	8
7	7	Hardy	Thomas	5	2
8	8	O'Leary	Rhea	4	9
9	9	Locario	Paulo	6	1

The Projects table

	PROJECT_NUMBER	EMPLOYEE_ID
1	P1011	8
2	P1011	4
3	P1012	3
4	P1012	1
5	P1012	5
6	P1013	6
7	P1013	9
8	P1014	10

Join the three tables using left outer joins

```
SELECT department_name, last_name,
    project_number AS proj_no
FROM departments d
    LEFT JOIN employees e
        ON d.department_number =
           e.department_number
    LEFT JOIN projects p
        ON e.employee_id =
           p.employee_id
ORDER BY department_name, last_name,
    project_number
```

	DEPARTMENT_NAME	LAST_NAME	PROJ_NO
1	Accounting	Hernandez	P1011
2	Maintenance	Hardy	(null)
3	Operations	(null)	(null)
4	Payroll	Aaronsen	P1012
5	Payroll	Simonian	P1012
6	Payroll	Smith	P1012
7	Personnel	Jones	(null)
8	Personnel	O'Leary	P1011

Join the three tables using full outer joins

```
SELECT department_name, last_name,
    project_number AS proj_no
FROM departments dpt
    FULL JOIN employees emp
        ON dpt.department_number =
           emp.department_number
    FULL JOIN projects prj
        ON emp.employee_id =
           prj.employee_id
ORDER BY department_name
```

	DEPARTMENT_NAME	LAST_NAME	PROJ_NO
1	Accounting	Hernandez	P1011
2	Maintenance	Hardy	(null)
3	Operations	(null)	(null)
4	Payroll	Simonian	P1012
5	Payroll	Aaronsen	P1012
6	Payroll	Smith	P1012
7	Personnel	Jones	(null)
8	Personnel	O'Leary	P1011
9	(null)	Locario	P1013
10	(null)	(null)	P1014
11	(null)	Watson	P1013

Description

- You can use outer joins to join multiple tables.
- You can combine inner and outer joins within a single SELECT statement.

Figure 4-9 Outer joins that join more than two tables

How to use the implicit outer join syntax

Like inner joins, outer joins also have an older, implicit syntax that you may come across in legacy code. The implicit syntax for an outer join uses the *join operator* ((+)) to indicate whether a join is left or right. For a left join, the operator is coded after the second table name, as seen in the first example of figure 4-10. For a right join, the operator is coded after the first table name. This is shown in the second example.

Note, however, that you can't perform a full outer join using the implicit syntax. As a result, if you need to perform a full outer join, you'll have to use the explicit outer join syntax.

The implicit syntax for an outer join

```
SELECT select_list
FROM table_1, table_2 [, table 3]...
WHERE table_1.column_name [(+)] table_2.column_name [(+)]
    [table_2.column_name [(+)] table_3.column_name [(+)]]...
```

Join two tables using a left outer join

```
SELECT department_name AS dept_name,
    dpt.department_number AS dept_no,
    last_name
FROM departments dpt, employees emp
WHERE dpt.department_number = emp.department_number (+)
ORDER BY department_name
```

The result set

	DEPT_NAME	DEPT_NO	LAST_NAME
1	Accounting	1	Hernandez
2	Maintenance	5	Hardy
3	Operations	3	(null)
4	Payroll	2	Smith
5	Payroll	2	Simonian
6	Payroll	2	Aaronsen
7	Personnel	4	Jones
8	Personnel	4	O'Leary

Join two tables using a right outer join

```
SELECT department_name AS dept_name,
    emp.department_number AS dept_no,
    last_name
FROM departments dpt, employees emp
WHERE dpt.department_number (+) = emp.department_number
ORDER BY department_name
```

The result set

	DEPT_NAME	DEPT_NO	LAST_NAME
1	Accounting	1	Hernandez
2	Maintenance	5	Hardy
3	Payroll	2	Smith
4	Payroll	2	Simonian
5	Payroll	2	Aaronsen
6	Personnel	4	O'Leary
7	Personnel	4	Jones
8	(null)	6	Watson
9	(null)	6	Locario

Description

- The implicit syntax for outer joins is an alternative to the SQL-92 standards.
- The *join operator* ((+)) indicates whether an outer join using the implicit syntax is a left join or right join.
- Full outer joins are not possible with the implicit syntax.

Figure 4-10 How to use the implicit outer join syntax

Other skills for working with joins

The figures that follow present other skills for working with joins. First, you'll learn how to use inner and outer joins in the same statement. Next, you'll learn how to join tables with the USING and NATURAL keywords. Then, you'll learn how to use another type of join, called a cross join.

How to combine inner and outer joins

Figure 4-11 shows how you can combine inner and outer joins. In this example, the Departments table is joined with the Employees table using an inner join and the Employees table is joined to the Projects table with a left outer join. The result set includes all of the departments that have employees assigned to them, all of the employees assigned to those departments, and the projects those employees are assigned to. Here, you can clearly see that two employees, Hardy and Jones, haven't been assigned projects.

The Departments table

	DEPARTMENT_NUMBER	DEPARTMENT_NAME
1	1	Accounting
2	2	Payroll
3	3	Operations
4	4	Personnel
5	5	Maintenance

The Employees table

	EMPLOYEE_ID	LAST_NAME	FIRST_NAME	DEPARTMENT_NUMBER	MANAGER_ID
1	1	Smith	Cindy	2	(null)
2	2	Jones	Elmer	4	1
3	3	Simonian	Ralph	2	2
4	4	Hernandez	Olivia	1	9
5	5	Aaronsen	Robert	2	4
6	6	Watson	Denise	6	8
7	7	Hardy	Thomas	5	2
8	8	O'Leary	Rhea	4	9
9	9	Locario	Paulo	6	1

The Projects table

	PROJECT_NUMBER	EMPLOYEE_ID
1	P1011	8
2	P1011	4
3	P1012	3
4	P1012	1
5	P1012	5
6	P1013	6
7	P1013	9
8	P1014	10

Combine an outer and inner join

```
SELECT department_name AS dept_name, last_name, project_number
FROM departments dpt
    JOIN employees emp
        ON dpt.department_number = emp.department_number
    LEFT JOIN projects prj
        ON emp.employee_id = prj.employee_id
ORDER BY department_name
```

The result set

	DEPT_NAME	LAST_NAME	PROJECT_NUMBER
1	Accounting	Hernandez	P1011
2	Maintenance	Hardy	(null)
3	Payroll	Simonian	P1012
4	Payroll	Smith	P1012
5	Payroll	Aaronsen	P1012
6	Personnel	O'Leary	P1011
7	Personnel	Jones	(null)

Description

- You can combine inner and outer joins within a single SELECT statement using either the explicit or the implicit join syntax.

Figure 4-11 How to combine inner and outer joins

How to join tables with the USING keyword

When you use the equal operator to join two tables on a common column, the join can be referred to as an *equijoin* (or an *equi-join*). When you code an equijoin, it's common for the columns that are being compared to have the same name. For joins like these, you can simplify the query with the USING keyword. You do this by coding a USING clause instead of an ON clause to specify the join as shown in figure 4-12.

Here, the first example shows how to join the Vendors and Invoices tables from the AP schema on the vendor_id column with a USING clause. This returns the same results as the query shown in figure 4-1 that uses the ON clause. Note that the USING clause only works because the vendor_id column exists and has the same name in both the Vendors and Invoices tables.

The second example shows how to join the Departments, Employees, and Projects tables from the EX schema with the USING keyword. Here, the first USING clause uses an inner join to join the Departments table to the Employees table on the department_number column. Then, the second USING clause uses a left join to join the Employees table to the Projects table on the employee_id column. This shows that you can use a USING clause for both inner and outer joins, and it returns the same result as the query shown in figure 4-11.

In some rare cases, you may want to join a table by multiple columns. To do that with a USING clause, you can code multiple column names within the parentheses, separating each column name with a comma. This yields the same result as coding two equijoins connected with the AND operator.

Since the USING clause is more concise than the ON clause, it can make your code easier to read and maintain. As a result, it often makes sense to use the USING clause when you're developing new statements. However, if you can't get the USING clause to work correctly because of the way your database is structured, you can always use an ON clause instead.

The syntax for a join that uses the USING keyword

```
SELECT select_list
FROM table_1
    [{LEFT|RIGHT|FULL} [OUTER]] JOIN table_2
        USING(join_column_1[, join_column_2]...)
   [[{LEFT|RIGHT|FULL} [OUTER]] JOIN table_3
        USING (join_column_2[, join_column_2]...)]...
```

Use the USING keyword to join two tables

```
SELECT invoice_number, vendor_name
FROM vendors
    JOIN invoices USING (vendor_id)
```

The result set

	INVOICE_NUMBER	VENDOR_NAME
1	QP58872	IBM
2	Q545443	IBM
3	547479217	Blue Cross
4	547480102	Blue Cross

(114 rows)

Use the USING keyword to join three tables

```
SELECT department_name AS dept_name, last_name, project_number
FROM departments
    JOIN employees USING (department_number)
    LEFT JOIN projects USING (employee_id)
ORDER BY department_name
```

The result set

	DEPT_NAME	LAST_NAME	PROJECT_NUMBER
1	Accounting	Hernandez	P1011
2	Maintenance	Hardy	(null)
3	Payroll	Simonian	P1012
4	Payroll	Smith	P1012
5	Payroll	Aaronsen	P1012
6	Personnel	O'Leary	P1011
7	Personnel	Jones	(null)

(7 rows)

Description

- You can use the USING keyword to simplify the syntax for joining tables.
- The join can be an inner join or an outer join.
- The tables must be joined by a column that has the same name in both tables.
- To include multiple columns, separate them with commas.
- The join must be an *equijoin*, which means that the equals operator is used to compare the two columns.

Figure 4-12 How to join tables with the USING keyword

How to join tables with the NATURAL keyword

Figure 4-13 shows how to use the NATURAL keyword to code a *natural join*. When you code a natural join, you don't specify the column that's used to join the two tables. Instead, the database automatically joins the two tables based on all columns in the two tables that have the same names. As a result, this type of join only works correctly if the database is designed in a certain way.

For instance, if you use a natural join to join the Vendors and Invoices tables as shown in the first example, the join works correctly because these tables only have one column in common: the vendor_id column. As a result, the database joins these two tables on the vendor_id column. However, if these tables had another column name in common, this query would attempt to join these tables on both columns and might yield unexpected results.

Although natural joins are easy to code, these joins don't explicitly specify the join column or conditions. As a result, they might not work correctly if the structure of the database changes later. Because of that, you'll usually want to avoid using natural joins for production code.

In addition, if you use natural joins, you may get unexpected results for more complex queries. In that case, you can use the USING or ON clauses to explicitly specify the join since these clauses give you more control over the join. If necessary, you can mix a natural join with the USING or ON clauses within a single SELECT statement. In this figure, for example, the second SELECT statement uses a natural join for the first join and a USING clause for the second join. The result is the same as the result for the second statement in figure 4-12.

The syntax for a join that uses the NATURAL keyword

```
SELECT select_list
FROM table_1
    NATURAL JOIN table_2
    [NATURAL JOIN table_3]...
```

Use the NATURAL keyword to join tables

```
SELECT invoice_number, vendor_name
FROM vendors
    NATURAL JOIN invoices
```

The result set

	INVOICE_NUMBER	VENDOR_NAME
1	QP58872	IBM
2	Q545443	IBM
3	547479217	Blue Cross
4	547480102	Blue Cross

(114 rows)

Use the NATURAL keyword to join three tables

```
SELECT department_name AS dept_name, last_name, project_number
FROM departments
    NATURAL JOIN employees
    LEFT JOIN projects USING (employee_id)
ORDER BY department_name
```

The result set

	DEPT_NAME	LAST_NAME	PROJECT_NUMBER
1	Accounting	Hernandez	P1011
2	Maintenance	Hardy	(null)
3	Payroll	Simonian	P1012
4	Payroll	Smith	P1012
5	Payroll	Aaronsen	P1012
6	Personnel	O'Leary	P1011
7	Personnel	Jones	(null)

(7 rows)

Description

- You can use the NATURAL keyword to create a *natural join* that joins two tables based on all columns in the two tables that have the same names.
- A natural join only works correctly for certain types of database structures, and often yields unexpected results for complex queries. As a result, it's not commonly used.

Figure 4-13 How to join tables with the NATURAL keyword

How to use cross joins

A *cross join* produces a result set that includes each row from the first table joined, or crossed, with each row from the second table. The result set is known as the *Cartesian product* of the tables. Figure 4-14 shows how to code a cross join using both the explicit and implicit syntax.

To use the explicit syntax, you include the CROSS JOIN keywords between the two tables in the FROM clause. Notice that because of the way a cross join works, you don't include a join condition. The same is true when you use the implicit syntax. In that case, you list the tables in the FROM clause and omit the join condition from the WHERE clause.

The two SELECT statements in this figure illustrate how cross joins work. Both of these statements combine data from the Departments and Employees tables from the EX schema. As you can see, the result is a table that includes 45 rows. That's each of the five rows in the Departments table combined with each of the nine rows in the Employees table. Although this result set is relatively small, you can imagine how large it would be if the tables included hundreds or thousands of rows.

Generally, cross joins have few practical uses in a relational database. As a result, you'll rarely need to use one. In fact, you're most likely to code a cross join by accident if you use the implicit join syntax and forget to code the join condition in the WHERE clause. That's one of the reasons why it's generally considered a good practice to use the explicit join syntax.

A cross join using the explicit syntax

The explicit syntax for a cross join

```
SELECT select_list
FROM table_1 CROSS JOIN table_2
```

A cross join that uses the explicit syntax

```
SELECT departments.department_number, department_name, employee_id,
    last_name
FROM departments CROSS JOIN employees
```

A cross join using the implicit syntax

The implicit syntax for a cross join

```
SELECT select_list
FROM table_1, table_2
```

A cross join that uses the implicit syntax

```
SELECT departments.department_number, department_name, employee_id,
    last_name
FROM departments, employees
```

The result set

	DEPARTMENT_NUMBER	DEPARTMENT_NAME	EMPLOYEE_ID	LAST_NAME
1	1	Accounting	1	Smith
2	1	Accounting	2	Jones
3	1	Accounting	3	Simonian
4	1	Accounting	4	Hernandez
5	1	Accounting	5	Aaronsen
6	1	Accounting	6	Watson
7	1	Accounting	7	Hardy
8	1	Accounting	8	O'Leary
9	1	Accounting	9	Locario
10	2	Payroll	1	Smith
11	2	Payroll	2	Jones
12	2	Payroll	3	Simonian

(45 rows)

Description

- A *cross join* joins each row from the first table combined, or crossed, with each row from the second table.
- The result set returned by a cross join is known as a *Cartesian product*.
- To code a cross join using the explicit syntax, use the CROSS JOIN keywords in the FROM clause.
- To code a cross join using the implicit syntax, list the tables in the FROM clause and omit a join condition from the WHERE clause.

Figure 4-14 How to use cross joins

How to work with unions

Like a join, a *union* combines data from two or more tables. Instead of combining columns from base tables, however, a union combines rows from two or more result sets.

The syntax of a union

Figure 4-15 shows how to code a union. As the syntax shows, you create a union by connecting two or more SELECT statements with the UNION keyword. For this to work, the result of each SELECT statement must have the same number of columns, and the data types of the corresponding columns in each set must be compatible.

In this figure, the SELECT statements that are connected by the UNION operator have been indented to make it easier to see how this statement works. However, in a production environment, it's common to see the SELECT statements and the UNION operator coded at the same level of indentation.

If you want to sort the result of a union operation, you can code an ORDER BY clause after the last SELECT statement. Note that the column names you use in this clause must be the same as those used in the first SELECT statement. That's because the column names you use in the first SELECT statement are the ones that are used in the result set.

By default, a union operation removes duplicate rows from the result set. If that's not what you want, you can include the ALL keyword. In most cases, though, you'll omit this keyword.

Unions that combine data from different tables

The example in this figure shows how to use a union to combine data from two different tables from the EX schema. In this case, the Active_Invoices table contains invoices with outstanding balances, and the Paid_Invoices table contains invoices that have been paid in full. Both of these tables have the same structure as the Invoices table you've seen in previous figures.

This union operation combines the rows in both tables that have an invoice date on or after June 1, 2024. Notice that the first SELECT statement includes a column named Source that contains the literal value "Active." The second SELECT statement includes a column by the same name, but it contains the literal value "Paid." This column is used to indicate which table each row in the result set came from.

Although this column is assigned the same name in both SELECT statements, that doesn't have to be the case. In fact, none of the columns have to have the same names. Corresponding columns have to have compatible data types, but the corresponding relationships are determined by the order in which the columns are coded in the SELECT clauses, not by their names. When you use column aliases, though, you'll typically assign the same name to corresponding columns so that the statement is easier to understand.

The syntax for a union operation

```
    SELECT_statement_1
UNION [ALL]
    SELECT_statement_2
[UNION [ALL]
    SELECT_statement_3]...
[ORDER BY order_by_list]
```

Combine result sets from two different tables

```
    SELECT 'Active' AS source, invoice_number, invoice_date, invoice_total
    FROM active_invoices
    WHERE invoice_date >= '01-JUN-2024'
UNION
    SELECT 'Paid' AS source, invoice_number, invoice_date, invoice_total
    FROM paid_invoices
    WHERE invoice_date >= '01-JUN-2024'
ORDER BY invoice_total DESC
```

The result set

	SOURCE	INVOICE_NUMBER	INVOICE_DATE	INVOICE_TOTAL
1	Active	40318	18-JUL-24	21842
2	Paid	P02-3772	03-JUN-24	7125.34
3	Paid	10843	04-JUN-24	4901.26
4	Paid	77290	04-JUN-24	1750
5	Paid	RTR-72-3662-X	04-JUN-24	1600
6	Paid	75C-90227	06-JUN-24	1367.5
7	Paid	P02-88D77S7	06-JUN-24	856.92
8	Active	I77271-O01	05-JUN-24	662
9	Active	9982771	03-JUN-24	503.2

(22 rows)

Description

- A *union* combines the result sets of two or more SELECT statements into one result set.
- Each result set must return the same number of columns, and the corresponding columns in each result set must have compatible data types.
- By default, a union eliminates duplicate rows. If you want to include duplicate rows, code the ALL keyword.
- The column names in the final result set are taken from the first SELECT clause. Column aliases assigned by the other SELECT clauses have no effect on the final result set.
- To sort the rows in the final result set, code an ORDER BY clause after the last SELECT statement. This clause must refer to the column names assigned in the first SELECT clause.

Figure 4-15 Unions that combine data from different tables

Unions that combine data from the same table

Figure 4-16 shows how to use unions to combine data from a single table. In the first example, rows from the Invoices table that have a balance due are combined with rows from the same table that are paid in full. As in the example in the previous figure, a column named Source is added at the beginning of each interim table. That way, the final result set indicates whether each invoice is active or paid.

The second example in this figure shows how you can use a union with data that's joined from two tables. Here, each SELECT statement joins data from the Invoices and Vendors tables. The first SELECT statement retrieves invoices with totals greater than $10,000. Then, it calculates a payment of 33% of the invoice total. The two other SELECT statements are similar. The second one retrieves invoices with totals between $500 and $10,000 and calculates a 50% payment, and the third one retrieves invoices with totals less than $500 and sets the payment amount at 100% of the total. Although this isn't the most practical example, it helps illustrate the flexibility of union operations.

In both of these examples, the same column aliases are assigned in each SELECT statement. Although the aliases in the second and third SELECT statements are optional, they make the query easier to read. In particular, they make it easy to see that the three SELECT statements have the same number and types of columns.

Combine data from the Invoices table

```
    SELECT 'Active' AS source, invoice_number, invoice_date, invoice_total
    FROM invoices
    WHERE (invoice_total - payment_total - credit_total) > 0
UNION
    SELECT 'Paid' AS source, invoice_number, invoice_date, invoice_total
    FROM invoices
    WHERE (invoice_total - payment_total - credit_total) <= 0
ORDER BY invoice_total DESC
```

The result set

	SOURCE	INVOICE_NUMBER	INVOICE_DATE	INVOICE_TOTAL
1	Paid	0-2058	08-MAY-24	37966.19
2	Paid	P-0259	16-APR-24	26881.4
3	Paid	0-2060	08-MAY-24	23517.58
4	Active	40318	18-JUL-24	21842
5	Active	P-0608	11-APR-24	20551.18
6	Active	0-2436	07-MAY-24	10976.06

(114 rows)

Combine payment data from the same joined tables

```
    SELECT invoice_number, vendor_name, '33% Payment' AS payment_type,
        invoice_total AS total, (invoice_total * 0.333) AS payment
    FROM invoices JOIN vendors
        ON invoices.vendor_id = vendors.vendor_id
    WHERE invoice_total > 10000
UNION
    SELECT invoice_number, vendor_name, '50% Payment' AS payment_type,
        invoice_total AS total, (invoice_total * 0.5) AS payment
    FROM invoices JOIN vendors
        ON invoices.vendor_id = vendors.vendor_id
    WHERE invoice_total BETWEEN 500 AND 10000
UNION
    SELECT invoice_number, vendor_name, 'Full amount' AS payment_type,
        invoice_total AS Total, invoice_total AS Payment
    FROM invoices JOIN vendors
        ON invoices.vendor_id = vendors.vendor_id
    WHERE invoice_total < 500
ORDER BY payment_type, vendor_name, invoice_number
```

The result set

	INVOICE_NUMBER	VENDOR_NAME	PAYMENT_TYPE	TOTAL	PAYMENT
1	40318	Data Reproductions Corp	33% Payment	21842	7273.386
2	0-2058	Malloy Lithographing Inc	33% Payment	37966.19	12642.74127
3	0-2060	Malloy Lithographing Inc	33% Payment	23517.58	7831.35414
4	0-2436	Malloy Lithographing Inc	33% Payment	10976.06	3655.02798
5	P-0259	Malloy Lithographing Inc	33% Payment	26881.4	8951.5062
6	P-0608	Malloy Lithographing Inc	33% Payment	20551.18	6843.54294
7	509786	Bertelsmann Industry Svcs. Inc	50% Payment	6940.25	3470.125

(114 rows)

Figure 4-16 Unions that combine data from the same table

How to use the MINUS and INTERSECT operators

Like the UNION operator, the MINUS and INTERSECT operators work with two or more result sets. Because of that, all three of these operators can be referred to as *set operators*. In addition, the MINUS and INTERSECT operators follow many of the same rules as the UNION operator.

The first query shown in figure 4-17 uses the MINUS operator to return the first and last names of all customers in the Customers table except any customers whose first and last names also exist in the Employees table. Since Thomas Hardy is the only name that's the same in both tables, this is the only record that's excluded from the result set for the query that comes before the MINUS operator. Using the MINUS operator this way can also be called an *antijoin*.

The second query shown in this figure uses the INTERSECT operator to return the first and last names of all customers in the Customers table whose first and last names also exist in the Employees table. Since Thomas Hardy is the only name that exists in both tables, this is the only record that's returned for the result set for this query. Using the INTERSECT operator this way can also be called a *semijoin*.

When you use the MINUS and INTERSECT operators, you must follow many of the same rules as for working with the UNION operator. To start, both of the statements that are connected by these operators must return the same number of columns. In addition, the data types for these columns must be compatible.

Finally, when two queries are joined by a MINUS or INTERSECT operator, the column names in the final result set are taken from the first query. When you code the ORDER BY clause, you use the column names from the first query.

The syntax for the MINUS and INTERSECT operations

```
    SELECT_statement_1
{MINUS | INTERSECT}
    SELECT_statement_2
[ORDER BY order_by_list]
```

The Customers table

	CUSTOMER_FIRST_NAME	CUSTOMER_LAST_NAME
1	Maria	Anders
2	Ana	Trujillo
3	Antonio	Moreno
4	Thomas	Hardy
5	Christina	Berglund
6	Hanna	Moos
7	Fred	Citeaux
8	Martin	Summer
9	Laurence	Lebihan

(24 rows)

The Employees table

	FIRST_NAME	LAST_NAME
1	Cindy	Smith
2	Elmer	Jones
3	Ralph	Simonian
4	Olivia	Hernandez
5	Robert	Aaronsen
6	Denise	Watson
7	Thomas	Hardy
8	Rhea	O'Leary
9	Paulo	Locario

(9 rows)

Exclude rows from the first query if they also occur in the second query

```
    SELECT customer_first_name, customer_last_name
    FROM customers
MINUS
    SELECT first_name, last_name
    FROM employees
ORDER BY customer_last_name
```

The result set

	CUSTOMER_FIRST_NAME	CUSTOMER_LAST_NAME
1	Maria	Anders
2	Christina	Berglund
3	Art	Braunschweiger
4	Donna	Chelan

(23 rows)

Include rows only if they occur in both queries

```
    SELECT customer_first_name, customer_last_name
    FROM customers
INTERSECT
    SELECT first_name, last_name
    FROM employees
```

The result set

	CUSTOMER_FIRST_NAME	CUSTOMER_LAST_NAME
1	Thomas	Hardy

Description

- To use MINUS or INTERSECT, the number of columns must be the same in both SELECT statements, and the data types for each column must be compatible.
- The column names in the final result set are taken from the first SELECT statement.

Figure 4-17 How to use the MINUS and INTERSECT operators

Perspective

In this chapter, you learned several techniques for combining data from two or more tables into a single result set. In particular, you learned how to use the explicit syntax for combining data using inner joins. Of all the techniques presented in this chapter, this is the one you'll use most often. So, you'll want to be sure you understand it thoroughly before you go on.

Terms

join
join condition
inner join
qualified column name
explicit syntax
table alias
self-join
implicit syntax
outer join
left outer join
right outer join
full outer join
join operator
equijoin
natural join
cross join
Cartesian product
union
set operators
antijoin
semijoin

Exercises

1. Write a SELECT statement that returns all columns from the Vendors table inner-joined with all columns from the Invoices table.
2. Write a SELECT statement that returns four columns:

vendor_name	vendor_name from the Vendors table
invoice_number	invoice_number from the Invoices table
invoice_date	invoice_date from the Invoices table
balance_due	invoice_total minus payment_total minus credit_total from the Invoices table

   The result set should have one row for each invoice with a non-zero balance.

   Sort the result set by vendor_name in ascending order.
3. Write a SELECT statement that returns three columns:

vendor_name	vendor_name from the Vendors table
default_account	default_account_number from the Vendors table
description	account_description from the General_Ledger_Accounts table

   The result set should have one row for each vendor.

   Sort the result set by account_description and then by vendor_name.

4. Write a SELECT statement that returns five columns from three tables:

vendor_name	vendor_name from the Vendors table
invoice_date	invoice_date from the Invoices table
invoice_number	invoice_number from the Invoices table
li_sequence	invoice_sequence from the Invoice_Line_Items table
li_amount	line_item_amt from the Invoice_Line_Items table

   Use these aliases for the tables: ven for the Vendors table, inv for the Invoices table, and li for the Invoice_Line_Items table.

   Sort the final result set by vendor_name, invoice_date, invoice_number, and invoice_sequence.

5. Write a SELECT statement that returns three columns:

vendor_id	vendor_id from the Vendors table
vendor_name	vendor_name from the Vendors table
contact_name	A concatenation of vendor_contact_first_name and vendor_contact_last_name with a space in between

   The result set should have one row for each vendor whose contact has the same last name as another vendor's contact. Hint: Use a self-join.

   Sort the result set by vendor_contact_last_name.

6. Write a SELECT statement that returns two columns from the General_Ledger_Accounts table: account_number and account_description. The result set should have one row for each account number that has never been used. Hint: Use an outer join to the Invoice_Line_Items table.

   Sort the final result set by account_number.

7. Use the UNION operator to generate a result set consisting of two columns from the Vendors table: vendor_name and vendor_state. If the vendor is in California, the vendor_state value should be "CA"; otherwise, the VendorState value should be "Outside CA."

   Sort the final result set by vendor_name.

5

How to code summary queries

In this chapter, you'll learn how to code queries that summarize data. For example, you can use summary queries to report sales totals by vendor or state, or to get a count of the number of invoices that were processed each day of the month. You'll also learn how to use aggregate and analytic functions, which allow you to do jobs like calculate averages, summarize totals, or find the highest value for a given column.

How to work with aggregate functions 128
How to code aggregate functions 128
Queries that use aggregate functions 130
How to group and summarize data 132
How to code the GROUP BY and HAVING clauses 132
Queries that use the GROUP BY and HAVING clauses 134
How the HAVING clause compares to the WHERE clause 136
How to code complex search conditions 138
How to summarize data using the ROLLUP CUBE clause 140
How to use ROLLUP 140
How to use CUBE 142
How to code analytic functions 144
How analytic functions work 144
How to code frames 146
Two more examples of frames 148
How to use named windows 150
How to use the ranking functions 152
Perspective 156

How to work with aggregate functions

An *aggregate function* operates on a series of values and returns a single summary value. Because aggregate functions typically operate on the values in columns, they are sometimes referred to as *column functions*. A query that contains one or more aggregate functions is typically referred to as a *summary query*.

How to code aggregate functions

Figure 5-1 presents the syntax of the most common aggregate functions. Most of these functions operate on an expression. Typically, the expression is just a column name. This is shown in the first example, which uses the MAX function on the invoice_total column to get the highest total from the Invoices table.

However, an expression can also be more complex. In the second example, the expression that's coded for the SUM function calculates the balance due of an invoice using the invoice_total, payment_total, and credit_total columns. Furthermore, the WHERE clause selects only those invoices with a balance due. The result is a single value that represents the total amount due for all the selected invoices.

When you use these functions, you can also code the ALL or DISTINCT keyword. The ALL keyword is the default, which means that all values except null values are included in the calculation. In other words, null values are excluded from these functions.

If you don't want duplicate values included, you can code the DISTINCT keyword. In most cases, you'll use DISTINCT only with the COUNT function. You won't use it with MIN or MAX because it has no effect on those functions. And it doesn't usually make sense to use it with the AVG and SUM functions.

Unlike the other aggregate functions, you can't use the ALL or DISTINCT keywords or an expression with COUNT(*). Instead, you code this function exactly as shown in the syntax. The value returned by this function is the number of rows in the base table that satisfy the search condition of the query, including rows with null values. In this figure, for example, the COUNT(*) function in the query indicates that the Invoices table contains 40 invoices with a balance due.

Finally, note that when you code an aggregate function in a SELECT statement, you usually can't include any non-aggregate columns in that same statement. If you do, Oracle will return an error as shown in the third example.

The syntax of common aggregate functions

Function syntax	Result
`AVG([ALL|DISTINCT] expression)`	The average of the non-null values in the expression.
`SUM([ALL|DISTINCT] expression)`	The total of the non-null values in the expression.
`MIN([ALL|DISTINCT] expression)`	The lowest non-null value in the expression.
`MAX([ALL|DISTINCT] expression)`	The highest non-null value in the expression.
`COUNT([ALL|DISTINCT] expression)`	The number of non-null values in the expression.
`COUNT(*)`	The number of rows selected by the query.

Get the highest invoice total

```
SELECT MAX(invoice_total)
FROM invoices
```

	MAX(INVOICE_TOTAL)
1	37966.19

Count unpaid invoices and calculate the total due

```
SELECT COUNT(*) AS number_of_invoices,
       SUM(invoice_total - payment_total - credit_total) AS total_due
FROM invoices
WHERE invoice_total - payment_total - credit_total > 0
```

	NUMBER_OF_INVOICES	TOTAL_DUE
1	40	66796.24

An error due to a non-aggregate column

```
SELECT MAX(invoice_total), invoice_id
FROM invoices
```

```
ORA-00937: not a single-group group function
00937. 00000 -  "not a single-group group function"
*Cause:
*Action:
Error at Line: 1 Column: 28
```

Description

- *Aggregate functions*, also called *column functions*, perform a calculation on the values in a set of selected rows.
- A *summary query* is a SELECT statement that includes one or more aggregate functions.
- The AVG and SUM functions must be used with numeric data.
- The MIN, MAX, and COUNT functions can be used with numeric, date, or string data.
- By default, all values are included in the calculation regardless of whether they're duplicated. To omit duplicate values, code the DISTINCT keyword. This keyword is typically used with the COUNT function.
- All of the aggregate functions except for COUNT(*) ignore null values.
- If you code an aggregate function in the SELECT clause, you usually can't also include non-aggregate columns.

Figure 5-1 How to code aggregate functions

Queries that use aggregate functions

Figure 5-2 presents five more queries that use aggregate functions. The first two queries use the COUNT(*) function to count the number of rows in the Invoices table that satisfy the search condition. In both cases, only those invoices with invoice dates after 1/1/2024 are included in the count, which happens to be all of the invoices in the table.

In addition, the first query uses the AVG function to calculate the average amount of those invoices and the SUM function to calculate the total amount of those invoices. In contrast, the second query uses the MIN and MAX functions to get the minimum and maximum invoice amounts.

Although the MIN, MAX, and COUNT functions are typically used on columns that contain numeric data, they can also be used on columns that contain character or date data. In the third query, for example, they're used on the vendor_name column in the Vendors table. Here, the MIN function returns the name of the vendor that's lowest in the sort sequence, the MAX function returns the name of the vendor that's highest in the sort sequence, and the COUNT function returns the total number of vendors. Note that since the vendor_name column can't contain null values, the COUNT(*) function would have returned the same result.

The fourth query illustrates how using the DISTINCT keyword can affect the result of a COUNT function. Here, the first COUNT function uses the DISTINCT keyword to count the number of vendors that have invoices dated 1/1/2024 or later in the Invoices table. To do that, it looks for distinct values in the vendor_id column.

In contrast, because the second COUNT function doesn't include the DISTINCT keyword, it counts every invoice that's dated 1/1/2024 or later. Of course, you could accomplish the same thing using the COUNT(*) function. COUNT(vendor_id) is used here only to illustrate the difference between coding and not coding the DISTINCT keyword.

Finally, the last example shows how to use the scalar function ROUND to round a numeric value to a specified number of places. ROUND is used with AVG throughout this chapter to make the results easier to read and understand. You'll learn more about scalar functions in chapter 8.

In addition, this example illustrates an exception to the rule that summary queries can contain only aggregate functions. If the column specification results in a literal value, then it may be included in a summary query. In this case, that's the string literal "All invoices".

Use the COUNT(*), AVG, and SUM functions

```
SELECT COUNT(*) AS number_of_invoices,
       AVG(invoice_total) AS avg_invoice_amt,
       SUM(invoice_total) AS total_invoice_amt
FROM invoices
WHERE invoice_date > '01-JAN-2024'
```

	NUMBER_OF_INVOICES	AVG_INVOICE_AMT	TOTAL_INVOICE_AMT
1	114	1879.741315789473684210526315789473684211	214290.51

Use the MIN and MAX functions

```
SELECT COUNT(*) AS number_of_invoices,
       MAX(invoice_total) AS highest_invoice_total,
       MIN(invoice_total) AS lowest_invoice_total
FROM invoices
```

	NUMBER_OF_INVOICES	HIGHEST_INVOICE_TOTAL	LOWEST_INVOICE_TOTAL
1	114	37966.19	6

Use the MIN and MAX functions with a non-numeric column

```
SELECT MIN(vendor_name) AS first_vendor,
       MAX(vendor_name) AS last_vendor,
       COUNT(vendor_name) AS number_of_vendors
FROM vendors
```

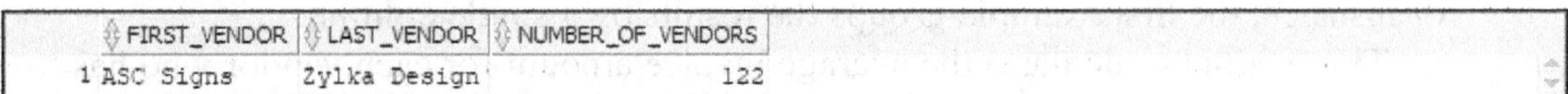

	FIRST_VENDOR	LAST_VENDOR	NUMBER_OF_VENDORS
1	ASC Signs	Zylka Design	122

Use the DISTINCT keyword

```
SELECT COUNT(DISTINCT vendor_id) AS number_of_vendors,
       COUNT(vendor_id) AS number_of_invoices,
       AVG(invoice_total) AS avg_invoice_amt,
       SUM(invoice_total) AS total_invoice_amt
FROM invoices
```

	NUMBER_OF_VENDORS	NUMBER_OF_INVOICES	AVG_INVOICE_AMT	TOTAL_INVOICE_AMT
1	34	114	1879.741315789473684210526315789473684211	214290.51

Use the ROUND function to round an average to 2 decimal places

```
SELECT 'All invoices' AS selected_invoices,
       ROUND(AVG(invoice_total), 2) AS avg_invoice_amt
FROM invoices
```

	SELECTED_INVOICES	AVG_INVOICE_AMT
1	All invoices	1879.74

Notes

- To count all of the selected rows, you typically use COUNT(*). Alternately, you can use the COUNT function with any column that doesn't contain null values.
- To count only the rows with unique values in a specified column, you can code the COUNT function with the DISTINCT keyword followed by the name of the column.
- You can use the ROUND function to round averages. The first parameter is the value to be rounded, and the second parameter is the number of decimal places to round to.

Figure 5-2 Queries that use aggregate functions

How to group and summarize data

Sometime you want to summarize only some of the data in a table, or group data before you analyze each individual group. Fortunately, that's easy to do with the GROUP BY and HAVING clauses of the SELECT statement.

How to code the GROUP BY and HAVING clauses

Figure 5-3 shows the syntax of the SELECT statement with the GROUP BY and HAVING clauses. The GROUP BY clause determines how the selected rows are grouped, and the HAVING clause determines which groups are included in the final results. These clauses are coded after the WHERE clause but before the ORDER BY clause. That makes sense because the WHERE clause is applied before the rows are grouped, and the ORDER BY clause is applied after the rows are grouped.

In the GROUP BY clause, you list one or more columns or expressions separated by commas. Then, the rows in the result set are grouped by those columns or expressions in ascending sequence. That means that a single row is returned for each unique set of values in the GROUP BY columns. In this figure, for instance, the first example groups the results by a single column.

This example calculates the average invoice amount for each vendor who has invoices in the Invoices table. To do that, it uses a GROUP BY clause to group the invoices by vendor_id. As a result, the AVG function calculates the average of the invoice_total column for each vendor rather than for the entire result set.

The example in this figure also includes a HAVING clause. The search condition in this clause specifies that only those vendors with invoice totals that average over $2,000 should be included. Note that this condition must be applied after the rows are grouped and the average for each group has been calculated.

In addition to the AVG function, the SELECT clause includes the vendor_id column. That's usually what you want since it shows which average goes with which group. However, you don't have to include the columns used in the GROUP BY clause in the SELECT clause if you don't want to.

In most cases, the SELECT clause for a statement that includes a GROUP BY clause will only include the columns that are used for grouping along with the aggregate functions. However, you can also include expressions that result in a literal value.

The syntax of a SELECT statement with GROUP BY and HAVING clauses

```
SELECT select_list
FROM table_source
[WHERE search_condition]
[GROUP BY group_by_list]
[HAVING search_condition]
[ORDER BY order_by_list]
```

Get the average invoice amount by vendor and display averages >2000

```
SELECT vendor_id, ROUND(AVG(invoice_total), 2) AS average_invoice_amount
FROM invoices
GROUP BY vendor_id
HAVING AVG(invoice_total) > 2000
ORDER BY average_invoice_amount DESC
```

	VENDOR_ID	AVERAGE_INVOICE_AMOUNT
1	110	23978.482
2	72	10963.655
3	104	7125.34
4	99	6940.25

(8 rows)

Description

- The GROUP BY clause groups the rows of a result set based on one or more columns or expressions. To include two or more columns or expressions, separate them by commas.
- If you include aggregate functions in the SELECT clause, the aggregate is calculated for each group specified by the GROUP BY clause.
- A group-by list typically consists of the names of one or more columns separated by commas. However, it can contain any expression except for those that contain aggregate functions.
- If you include two or more columns or expressions in the GROUP BY clause, the results are grouped by the first column, then by the second column if there are any ties, and so on.
- When a SELECT statement includes a GROUP BY clause, the SELECT clause can include aggregate functions, the columns used for grouping, and expressions that result in a literal value.
- The HAVING clause specifies a search condition for a group or an aggregate. This condition is applied after the rows that satisfy the search condition in a WHERE clause are grouped.

Figure 5-3 How to code the GROUP BY and HAVING clauses

Queries that use the GROUP BY and HAVING clauses

Figure 5-4 presents three more queries that group data. The first query groups the rows in the Invoices table by vendor_id and returns a count of the number of invoices for each vendor.

The second query shows how you can group by more than one column. Here, a join is used to combine the vendor_state and vendor_city columns from the Vendors table with a count and average of the invoice totals in the Invoices table. Because the rows are grouped by both state and city, a row is returned for each state and city combination. Then, the ORDER BY clause sorts the rows by city within state. Without this clause, the rows would be returned in no particular sequence.

The third query is identical to the second query except that it includes a HAVING clause. This clause uses the COUNT function to limit the state and city groups that are included in the result set to those that have two or more invoices. In other words, it excludes groups that have only one invoice.

Count the number of invoices by vendor

```
SELECT vendor_id, COUNT(*) AS invoice_qty
FROM invoices
GROUP BY vendor_id
ORDER BY vendor_id
```

	VENDOR_ID	INVOICE_QTY
1	34	2
2	37	3
3	48	1
4	72	2
5	80	2

(34 rows)

Calculate the number of invoices and average total for each state and city

```
SELECT vendor_state, vendor_city, COUNT(*) AS invoice_qty,
       ROUND(AVG(invoice_total), 2) AS invoice_avg
FROM invoices JOIN vendors
    ON invoices.vendor_id = vendors.vendor_id
GROUP BY vendor_state, vendor_city
ORDER BY vendor_state, vendor_city
```

	VENDOR_STATE	VENDOR_CITY	INVOICE_QTY	INVOICE_AVG
1	AZ	Phoenix	1	662
2	CA	Fresno	19	1208.75
3	CA	Los Angeles	1	503.2
4	CA	Oxnard	3	188
5	CA	Pasadena	5	196.12

(20 rows)

Limit the groups to those with two or more invoices

```
SELECT vendor_state, vendor_city, COUNT(*) AS invoice_qty,
       ROUND(AVG(invoice_total), 2) AS invoice_avg
FROM invoices JOIN vendors
    ON invoices.vendor_id = vendors.vendor_id
GROUP BY vendor_state, vendor_city
HAVING COUNT(*) >= 2
ORDER BY vendor_state, vendor_city
```

	VENDOR_STATE	VENDOR_CITY	INVOICE_QTY	INVOICE_AVG
1	CA	Fresno	19	1208.75
2	CA	Oxnard	3	188
3	CA	Pasadena	5	196.12
4	CA	Sacramento	7	253
5	CA	San Francisco	3	1211.04

(12 rows)

Note

- You can use a join with a summary query to group and summarize the data in two or more tables.

Figure 5-4 Queries that use the GROUP BY and HAVING clauses

How the HAVING clause compares to the WHERE clause

You can limit the groups included in a result set by using both the WHERE and the HAVING clauses. So, what's the difference? To apply a search condition to each row before it's included in a group, you use the WHERE clause. To apply a condition to a group, you use the HAVING clause. Figure 5-5 presents two examples illustrating this.

In the first example, the invoices in the Invoices table are grouped by vendor name and a count and average invoice total are calculated for each group. Then, the HAVING clause limits the groups in the result set to those that have an average invoice total greater than $500.

In contrast, the second example includes a search condition in the WHERE clause that limits the invoices included in the groups to those that have an invoice total greater than $500. In other words, the search condition in this example is applied to every row. In the previous example, it was applied to each group of rows.

There are two other main differences between the WHERE and HAVING clauses. First, the HAVING clause can include aggregate functions, but the WHERE clause can't. That's because the search condition in a WHERE clause is applied before the rows are grouped. Second, although the WHERE clause can refer to any column in the base tables, the HAVING clause can only refer to columns included in the SELECT clause. That's because it filters the groups, not the rows in the base tables.

A search condition in the HAVING clause

```
SELECT vendor_name, COUNT(*) AS invoice_qty,
       ROUND(AVG(invoice_total), 2) AS invoice_avg
FROM vendors JOIN invoices
    ON vendors.vendor_id = invoices.vendor_id
GROUP BY vendor_name
HAVING AVG(invoice_total) > 500
ORDER BY invoice_qty DESC
```

	VENDOR_NAME	INVOICE_QTY	INVOICE_AVG
1	United Parcel Service	9	2575.33
2	Zylka Design	8	867.53
3	Malloy Lithographing Inc	5	23978.48
4	IBM	2	600.06
5	Data Reproductions Corp	2	10963.66

(19 rows)

A search condition in the WHERE clause

```
SELECT vendor_name, COUNT(*) AS invoice_qty,
       ROUND(AVG(invoice_total), 2) AS invoice_avg
FROM vendors JOIN invoices
    ON vendors.vendor_id = invoices.vendor_id
WHERE invoice_total > 500
GROUP BY vendor_name
ORDER BY invoice_qty DESC
```

	VENDOR_NAME	INVOICE_QTY	INVOICE_AVG
1	United Parcel Service	9	2575.33
2	Zylka Design	7	946.67
3	Malloy Lithographing Inc	5	23978.48
4	Ingram	2	1077.21
5	Digital Dreamworks	1	7125.34

(20 rows)

Description

- When you include a WHERE clause in a SELECT statement that uses grouping and aggregates, the search condition is applied before the rows are grouped and the aggregates are calculated.
- When you include a HAVING clause in a SELECT statement that uses grouping and aggregates, the search condition is applied after the rows are grouped and the aggregates are calculated.
- A HAVING clause can only refer to a column included in the SELECT clause.
- A WHERE clause can refer to any column in the base tables.
- A HAVING clause can contain aggregate functions.
- A WHERE clause can't contain aggregate functions.

Figure 5-5 How the HAVING clause compares to the WHERE clause

How to code complex search conditions

You can code compound search conditions in a HAVING clause just as you can in a WHERE clause. The first example in figure 5-6 shows how this works. This query groups invoices by invoice date and calculates a count of the invoices and the sum of the invoice totals for each date.

In addition, the HAVING clause specifies three conditions. First, the invoice date must be between 5/1/2024 and 5/31/2024. Second, the invoice count for each date must be greater than 1. And third, the sum of the invoice totals for each date must be greater than $100.

In the HAVING clause of this query, the second and third conditions include aggregate functions. As a result, they must be coded in the HAVING clause. The first condition, however, doesn't include an aggregate function, so it can be coded in either the HAVING or WHERE clause. The second example shows this condition coded in the WHERE clause. Either way, both queries return the same result set.

Since a search condition in the WHERE clause is applied before the rows are grouped while a search condition in the HAVING clause isn't applied until after the grouping, you might expect a performance advantage by coding all search conditions in the HAVING clause. However, Oracle takes care of this performance issue for you when it optimizes the query. To do that, it automatically moves search conditions to whichever clause will result in the best performance, as long as that doesn't change the logic of your query. As a result, you can code search conditions wherever they result in the most readable code without worrying about system performance.

A compound condition in the HAVING clause

```
SELECT
    invoice_date,
    COUNT(*) AS invoice_qty,
    SUM(invoice_total) AS invoice_sum
FROM invoices
GROUP BY invoice_date
HAVING invoice_date BETWEEN '01-MAY-2024' AND '31-MAY-2024'
    AND COUNT(*) > 1
    AND SUM(invoice_total) > 100
ORDER BY invoice_date DESC
```

The same query coded with a WHERE clause

```
SELECT
    invoice_date,
    COUNT(*) AS invoice_qty,
    SUM(invoice_total) AS invoice_sum
FROM invoices
WHERE invoice_date BETWEEN '01-MAY-2024' AND '31-MAY-2024'
GROUP BY invoice_date
HAVING COUNT(*) > 1
    AND SUM(invoice_total) > 100
ORDER BY invoice_date DESC
```

The result set returned by both queries

	INVOICE_DATE	INVOICE_QTY	INVOICE_SUM
1	31-MAY-24	3	11557.75
2	23-MAY-24	6	2761.17
3	22-MAY-24	2	442.5
4	20-MAY-24	3	308.64

(15 rows)

Description

- You can use the AND and OR operators to code compound search conditions in a HAVING clause just as you can in a WHERE clause.
- If a search condition includes an aggregate function, it must be coded in the HAVING clause. Otherwise, it can be coded in either the HAVING or the WHERE clause.

How to summarize data using the ROLLUP CUBE clause

So far, this chapter has shown only standard SQL. That means the code presented so far will work the same way in almost any SQL dialect. However, there are some keywords, functions, and clauses that are useful but are not part of standard SQL. One of these is the ROLLUP CUBE clause.

Although ROLLUP functionality is provided by many SQL dialects, the syntax differs depending on how the dialect implemented it. So, that's something you should be aware of if you need to convert your code from Oracle to another SQL dialect. Fortunately, converting usually isn't difficult as long as you understand the underlying concept.

How to use ROLLUP

The ROLLUP CUBE clause is a subclause of the GROUP BY clause. It has a simple syntax as shown at the top of figure 5-7. You code either the ROLLUP or CUBE keyword and then a list of groups in parentheses, separated by commas. Which keyword you use depends on how you want to summarize your data.

The ROLLUP keyword adds one or more summary rows to a result set that uses grouping and aggregates. If the GROUP BY clause has just one column, it adds a final summary row. If the clause lists more than one column, it adds a summary row for each group except the rightmost group, plus a final summary row. This is because the rightmost group is already summarized by the aggregate functions. The two examples illustrate how this works.

The first example shows how the ROLLUP keyword works when you group by a single column. Here, the invoices in the Invoices table are grouped by vendor_id, and an invoice count and total are calculated for each vendor. Because the ROLLUP keyword is included, a summary row is added at the end of the result set. This row summarizes all the aggregate columns in the result set. In this case, it summarizes the invoice_count and invoice_total columns. Because the vendor_id column can't be summarized, it's assigned a null value.

The second query in this figure shows how the ROLLUP keyword works when you group by two columns. This query groups the vendors in the Vendors table by state and city and counts the number of vendors in each group. Because there are two columns named in the subclause, summary rows are included for each state in addition to a summary row at the end of the result set. The reason the state column is summarized but not the city column is because vendor_city is the rightmost column.

When you use an ORDER BY clause with the ROLLUP operator, the rows are sorted after the summary rows are added. Then, because null values come after other values in the Oracle sort sequence, the summary rows come after the rows that they summarize.

The syntax of the ROLLUP CUBE clause

```
{ROLLUP|CUBE} (group_by_list)
```

Add a summary row to the end of the result set

```
SELECT vendor_id, COUNT(*) AS invoice_count,
       SUM(invoice_total) AS invoice_total
FROM invoices
GROUP BY ROLLUP(vendor_id)
```

	VENDOR_ID	INVOICE_COUNT	INVOICE_TOTAL
32	121	8	6940.25
33	122	9	23177.96
34	123	47	4378.02
35	(null)	114	214290.51

Summary row

(35 rows)

Add a summary row for each grouping level

```
SELECT vendor_state, vendor_city, COUNT(*) AS qty_vendors
FROM vendors
WHERE vendor_state IN ('IA', 'NJ')
GROUP BY ROLLUP (vendor_state, vendor_city)
ORDER BY vendor_state, vendor_city
```

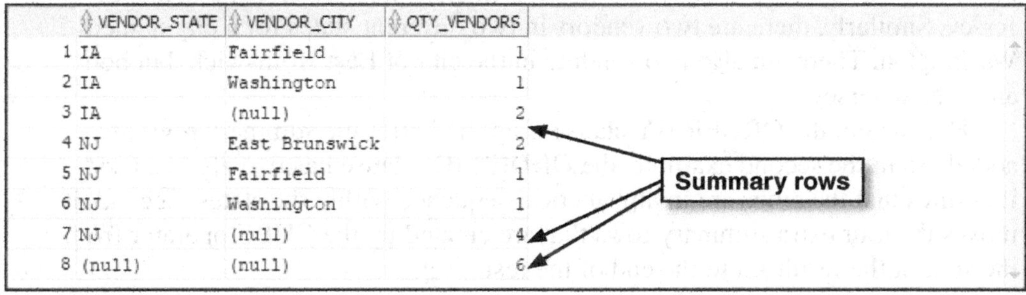

	VENDOR_STATE	VENDOR_CITY	QTY_VENDORS
1	IA	Fairfield	1
2	IA	Washington	1
3	IA	(null)	2
4	NJ	East Brunswick	2
5	NJ	Fairfield	1
6	NJ	Washington	1
7	NJ	(null)	4
8	(null)	(null)	6

Description

- The ROLLUP CUBE clause is used with the GROUP BY clause to add summary rows to the final result set.
- The ROLLUP keyword adds a summary row for each group specified in the GROUP BY clause except for the rightmost group, which is summarized by the aggregate functions.
- The ROLLUP keyword also adds a summary row to the end of the result set that summarizes the entire result set.
- If the GROUP BY clause specifies a single group, only the final summary row is added.
- The sort sequence in the ORDER BY clause is applied after the summary rows are added.

Figure 5-7 How to use ROLLUP

How to use CUBE

Figure 5-8 shows you how to use the CUBE keyword with the GROUP BY clause. This works similarly to ROLLUP, except that it adds summary rows for every combination of groups. This is illustrated by the two examples in this figure. As you can see, these examples are the same as the ones in figure 5-7 except that they use CUBE instead of ROLLUP.

In the first example, the result set is grouped by a single column. In this case, a single row is added to the result set that summarizes all the groups. In other words, this works the same as it does with ROLLUP. The only difference is that the summary row is at the start of the result set instead of the end of the result set unless you use the ORDER BY clause.

In the second example, you can see how CUBE differs from ROLLUP when you group by two or more columns. In this case, the result set includes a summary row for each state just as it did when ROLLUP was used. For instance, the third row in this example indicates that there are two vendors in the state of Iowa. In addition, though, CUBE includes a summary row for each city. For instance, the ninth row in this example indicates that there are two vendors in a city named Fairfield. But if you look at the first and fifth rows in the result set, you'll see that one of those vendors is in Iowa and one is in New Jersey. Similarly, there are two vendors in two different states for a city named Washington. There are also two vendors in the city of East Brunswick, but both are in New Jersey.

Here again, the ORDER BY clause is applied after the summary rows are added. So in the second example, the ORDER BY clause has two effects. First, it ensures that the cities are in alphabetical sequence within the states. Second, it moves the four extra summary rows that are created by the CUBE operator from the start of the result set to the end of the result set.

Add a summary row to the beginning of the result set

```
SELECT vendor_id, COUNT(*) AS invoice_count,
       SUM(invoice_total) AS invoice_total
FROM invoices
GROUP BY CUBE(vendor_id)
```

	VENDOR_ID	INVOICE_COUNT	INVOICE_TOTAL
1	(null)	114	214290.51
2	34	2	1200.12
3	37	3	564
4	48	1	856.92

Summary row

(35 rows)

Add a summary row for each set of groups

```
SELECT vendor_state, vendor_city, COUNT(*) AS qty_vendors
FROM vendors
WHERE vendor_state IN ('IA', 'NJ')
GROUP BY CUBE(vendor_state, vendor_city)
ORDER BY vendor_state, vendor_city
```

	VENDOR_STATE	VENDOR_CITY	QTY_VENDORS
1	IA	Fairfield	1
2	IA	Washington	1
3	IA	(null)	2
4	NJ	East Brunswick	2
5	NJ	Fairfield	1
6	NJ	Washington	1
7	NJ	(null)	4
8	(null)	East Brunswick	2
9	(null)	Fairfield	2
10	(null)	Washington	2
11	(null)	(null)	6

Description

- The CUBE keyword adds a summary row for each group specified in the GROUP BY clause.
- The CUBE keyword also adds a summary row to the beginning of the result set that summarizes the entire result set.
- The sort sequence in the ORDER BY clause is applied after the summary rows are added.

Figure 5-8 How to use CUBE

How to code analytic functions

The final section in this chapter presents some useful techniques for further analyzing and partitioning data by using analytic functions. You'll also learn how to use the ranking functions to rank ordered data.

How analytic functions work

When you code an aggregate function, it returns a single value as shown in the first example in figure 5-9. When you use GROUP BY with an aggregate function, it returns a value for each group. And when you code an aggregate function as an *analytic function*, it returns a value for each row.

An analytic function can be almost any function that takes a column as a parameter and returns a single value. For example, all of the aggregate functions you've seen in this chapter can also be used as analytic functions. You can even create your own functions and use them as analytic functions.

To indicate that you're using a function as an analytic function, you code the function and any parameters as usual, and then the OVER keyword followed by a set of parentheses. Within the parentheses you can optionally include an *analytic clause* which specifies how the data should be partitioned and ordered.

A *partition* is simply a group of rows. Partitions in turn are used to define *windows*. A window is a group of rows that are analyzed together by a function. When used as seen in this figure, each partition is one window.

If you code an empty set of parentheses after OVER, it's called an *empty analytic clause*. The entire result set is then considered to be one window, and the same value is returned for each row. This is illustrated by the second example in this figure, which uses SUM as an analytic function.

Notice two main differences between this example and the first. One, instead of returning just one row, all rows in the table are returned. Two, because all rows are returned, columns besides aggregate functions and literals can be included in the SELECT statement. In this case, the invoice_id and vendor_id columns have been selected. This wouldn't be possible with the first example.

The third example codes an analytic clause in the parentheses after OVER to specify how the rows should be partitioned before being analyzed. In this case, the clause uses the PARTITION BY keywords to partition the data by vendor ID. So, SUM now returns the sum of the invoice_total column for each unique vendor ID instead of the overall sum.

The simplified syntax of an analytic function

```
function_name([parameters]) OVER(analytic_clause)
```

The simplified syntax of an analytic clause

```
[PARTITION BY {expr[, expr]...}]
[order_by_clause [windowing_clause]]
```

An aggregate function

```
SELECT SUM(invoice_total)
FROM invoices
```

	SUM(INVOICE_TOTAL)
1	214290.51

An analytic function

```
-- can select non-aggregate columns with analytic functions
SELECT invoice_id, vendor_id, SUM(invoice_total) OVER()
FROM invoices
```

	INVOICE_ID	VENDOR_ID	SUM(INVOICE_TOTAL)OVER()
1	1	34	214290.51
2	2	34	214290.51
3	3	110	214290.51

An analytic function with PARTITION BY

```
SELECT invoice_id, vendor_id,
       SUM(invoice_total) OVER(PARTITION BY vendor_id)
FROM invoices
```

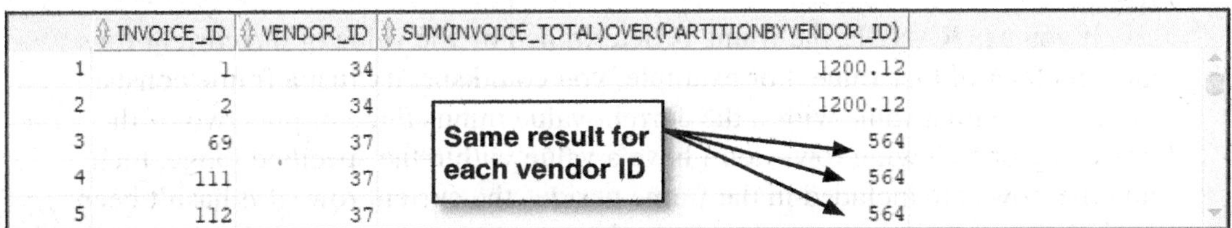

	INVOICE_ID	VENDOR_ID	SUM(INVOICE_TOTAL)OVER(PARTITIONBYVENDOR_ID)
1	1	34	1200.12
2	2	34	1200.12
3	69	37	564
4	111	37	564
5	112	37	564

Description

- A *partition* is an expression that specifies a group of rows.
- The group of rows analyzed by an analytic function is called a *window*.
- An *analytic function* is indicated by use of the OVER keyword after a function and computes a value for each specified window. It returns that value for each row.
- Almost any function that accepts a column as a parameter and returns a single value can be used as an analytic function, including all of the aggregate functions.
- You can optionally code an *analytic clause* that specifies how the data should be partitioned and ordered within the parentheses after the OVER keyword. If the
- If you code an *empty analytic clause*, the entire result set is treated as a single window.

Figure 5-9 How analytic functions work

How to code frames

In the previous figure, you saw windows defined by partitions. However, you can also define a window with a *frame*.

Just as a window frame outlines a window in a house, you can think of a window frame as outlining a window of data to be analyzed. So, instead of windows that are the same as the partitions, the window frame allows you to "move" the window to analyze different groups of data within or across partitions. The definition of a frame is always relative to the current row. That makes it easy to calculate cumulative totals and moving averages.

Figure 5-10 begins by showing the simplified syntax for the windowing clause, which is used to define a frame. To code a windowing clause, you must precede it with an ORDER BY clause in the analytic clause.

After the ORDER BY clause, you code either the ROWS, RANGE, or GROUPS keyword. Then, you specify the starting and ending points for your frame using BETWEEN and AND. Or, if you only want to specify a starting row or value, then the ending point is the current row by default.

Finally, you can choose to exclude the current row, the current row and any peers (GROUPS), peers of the current row (TIES), or no others. NO OTHERS does not exclude anything and is the default. A *peer* is a row that has the same sort sequence as another row in the partition. A *peer group* is all rows with the same value in a partition.

If you use ROWS, the frame is determined by the number of rows before and after the current row. For example, you could specify a frame that contains the five rows before the current row, the current row, and all rows after the current row.

If you use RANGE, the frame is determined by the value of the current row and an offset of that value. For example, you could specify that a frame consist of all rows with a value within the current value minus three or plus two. If the preceding or following rows don't have a value within the specified range, then no other rows are included in the frame besides the current row (if it hasn't been excluded).

If you use GROUPS, the frame is determined by the number of peer groups with the same value before and after the current peer group. For example, you could specify the current row's peer group, all rows in the two peer groups before the current row, and all rows in the following peer group.

When you use ROWS or GROUPS, you can partition the data with ORDER BY using as many columns as you like. If you use RANGE, however, you usually can only partition by one column.

In this figure's first example, the rows have been ordered by invoice date and the frame covers all rows from the beginning of the data set to the current row. This allows you to calculate a cumulative total, in this case partitioned by vendor. The second example returns the same results but uses BETWEEN and AND to define the frame.

The third example uses the EXCLUDE clause of the windowing clause to exclude the current row. So, the frame is only the preceding row and the following row, which it then averages.

The simplified syntax of the windowing clause

```
{ROWS|RANGE|GROUPS} {start_row | BETWEEN start_row AND end_row}
[EXCLUDE {CURRENT ROW|GROUPS|TIES|NO OTHERS}]
```

ROWS, RANGE, and GROUPS

Value	Description
ROWS	The frame is determined based on an offset number of rows from the current row. Allows any number of columns with ORDER BY.
RANGE	The frame is determined based on a logical offset from a value from the current row. Usually allows only one column with ORDER BY.
GROUPS	The frame is determined based on an offset number of groups from the current group. Allows any number of columns with ORDER BY.

Allowed values for start_row and end_row

Value	Description
CURRENT ROW	The frame starts or ends with the current row.
UNBOUNDED PRECEDING	The frame starts with the first row in the partition.
UNBOUNDED FOLLOWING	The frame ends with the last row in the partition.
expr PRECEDING	The frame starts expr rows, values, or groups before the current row.
expr FOLLOWING	The frame ends expr rows, values, or groups after the current row.

A cumulative total for invoices by vendor

```
SUM(invoice_total) OVER(PARTITION BY vendor_id ORDER BY invoice_date
                   ROWS UNBOUNDED PRECEDING)
```

The same result but with BETWEEN and AND

```
SUM(invoice_total) OVER(PARTITION BY vendor_id ORDER BY invoice_date
                   ROWS BETWEEN UNBOUNDED PRECEDING AND CURRENT ROW)
```

The average of the previous and next invoice totals

```
AVG(invoice_total) OVER(ORDER BY invoice_date
                   ROWS BETWEEN 1 PRECEDING AND 1 FOLLOWING
                   EXCLUDE CURRENT ROW)
```

The average of the invoice totals for the last three months

```
AVG(invoice_total) OVER(ORDER BY EXTRACT(MONTH FROM invoice_date)
                   RANGE 2 PRECEDING)
```

The highest invoice total from the current day, previous day, and day after

```
MAX(invoice_total) OVER(ORDER BY invoice_date
                   RANGE BETWEEN 1 PRECEDING AND 1 FOLLOWING)
```

The highest invoice total from the current day, previous day with an entry, and day after with an entry

```
MAX(invoice_total) OVER(ORDER BY invoice_date
                   GROUPS BETWEEN 1 PRECEDING AND 1 FOLLOWING)
```

Figure 5-10 How to code frames

The fourth example uses RANGE to get all rows that share the current row's value as well as all rows with the preceding two values. Since it's ordered by invoice month, that means it gets all invoice totals from the current month and the preceding two months, which are then averaged.

The fifth example gets the highest invoice total from the current, previous, and next date. And the last example gets the highest invoice total from the current date, previous date that has an entry, and next date that has an entry.

Two more examples of frames

Figure 5-11 presents two more examples to help you understand analytic functions and frames. The first compares ROWS, RANGE, and GROUPS when you order the Invoices table by invoice_date. Each gets a sum of invoice totals for the current row plus some preceding and following rows.

The column named Rows_Example gets the sum of the current row, the immediately previous row, and the immediately following row. So, for example, the value for row 74 is equal to the sum of the values for rows 73, 74, and 75, even though row 73 has a date of May 19 and rows 74 and 75 have a date of May 20. All that matters is the order of the rows.

The column named Range_Example gets the sum for the current row and any peers, plus any rows with values immediately previous or following. Row 74 has an invoice date of May 20, so this means it gets the sum of rows with the dates May 19, 20, and 21. There are no rows with a date of May 21, so it calculates the total for rows 72-76.

The column named Groups_Example gets the sum for the current, preceding, and following peer groups. For row 74, the current peer group is rows with an invoice date of May 20, the previous peer group has a date of May 19, and the following peer group has a date of May 22. So, it calculates the total for rows 72-78.

The second example in this figure illustrates a common use for frames. Here, a moving average is calculated for the invoice totals for each month. A moving average is an average that's calculated on the current row plus a specified number of rows before and/or after the current row. It's particularly useful when working with data over a period of time to eliminate short-term fluctuations so long-term trends become more obvious.

In this example, a three-month average is calculated for the sum of invoice totals for each vendor. To do that, the RANGE keyword is coded followed by 2 PRECEDING to indicate that the invoice total for the current month and the two months prior should be used to calculate the average. The three-month average for month 5, for example, is calculated by adding the values in the invoice_total column for months 3, 4, and 5 and dividing by 3.

Note that this example includes a GROUP BY clause. Because of that, it can't include any non-aggregate columns that aren't used for grouping. It also uses the ROUND function, which must surround the entire analytic function, including the analytic clause, in order to work correctly.

A comparison of ROWS, RANGE, and GROUPS

```
SELECT vendor_id, invoice_date, invoice_total,
SUM(invoice_total) OVER(ORDER BY invoice_date
          ROWS BETWEEN 1 PRECEDING AND 1 FOLLOWING) as rows_example,
SUM(invoice_total) OVER(ORDER BY invoice_date
          RANGE BETWEEN 1 PRECEDING AND 1 FOLLOWING) as range_example,
SUM(invoice_total) OVER(ORDER BY invoice_date
          GROUPS BETWEEN 1 PRECEDING AND 1 FOLLOWING) as groups_example
FROM invoices;
```

The result set

	VENDOR_ID	INVOICE_DATE	INVOICE_TOTAL	ROWS_EXAMPLE	RANGE_EXAMPLE	GROUPS_EXAMPLE
72	37	19-MAY-24	224	334.59	783.97	783.97
73	123	19-MAY-24	42.67	490.67	783.97	783.97
74	37	20-MAY-24	224	306.87	575.31	1017.81
75	123	20-MAY-24	40.2	308.64	575.31	1017.81
76	123	20-MAY-24	44.44	127.14	575.31	1017.81
77	123	22-MAY-24	[illegible]	486.94	3203.67	3512.31
78	123	22-MAY-24	[illegible]	1021.92	3203.67	3512.31
79	83	23-MAY-24	579.42	2554.42	3209.67	3209.67
80	83	23-MAY-24	1575	2374.42	3209.67	3209.67
81	105	23-MAY-24	220	1967.5	3209.67	3209.67
82	123	23-MAY-24	172.5	539.75	3209.67	3209.67

Peer group

Get a moving three-month average for each vendor

```
-- can only include aggregate and analytic functions since using GROUP BY
SELECT EXTRACT(MONTH FROM invoice_date) AS month,
    SUM(invoice_total) AS total_invoices,
    ROUND(AVG(SUM(invoice_total))
                   OVER(ORDER BY EXTRACT(MONTH FROM invoice_date)
                        RANGE 2 PRECEDING), 2) as moving_avg
FROM invoices
GROUP BY EXTRACT(MONTH FROM invoice_date);
```

The result set

	MONTH	TOTAL_INVOICES	MOVING_AVG
1	2	116.54	116.54
2	3	1083.58	600.06
3	4	77878.59	26359.57
4	5	93283.48	57415.22
5	6	20086.32	63749.46
6	7	21842	45070.6

Description

- A *window frame* defines a window that can "move" based on the current row.
- To use a frame, you must include an ORDER BY clause in the analytic clause.
- If you include ORDER BY but omit a windowing clause, the analytic function defaults to RANGE UNBOUNDED PRECEDING.
- A *peer* is a row that has the same sort sequence as another row in the partition.
- A *peer group* is all rows with the same value in a partition.

Figure 5-11 Two more examples of frames

How to use named windows

In some cases, you'll need to code a SELECT statement with two or more aggregate functions that use the same window. Then, you may want to use a *named window* so you don't have to repeat the definition for the window for each function. Figure 5-12 shows how.

The first example in this figure shows a SELECT statement that includes two aggregate functions. Both functions have an OVER clause that partitions the rows in the result set by the vendor_id column. Because of that, the PARTITION BY clause is repeated on each OVER clause.

An easier way to do this is to name the window by coding a WINDOW clause as shown in the second example. Here, the window is named vw and is defined after the AS keyword. Then, the two aggregate functions include just the window name in the OVER clause. In other words, they don't have to repeat the PARTITION BY clause.

Using a WINDOW clause not only makes your code shorter, but also makes it easier to maintain. If you later need to change the window definition, you only need to do so in one place.

The third example in this figure shows how you can modify a window definition when you use it. To do that, you can add a PARTITION BY cause, ORDER BY clause, and/or a frame definition so long as the clause you're adding isn't already present.

In this example, the named window partitions the rows in the result set by the vendor_id column just like the second example. However, the analytic clause for the vendor_total column adds a frame to the window to create a cumulative total ordered by date.

When you use named windows, you can't modify any of the clauses that are in the window definition. For example, because vw has a PARTITION BY clause, you can't include that clause on an OVER clause that uses the named window. Instead, you can only add to the window definition.

The syntax for naming a window

```
WINDOW window_name AS (analytic_clause)
```

Two functions that use the same window

```
SELECT vendor_id, invoice_date, invoice_total,
       SUM(invoice_total) OVER(PARTITION BY vendor_id) AS vendor_total,
       MAX(invoice_total) OVER(PARTITION BY vendor_id) AS vendor_max
FROM invoices
```

Using a named window

```
SELECT vendor_id, invoice_date, invoice_total,
       SUM(invoice_total) OVER (vw) AS vendor_total,
       MAX(invoice_total) OVER (vw) AS vendor_max
FROM invoices
WINDOW vw AS (PARTITION BY vendor_id)
```

The result set for both statements

	VENDOR_ID	INVOICE_DATE	INVOICE_TOTAL	VENDOR_TOTAL	VENDOR_MAX
1	34	25-FEB-24	116.54	1200.12	1083.58
2	34	14-MAR-24	1083.58	1200.12	1083.58
3	37	17-MAY-24	116	564	224
4	37	19-MAY-24	224	564	224
5	37	20-MAY-24	224	564	224

Add to the specification for a named window

```
SELECT vendor_id, invoice_date, invoice_total,
       SUM(invoice_total) OVER (vw ORDER BY invoice_date
                          ROWS UNBOUNDED PRECEDING) AS vendor_total,
       MAX(invoice_total) OVER vw AS vendor_max
FROM invoices
WINDOW vw AS (PARTITION BY vendor_id)
```

Description

- To define a *named window*, code a WINDOW clause. This clause should be coded after the HAVING clause and before the ORDER BY clause, if those clauses are included.
- To use a named window, code its name in the analytic clause.
- You can add clauses to a named window when you use it.

Figure 5-12 How to use named windows

How to use the ranking functions

Figure 5-13 shows how to use four of Oracle's *ranking functions*. These functions provide ways to rank the rows returned by a result set. All four of these functions work similarly when used as analytic functions.

Like the previous analytic functions, you first code the name of the ranking function, followed by a set of parentheses, followed by the OVER keyword and a second set of parentheses. Within the second set of parentheses you must code an ORDER BY clause that specifies the sort order. Optionally, you can precede the ORDER BY clause with a PARTITION BY clause, but you can't include a windowing clause to create frames.

The first example shows how to use the ROW_NUMBER function to rank the vendors in the Vendors table alphabetically by name. As you can see, ASC Signs is ranked 1, AT&T is ranked 2, and so on. Note that because the sort is case-sensitive "ASC" has a sort order before "Abbey". If that's not what you want, you can use the LOWER function within the ORDER BY clause to make the sort case-insensitive as shown in the second example.

The second example uses the optional PARTITION BY clause to partition the rows before they are ranked. Here, the vendors are partitioned by state so that they are sorted and ranked within each state.

In addition to the ORDER BY clause in the analytic function, you can also code a regular ORDER BY clause that applies to the entire result set as you would any other SELECT statement. For example, you could modify the second example like this:

```
SELECT vendor_name, vendor_state,
       ROW_NUMBER() OVER(PARTITION BY vendor_state ORDER BY vendor_name)
         AS row_number
FROM vendors
ORDER BY vendor_name
```

In this case, the ORDER BY clause within the ranking function is used to sort the rows so they can be ranked by vendor name within each state. Then, the regular ORDER BY clause sorts the rows by name regardless of state.

The syntax for the ranking functions

```
ROW_NUMBER()                 OVER ([partition_by_clause] order_by_clause)
RANK()                       OVER ([partition_by_clause] order_by_clause)
DENSE_RANK()                 OVER ([partition_by_clause] order_by_clause)
NTILE(integer_expression)    OVER ([partition_by_clause] order_by_clause)
```

Rank vendors by name

```
SELECT vendor_name,
       ROW_NUMBER() OVER(ORDER BY vendor_name) AS row_number
FROM vendors
```

The result set

	VENDOR_NAME	ROW_NU...
1	ASC Signs	1
2	AT&T	2
3	Abbey Office Furnishings	3
4	American Booksellers Assoc	4
5	American Express	5
6	Ascom Hasler Mailing Systems	6

Rank vendors in each state by name (case-insensitive)

```
SELECT vendor_name, vendor_state,
       ROW_NUMBER() OVER(PARTITION BY vendor_state
                         ORDER BY LOWER(vendor_name))
       AS row_number
FROM vendors
```

The result set

	VENDOR_NAME	VENDOR_STATE	ROW_NUMBER
1	AT&T	AZ	1
2	Computer Library	AZ	2
3	Wells Fargo Bank	AZ	3
4	Abbey Office Furnishings	CA	1
5	American Express	CA	2
6	ASC Signs	CA	3

Description

- The ROW_NUMBER, RANK, DENSE_RANK, and NTILE functions are known as *ranking functions*.
- The ROW_NUMBER function returns the sequential number of a row within a partition of a result set, starting at 1 for the first row in each partition.
- The ORDER BY clause of a ranking function specifies the sort order in which the ranking function is applied.
- The optional PARTITION BY clause of a ranking function specifies the column that's used to divide the result set into groups.
- You can't include a windowing clause or use frames when coding a ranking function.
- The above syntax is for the ranking functions used as analytic functions. RANK and DENSE_RANK may also be used as aggregate functions with a different syntax.

Figure 5-13 How to use the ranking functions (part 1 of 2)

The third example shows how the RANK and DENSE_RANK functions work. Like ROW_NUMBER, they assign ranks to rows according to the specified sort order. Unlike ROW_NUMBER, however, they assign the same rank to rows that "tie", or have the same rank in the sort order.

In the example, the rows are sorted by the invoice_total column for both RANK and DENSE_RANK. Since the first three rows in this order have the same invoice total, both functions give these three rows the same rank, 1. However, the fourth row has a different value. To calculate the value for this row, the RANK function adds 1 to the total number of previous rows. Since the first three rows are tied for first place, the fourth row gets fourth place and is assigned a rank of 4.

The DENSE_RANK function, on the other hand, calculates the value for the fourth row by adding 1 to the rank of the previous row. As a result, this function assigns a rank of 2 to the fourth row. In other words, since the first three rows are tied for first place, the fourth row gets second place.

The fourth example shows how the NTILE function works. You use this function to divide rows into a specified number of groups called *buckets*. For instance, if a result set returns 100 rows, NTILE(10) would assign a rank of 1 to the first 10 rows according to the sort order, a rank of 2 to the next 10 rows, and so on. If the number of rows can't be divided evenly by the specified number of buckets, then the first groups get an extra row. There is never a difference of more than one row in bucket size.

In this example, the NTILE function is used to divide a small result set that contains 5 rows. The first NTILE function divides the rows into 2 buckets, with the first having 3 rows and the second having 2 rows. The second NTILE function divides the rows into 3 buckets, with the first and second having 2 rows each and the third just 1 row. And so on.

Now that you know the differences between the ranking functions, which should you use? It depends on your application. No matter which you choose, however, it's a good idea to test that the ranks are being calculated the way you want and expect before using these statements in a production environment.

RANK compared to DENSE_RANK

```
SELECT RANK() OVER (ORDER BY invoice_total) AS rank,
       DENSE_RANK() OVER (ORDER BY invoice_total) AS dense_rank,
       invoice_total, invoice_number
FROM invoices
```

The result set

	RANK	DENSE_RANK	INVOICE_TOTAL	INVOICE_NUMBER
1	1	1	6	25022117
2	1	1	6	24863706
3	1	1	6	24780512
4	4	2	9.95	21-4748363
5	4	2	9.95	21-4923721
6	6	3	10	4-342-8069

Description

- The RANK and DENSE_RANK functions both return the rank of each row within the partition of a result set.
- If there is a tie, both functions give the same rank to all rows that are tied.
- To determine the rank for the next distinct row, RANK adds 1 to the number of previous rows, while DENSE_RANK adds 1 to the rank for the previous row.

Divide rows into a specified number of groups

```
SELECT terms_description,
    NTILE(2) OVER (ORDER BY terms_id) AS tile2,
    NTILE(3) OVER (ORDER BY terms_id) AS tile3,
    NTILE(4) OVER (ORDER BY terms_id) AS tile4
FROM terms
```

The result set

	TERMS_DESCRIPTION	TILE2	TILE3	TILE4
1	Net due 10 days	1	1	1
2	Net due 20 days	1	1	1
3	Net due 30 days	1	2	2
4	Net due 60 days	2	2	3
5	Net due 90 days	2	3	4

Description

- The NTILE function divides the rows in a partition into the specified number of groups. These groups are sometimes called *buckets*.
- If the rows can't be evenly divided into buckets, the later buckets may have one less row than the earlier buckets.

Figure 5-13 How to use the ranking functions (part 2 of 2)

Perspective

In this chapter, you learned how to code queries that group and summarize data. You also learned how to further partition, sort, and analyze that data with a variety of useful functions. In most cases, you'll be able to use the techniques presented here to get all the summary information you need to create useful reports.

Terms

aggregate function
column function
summary query
partition
window
analytic function
analytic clause
empty analytic clause
window frame
peer
peer group
named window
ranking function
bucket

Exercises

1. Write a SELECT statement that returns one row for each vendor and contains these columns from the Invoices table:

    The vendor_id column

    The sum of the invoice_total column for that vendor

    The average of the invoice_total column for that vendor

    The result set should be sorted by vendor_id.

2. Write a SELECT statement that returns one row for each vendor and contains these columns:

    The vendor_name column from the Vendors table

    The count of the invoices for each vendor in the Invoices table

    The sum of the payment_total column for the vendor from the Invoices table

    The result set should be sorted in descending sequence by the payment total sum for each vendor.

3. Write a SELECT statement that returns one row for each general ledger account number and contains these columns:

    The account_description column from the General_Ledger_Accounts table

    The count of the entries in the Invoice_Line_Items table for that account_number

    The sum of the line item amounts in the Invoice_Line_Items table that have that account-number

Filter the result set to include only those rows with a count greater than 1; group the result set by account description; and sort the result set in descending sequence by the sum of the line item amounts. When you join the tables, use an alias for each table.

4. Modify your solution for exercise 3 to filter for only invoices dated in the second quarter of 2024 (April 1, 2024 to June 30, 2024).

   Hint: Join to the Invoices table to code a search condition based on invoice_date.

5. Write a SELECT statement that answers this question: What is the total amount invoiced for each general ledger account number? Use ROLLUP to include a row that gives the grand total.

   Hint: Use the line_item_amt column of the Invoice_Line_Items table.

6. Write a SELECT statement that answers this question: Which vendors are being paid from more than one account? Return two columns: the vendor name and the total number of accounts that apply to that vendor's invoices. You will need to use two joins.

   Hint: Use the DISTINCT keyword to count the account_number column in the Invoice_Line_Items table.

7. Write a SELECT statement that uses an analytic function to generate a four-week average for payments received in the Paid Invoices table in the EX schema. Include the following columns:

   Payment date

   Payment total

   The 4-week average for the payment total column

   Hint: Use RANGE with an offset of 28 days.

8. Modify your solution for exercise 7 to add another column that displays the cumulative total for payments up to the date of the current row. Then, add a WINDOW clause that names a window and use this window for both calculated columns.

   Hint: Only include the clause the two analytic functions have in common in the named window.

9. Write a SELECT statement that gets the following columns using the Active Invoices table in the EX schema:

   Invoice ID

   Invoice total

   Rank for each invoice total calculated using ROW_NUMBER

   Rank for each invoice total calculated using RANK

   Rank for each invoice total calculated using NTILE(5)

6

How to code subqueries

A subquery is a query coded inside of another query. As a result, you can use subqueries to build queries that would be difficult or impossible to do otherwise. In this chapter, you'll learn how to use subqueries within SELECT statements. Then, in the next chapter, you'll learn how to use them when you code INSERT, UPDATE, and DELETE statements.

An introduction to subqueries 160
How to use subqueries 160
How subqueries compare to joins 162
How to code subqueries in search conditions 164
How to use subqueries with the IN operator 164
How to compare a subquery with an expression 166
How to use the ALL keyword 168
How to use the ANY and SOME keywords 170
How to code correlated subqueries 172
How to use the EXISTS operator 174
Other ways to use subqueries 176
How to code subqueries in the FROM clause 176
How to code subqueries in the SELECT clause 178
Guidelines for working with complex queries 180
A complex query that uses subqueries 180
A procedure for building complex queries 182
How to use subquery factoring 184
How to use the WITH clause 184
How to code a recursive query 186
How to use the hierarchical query clause 188
Perspective 190

An introduction to subqueries

A *subquery* is a SELECT statement that's coded within another SQL statement. Since you already know how to code SELECT statements, you already know how to code subqueries. Now you just need to learn where you can code them and when you should use them.

How to use subqueries

A subquery can be coded, or *introduced*, in almost any clause of a SELECT statement. However, you're most likely to use one in the SELECT or FROM clauses to specify a column or table or in the WHERE or HAVING clauses as a search condition. Figure 6-1 shows how this works.

The first example in this figure is a simple SELECT statement that uses the AVG aggregate function to get the average invoice total from the Invoices database. The result is a value of approximately 1879.74.

The second example in this figure retrieves all the invoices from the Invoices table that have invoice totals greater than the average of all the invoices. It does this by coding the SELECT statement from the first example as a subquery in the WHERE clause. This is equivalent to coding the WHERE clause with a literal value like this:

```
WHERE invoice_total > 1879.74
```

However, using a subquery has several advantages over a literal value in this case. For one, you can run just one SELECT statement, not two. Second, you don't need to re-calculate the literal value if the data in the table changes. And third, it's clear to anyone reviewing your code just where the value you're using comes from and how it is calculated.

When a subquery returns a single value as it does in this example, you can use it anywhere you would normally use a single value. However, a subquery can also return a list of values, such as a result set that has one column. In that case, you can use the subquery in place of a list of values, such as the list for an IN operator. Or, a subquery can return multiple columns. In that case, you can use the subquery in the FROM clause in place of a table.

Finally, you can code a subquery within another subquery. In that case, the subqueries are said to be nested. Because *nested subqueries* can be difficult to read, you should use them only when necessary.

Four ways to introduce a subquery in a SELECT statement

1. In the SELECT clause as a column specification
2. In the FROM clause as a table specification
3. In the WHERE clause as a search condition
4. In the HAVING clause as a search condition

Get the average invoice total

```
SELECT AVG(invoice_total)
FROM invoices
```

The result set

	AVG(INVOICE_TOTAL)
1	1879.7413157894736842...

Get the average invoice total as a subquery in the WHERE clause

```
SELECT invoice_number, invoice_date, invoice_total
FROM invoices
WHERE invoice_total >
    (SELECT AVG(invoice_total)
     FROM invoices)
ORDER BY invoice_total
```

The result set

	INVOICE_NUMBER	INVOICE_DATE	INVOICE_TOTAL
1	989319-487	18-APR-24	1927.54
2	97/522	30-APR-24	1962.13
3	989319-417	26-APR-24	2051.59
4	989319-427	25-APR-24	2115.81

(21 rows)

Description

- A *subquery* is a SELECT statement that's coded within another SQL statement.
- A subquery can return a single value, a result set that contains a single column, or a result set that contains one or more columns.
- A subquery that returns a single value can be coded, or *introduced*, anywhere an expression is allowed.
- The syntax for a subquery is the same as for a standard SELECT statement. However, a subquery doesn't typically include the GROUP BY or HAVING clauses.
- Subqueries can be *nested* within other subqueries.

Figure 6-1 How to use subqueries

How subqueries compare to joins

In the last figure, you saw an example of a subquery that returns an aggregate value that's used in the search condition of a WHERE clause. This type of subquery provides for processing that can't be easily done any other way. However, most subqueries can be restated as joins, and most joins can be restated as subqueries as shown by the SELECT statements in figure 6-2.

Both SELECT statements in this figure return a result set that consists of selected rows and columns from the Invoices table. In this case, only the invoices for vendors in California are returned.

The first statement uses a join to combine the Vendors and Invoices tables so the vendor_state column can be tested for each invoice. In contrast, the second statement uses a subquery to return a result set that consists of the vendor_id column for each vendor in California. Then, that result set is used with the IN operator in the search condition so only invoices with a vendor_id in that result set are included in the final result set.

If you have a choice, which technique should you use? In general, it's best to use the technique that results in the most readable code. For example, a join tends to be more intuitive than a subquery when it uses an existing relationship between two tables. That's the case with the Vendors and Invoices tables in this figure. On the other hand, a subquery tends to be more intuitive when it uses an ad hoc relationship.

You should also realize that when you use a subquery in a search condition, its results can't be included in the final result set. For instance, the second example in this figure can't be changed to include the vendor_name column from the Vendors table. That's because the Vendors table isn't named in the FROM clause of the outer query. So if you need to include information from both tables in the result set, you need to use a join.

A query that uses an inner join

```
SELECT invoice_number, invoice_date, invoice_total
FROM invoices JOIN vendors
    ON invoices.vendor_id = vendors.vendor_id
WHERE vendor_state = 'CA'
ORDER BY invoice_date
```

The same query restated with a subquery

```
SELECT invoice_number, invoice_date, invoice_total
FROM invoices
WHERE vendor_id IN
    (SELECT vendor_id
    FROM vendors
    WHERE vendor_state = 'CA')
ORDER BY invoice_date
```

The result set returned by both queries

	INVOICE_NUMBER	INVOICE_DATE	INVOICE_TOTAL
1	QP58872	25-FEB-24	116.54
2	Q545443	14-MAR-24	1083.58
3	MAB01489	16-APR-24	936.93
4	97/553B	26-APR-24	313.55

(40 rows)

Advantages of joins

- The SELECT clause of a join can include columns from both tables.
- A join tends to be more intuitive when it uses an existing relationship between the two tables, such as a primary key to foreign key relationship.

Advantages of subqueries

- You can use a subquery to pass an aggregate value to the outer query.
- A subquery tends to be more intuitive when it uses an ad hoc relationship between the two tables.
- Long, complex queries can sometimes be easier to code using subqueries.

Description

- Like a join, a subquery can be used to code queries that work with two or more tables.
- Most subqueries can be restated as joins and most joins can be restated as subqueries.

Figure 6-2 How subqueries compare to joins

How to code subqueries in search conditions

You can use a variety of techniques to work with a subquery in a search condition. As you read this next section, keep in mind that although the examples illustrate the use of subqueries in a WHERE clause, all of this information applies to the HAVING clause as well.

How to use subqueries with the IN operator

The IN operator tests whether an expression is or isn't contained in a list of values. One way to provide that list of values is to use a subquery. This is illustrated in figure 6-3.

The example in this figure retrieves the vendors from the Vendors table that don't have invoices in the Invoices table. To do that, it uses a subquery to retrieve the vendor_id of each vendor in the Invoices table, which is returned as a result set with single column. Then, this result set is used to filter the vendors that are included in the final result set.

It's important that this subquery returns just one column because that's a requirement when a subquery is used with the IN operator. Note also that the subquery includes the DISTINCT keyword. That means that even if more than one invoice exists for a vendor, the vendor_id for that vendor will be included only once in the subquery's result set. The keyword DISTINCT is optional, but it reduces the values in the subquery's result set to be checked against. The final result set will be the same if you include it or omit it.

In the previous figure, you saw that a query that uses a subquery with the IN operator can be restated using an inner join. Similarly, a query that uses a subquery with the NOT IN operator can typically be restated using an outer join. For example, the first query shown in this figure can be restated as shown in the second query. In this case, though, the query with the subquery is more readable.

The syntax of a WHERE clause that uses an IN phrase with a subquery

```
WHERE test_expression [NOT] IN (subquery)
```

Return vendors without invoices

```
SELECT vendor_id, vendor_name, vendor_state
FROM vendors
WHERE vendor_id NOT IN
    (SELECT DISTINCT vendor_id
     FROM invoices)
ORDER BY vendor_id
```

The result of the subquery

	VENDOR_ID
1	34
2	37
3	48
4	72

(34 rows)

The result set

	VENDOR_ID	VENDOR_NAME	VENDOR_STATE
1	1	US Postal Service	WI
2	2	National Information Data Ctr	DC
3	3	Register of Copyrights	DC
4	4	Jobtrak	CA

(88 rows)

The query restated without a subquery

```
SELECT v.vendor_id, vendor_name, vendor_state
FROM vendors v LEFT JOIN invoices i
    ON v.vendor_id = i.vendor_id
WHERE i.vendor_id IS NULL
ORDER BY v.vendor_id
```

Description

- The IN operator allows you to test if a value is in a list of values returned by a subquery.
- When you use the IN operator, the subquery must return a single column of values.
- A query that uses the NOT IN operator with a subquery can typically be restated using an outer join.

Figure 6-3 How to use subqueries with the IN operator

How to compare a subquery with an expression

Figure 6-4 illustrates how you can use comparison operators like equal, less than, and greater than to compare an expression with the result of a subquery. In this example, the subquery returns the average balance due for the invoices in the Invoices table. Then, it uses that value to retrieve all invoices that have a balance due that's less than the average.

When you use a comparison operator as shown in this figure, the subquery must return a single value. In most cases, that means that the subquery uses an aggregate function. However, you can also use the comparison operators with subqueries that return two or more values. To do that, you use the SOME, ANY, or ALL keywords as shown in the next figures.

The syntax of a WHERE clause comparing an expression with a subquery

```
WHERE expression comparison_operator [SOME|ANY|ALL] (subquery)
```

Return invoices with a balance due that's less than the average

```
SELECT invoice_number, invoice_date,
    invoice_total - payment_total - credit_total AS balance_due
FROM invoices
WHERE invoice_total - payment_total - credit_total  > 0
   AND invoice_total - payment_total - credit_total <
    (
      SELECT AVG(invoice_total - payment_total - credit_total)
      FROM invoices
      WHERE invoice_total - payment_total - credit_total > 0
    )
ORDER BY invoice_total DESC
```

The value returned by the subquery

```
1669.906
```

The result set

	INVOICE_NUMBER	INVOICE_DATE	BALANCE_DUE
1	31359783	23-MAY-24	1575
2	97/553	27-APR-24	904.14
3	I77271-O01	05-JUN-24	662
4	31361833	23-MAY-24	579.42

(33 rows)

Description

- You can use a comparison operator in a WHERE clause to compare an expression with the results of a subquery.
- If you code a search condition without one of the ANY, SOME, or ALL keywords, the subquery must return a single value.
- If you include ANY, SOME, or ALL, the subquery can return a list of values.

Figure 6-4 How to compare a subquery with an expression

How to use the ALL keyword

Figure 6-5 shows you how to use the ALL keyword. This keyword modifies the comparison operator so the condition must be true for all of the values returned by a subquery. This is equivalent to coding a series of conditions connected by AND operators. The table at the top of this figure describes how this works for some of the common comparison operators.

If you use the greater than operator (>), the expression must be greater than the maximum value returned by the subquery. Conversely, if you use the less than operator (<), the expression must be less than the minimum value returned by the subquery. If you use the equal operator (=), the expression must be equal to all of the values returned by the subquery. And if you use the not equal operator (<>), the expression must not be equal to any of the values returned by the subquery. Note that a not equal condition could be restated using a NOT IN condition.

The query in this figure illustrates the use of the greater than operator with the ALL keyword. Here, the subquery selects the invoice_total column for all the invoices with a vendor_id value of 34. This results in a table with two rows. Then, the outer query retrieves the rows from the Invoices table that have invoice totals greater than all of the values returned by the subquery. This is equivalent to coding the WHERE clause as follows:

```
WHERE invoice_total > 116.54 AND invoice_total > 1083.58
```

Since any invoice total greater than 1083.58 will also be greater than 116.54, this clause effectively tests only to see if an invoice total is greater than 1083.58 since that's the larger number. In other words, this query returns all the invoices that have totals greater than the largest invoice for vendor number 34.

When you use the ALL operator, you should realize that if the subquery doesn't return any rows, the comparison operation will always be true. In contrast, if the subquery returns only null values, the comparison operation will always be false.

In many cases, a search condition with the ALL keyword can be rewritten using a function so it's easier to read and maintain. For example, the condition in the query in this figure could be rewritten to use the MAX function like this:

```
WHERE invoice_total >
    (SELECT MAX(invoice_total)
     FROM invoices
     WHERE vendor_id = 34)
```

As a result, consider replacing the ALL keyword with an equivalent condition whenever it makes the query easier to read.

How the ALL keyword works

Condition	Equivalent expression	Description
`x > ALL (1, 2)`	`x > 2`	Evaluates to true if x is greater than the largest value returned by the subquery.
`x < ALL (1, 2)`	`x < 1`	Evaluates to true if x is less than the smallest value returned by the subquery.
`x = ALL (1, 2)`	`(x = 1) AND (x = 2)`	Evaluates to true if the subquery returns a single value that's equal to x or if the subquery returns multiple values that are the same and these values are all equal to x.
`x <> ALL (1, 2)`	`(x <> 1) AND (x <> 2)`	Evaluates to true if x is not one of the values returned by the subquery.

Get invoices greater than the largest invoice for vendor 34

```
SELECT vendor_name, invoice_number, invoice_total
FROM invoices i JOIN vendors v ON i.vendor_id = v.vendor_id
WHERE invoice_total > ALL
    (SELECT invoice_total
     FROM invoices
     WHERE vendor_id = 34)
ORDER BY vendor_name
```

The result of the subquery

	INVOICE_TOTAL
1	116.54
2	1083.58

The result set

	VENDOR_NAME	INVOICE_NUMBER	INVOICE_TOTAL
1	Bertelsmann Industry Svcs. Inc	509786	6940.25
2	Cahners Publishing Company	587056	2184.5
3	Computerworld	367447	2433
4	Data Reproductions Corp	40318	21842

(25 rows)

Description

- You can use the ALL keyword to test if a condition is true for all of the values returned by a subquery.
- If no rows are returned by the subquery, a comparison that uses the ALL keyword is always true.
- If all of the rows returned by the subquery contain a null value, a comparison that uses the ALL keyword is always false.

Figure 6-5 How to use the ALL keyword

How to use the ANY and SOME keywords

Figure 6-6 shows how to use the ANY and SOME keywords to test whether a comparison is true for any of the values returned by a subquery. Since both of these keywords work the same way, you can use whichever one you prefer.

You use the ANY and SOME keywords to test if a comparison is true for any (or some) of the values returned by a subquery. This is equivalent to coding a series of conditions connected with OR operators. The table at the top of this figure describes how these keywords work with some of the common comparison operators.

The example in this figure shows how you can use the ANY keyword with the less than operator. This statement is similar to the one you saw in the previous figure, except that it retrieves invoices with invoice totals that are less than at least one of the invoice totals for a given vendor. Like the statement in the previous figure, this condition could be rewritten using the MAX function as follows:

```
WHERE invoice_total <
    (SELECT MAX(invoice_total)
     FROM invoices
     WHERE vendor_id = 115)
```

Because you can usually replace an ANY condition with an equivalent condition that's more readable, you probably won't use ANY often.

How the ANY and SOME keywords work

Condition	Equivalent expression	Description
`x > ANY (1, 2)`	`x > 1`	Evaluates to true if x is greater than the minimum value returned by the subquery.
`x < ANY (1, 2)`	`x < 2`	Evaluates to true if x is less than the maximum value returned by the subquery.
`x = ANY (1, 2)`	`(x = 1) OR (x = 2)`	Evaluates to true if x is equal to any of the values returned by the subquery.
`x <> ANY (1, 2)`	`(x <> 1) OR (x <> 2)`	Evaluates to true if x is not equal to at least one of the values returned by the subquery.

Get invoices smaller than the largest invoice for vendor 115

```
SELECT vendor_name, invoice_number, invoice_total
FROM vendors JOIN invoices ON vendors.vendor_id = invoices.invoice_id
WHERE invoice_total < ANY
     (SELECT invoice_total
      FROM invoices
      WHERE vendor_id = 115)
```

The result of the subquery

	INVOICE_TOTAL
1	6
2	25.67
3	6
4	6

The result set

	VENDOR_NAME	INVOICE_NUMBER	INVOICE_TOTAL
1	Reiter's Scientific & Pro Books	25022117	6
2	Boucher Communications Inc	24863706	6
3	Champion Printing Company	24780512	6
4	Opamp Technical Books	21-4748363	9.95

(17 rows)

Description

- You can use the ANY keyword to test if a condition is true for at least one of the values returned by a subquery.
- If the subquery doesn't return any values, or if it only returns null values, a comparison that uses the ANY keyword evaluates to false.
- The SOME keyword works the same as the ANY keyword.

Figure 6-6 How to use the ANY and SOME keywords

How to code correlated subqueries

So far, all of the subqueries in this chapter have been noncorrelated subqueries. A *noncorrelated subquery* is executed only once for the entire query. However, you can also code a *correlated subquery* that's executed once for each row that's processed by the main query. This type of query is similar to using a loop to do repetitive processing in a procedural programming language like PHP or Java.

Figure 6-7 illustrates how correlated subqueries work. The example in this figure retrieves rows from the Invoices table for those invoices that have an invoice total that's greater than the average of all the invoices for the same vendor. To do that, the search condition in the WHERE clause of the subquery refers to the vendor_id value of the current invoice. That way, only the invoices for the current vendor will be included in the calculated average.

Each time a row in the outer query is processed, the value in the vendor_id column for that row is substituted for the column reference in the subquery. Then, the subquery is executed based on the current value. If the vendor_id for a row in the Invoices table is 95, for example, this subquery will be executed:

```
SELECT AVG(invoice_total)
FROM invoices
WHERE vendor_id = 95
```

The value this subquery returns is used to determine whether the current row from the Invoices table is included in the result set. For example, the value returned by the subquery for vendor 95 is approximately 28.5016. Then, that value is compared with the invoice total of the current invoice. If the invoice total is greater than that value, the invoice is included in the result set. Otherwise, it's not. This process is repeated until each of the invoices in the Invoices table has been processed.

In this figure, the WHERE clause of the subquery qualifies the vendor_id column from the main query with the alias that's assigned to the Invoices table in that query. This is necessary because this statement uses the same table in both the sub and main queries. So, the use of a table alias avoids ambiguity. However, if a subquery uses a different table than the main query, a table alias isn't necessary.

Since correlated subqueries can be difficult to code, you may want to test a subquery separately before using it within a SELECT statement. To do that, however, you'll need to substitute a constant value for the variable that refers to a column in the outer query. That's what this example does to get the average invoice total for vendor 95. Once you're sure that the subquery works on its own, you can replace the constant value with a reference to the outer query so you can use it within a SELECT statement.

Get each invoice amount that's higher than the average for each vendor

```
SELECT vendor_id, invoice_number, invoice_total
FROM invoices i
WHERE invoice_total >
    (SELECT AVG(invoice_total)
     FROM invoices
     WHERE vendor_id = i.vendor_id)
ORDER BY vendor_id, invoice_total
```

The value returned by the subquery for vendor 95

```
28.50166...
```

The result set

	VENDOR_ID	INVOICE_NUMBER	INVOICE_TOTAL
6	83	31359783	1575
7	95	111-92R-10095	32.7
8	95	111-92R-10093	39.77
9	95	111-92R-10092	46.21

```
(36 rows)
```

Description

- A *correlated subquery* is a subquery that is executed once for each row in the main, or outer, query.
- A *noncorrelated subquery* (or *uncorrelated subquery*) is executed only once. All of the subqueries in the previous figures were noncorrelated subqueries.
- A correlated subquery refers to a value provided by a column in the outer query. Because that value varies depending on each row, each execution of the subquery returns a different result.
- To refer to a column in the main query, you can qualify the column with a table name or alias.
- If a correlated subquery uses the same table as the main query, you can use table aliases to remove ambiguity.

Note

- Because a correlated subquery is executed for each row processed by the outer query, a query with a correlated subquery typically takes longer to run than a query with a noncorrelated subquery.

How to use the EXISTS operator

Figure 6-8 shows you how to use the EXISTS operator with a subquery. This operator tests whether or not the subquery returns a result set. In other words, it tests whether a result set with at least one row exists. When you use this operator, the subquery doesn't actually return a result set to the outer query. Instead, it returns an indication of whether any rows satisfy the search condition of the subquery. Because of that, queries that use this operator execute quickly.

The query in this example retrieves all the vendors in the Vendors table that don't have invoices in the Invoices table. To do that, the correlated subquery tests each row to see if its vendor_id value is in the Invoices table. If it isn't, the result set for the subquery doesn't contain any rows, so NOT EXISTS returns true.

Because a subquery used with EXISTS doesn't actually return a result set, it doesn't matter what columns are included in the SELECT clause. So, it's customary to just code an asterisk.

You typically use the EXISTS operator with a correlated subquery, as illustrated in this figure. That's because noncorrelated subqueries with EXISTS are usually simpler to code and understand as joins. With correlated subqueries, however, EXISTS usually executes faster than the equivalent join. For example, notice that the query in this figure returns the same vendors as the two queries you saw in figure 6-3. Those queries use the IN operator with a subquery and an outer join. However, the query in this figure executes more quickly than either of them.

The syntax of a subquery that uses the EXISTS operator

```
WHERE [NOT] EXISTS (subquery)
```

Get all vendors that don't have invoices

```
SELECT vendor_id, vendor_name, vendor_state
FROM vendors
WHERE NOT EXISTS
    (SELECT *
     FROM invoices
     WHERE vendor_id = vendors.vendor_id)
```

The result set

	VENDOR_ID	VENDOR_NAME	VENDOR_STATE
1	1	US Postal Service	WI
2	2	National Information Data Ctr	DC
3	3	Register of Copyrights	DC
4	4	Jobtrak	CA

```
(88 rows)
```

Description

- You can use the EXISTS operator to test if at least one row is returned by a subquery.
- You can use NOT EXISTS to test if no rows are returned by a subquery.
- When you use these operators with a subquery, it doesn't matter what columns you specify in the SELECT clause. As a result, you typically just code an asterisk (*).
- It's usually easier to use a join than a noncorrelated subquery with EXISTS.

Figure 6-8 How to use the EXISTS operator

Other ways to use subqueries

Although you'll typically use subqueries in the WHERE or HAVING clause of a SELECT statement, you can also use them in the FROM and SELECT clauses. You'll learn how to do that in the figures that follow.

How to code subqueries in the FROM clause

Figure 6-9 shows how to code a subquery in the FROM clause. To do that, you code a subquery in place of a table specification. When you code a subquery in the FROM clause, it can return a result set that contains any number of rows and columns. This result set is sometimes referred to as an *inline view*.

Subqueries are typically used in the FROM clause to create inline views that provide summarized data to another summary query. In this figure, for example, the subquery returns a result set that contains the vendor state, vendor name, and sum of invoice totals for each vendor. To do that, it groups the vendors by state and name and calculates the sum of the invoices for each vendor. Once the subquery returns this result set, the main query groups the result set by vendor state and gets the largest sum of invoice totals for a vendor in each state.

When you code a subquery in the FROM clause, you must assign a table alias to the subquery. This is required even if you don't use the table alias in the main query. In this figure, for example, the query assigns a table alias of *t* (for temporary table) to the subquery.

In addition, you should assign a column alias to all calculated columns in a subquery. In this figure, for example, the subquery assigns a column alias of sum_of_invoices to the result of the SUM function. That makes it easier to refer to these columns from other clauses in the subquery if you need to do that. It also makes it possible to refer to the column from the main query. In this example, for instance, the sum_of_invoices column is referred to by the SELECT clause within the MAX function.

Get the largest sum of invoice totals for a vendor in each state

```
SELECT vendor_state, MAX(sum_of_invoices)
FROM
(
     SELECT vendor_state, vendor_name, SUM(invoice_total) AS sum_of_invoices
     FROM vendors v JOIN invoices i
         ON v.vendor_id = i.vendor_id
     GROUP BY vendor_state, vendor_name
) t
GROUP BY vendor_state
ORDER BY vendor_state
```

The result of the subquery (an inline view)

	VENDOR_STATE	VENDOR_NAME	SUM_OF_INVOICES
1	CA	IBM	1200.12
2	CA	Blue Cross	564
3	CA	Fresno County Tax Collector	856.92
4	MI	Data Reproductions Corp	21927.31

(34 rows)

The result set

	VENDOR_STATE	MAX(SUM_OF_INVOICES)
1	AZ	662
2	CA	7125.34
3	DC	600
4	MA	1367.5

(10 rows)

Description

- A subquery that's coded in the FROM clause returns a result set that can be referred to as an *inline view*.
- When you code a subquery in the FROM clause, you must assign an alias to it. Then, you can use that alias just as you would any other table name or alias.
- When you code a subquery in the FROM clause, you must assign names to any calculated values used in the result set.
- Inline views are most useful when you need to further summarize the results of a summary query.

How to code subqueries in the SELECT clause

Figure 6-10 shows you how to use subqueries in the SELECT clause. As you can see, you can use a subquery in place of a column specification. Because of that, the subquery must return a single value per row.

In most cases, the subqueries you use in the SELECT clause will be correlated subqueries. The subquery in this figure, for example, calculates the maximum, or most recent, invoice date for each vendor in the Vendors table. To do that, it refers to the vendor_id column from the Invoices table in the outer query.

Because subqueries coded in the SELECT clause can be difficult to read, you generally should avoid them if you can find a more readable solution. In most cases, the subquery can be replaced with a join. For example, the first query shown in this figure could be restated as shown in the second query. This query joins the Vendors and Invoices tables, groups the rows by vendor_name, and then uses the MAX function to calculate the maximum invoice date for each vendor. As you can see, this query is easier to read than the one with the subquery.

Get the most recent invoice date for each vendor

```
SELECT vendor_name,
    (
     SELECT MAX(invoice_date) FROM invoices i
     WHERE i.vendor_id = v.vendor_id
    ) AS latest_inv
FROM vendors v
ORDER BY latest_inv
```

The result set

	VENDOR_NAME	LATEST_INV
1	IBM	14-MAR-24
2	Wang Laboratories, Inc.	16-APR-24
3	Reiter's Scientific & Pro Books	17-APR-24
4	United Parcel Service	26-APR-24

```
(122 rows)
```

The same query restated using a join

```
SELECT vendor_name, MAX(invoice_date) AS latest_inv
FROM vendors v
    LEFT JOIN invoices i ON v.vendor_id = i.vendor_id
GROUP BY vendor_name
ORDER BY latest_inv
```

Description

- When you code a subquery for a column specification in the SELECT clause, the subquery must return a single value per row.
- A subquery that's coded within a SELECT clause is typically a correlated subquery.
- A query with a subquery in its SELECT clause can typically be restated using a join.

Figure 6-10 How to code subqueries in the SELECT clause

Guidelines for working with complex queries

So far, the examples you've seen of queries that use subqueries have been relatively simple. However, these types of queries can get complicated in a hurry, particularly if the subqueries are nested. Because of that, you'll want to be sure that you plan and test these queries carefully. To illustrate how to do that, this section presents an example of a complex query and then breaks down the steps for creating it.

A complex query that uses subqueries

Figure 6-11 presents a complex query that uses multiple subqueries. The first subquery is used in the FROM clause of the outer query to create a result set that contains the state, name, and total invoice amount for each vendor in the Vendors table. This is the same subquery that was described in figure 6-9.

The second subquery is also used in the FROM clause of the outer query to create a result set that's joined with the first result set. This result set contains the state and the sum of invoice totals for the vendor in each state that has the largest sum of invoice totals. To create this result set, a third subquery is nested within the FROM clause of the subquery. This subquery is identical to the first subquery.

After this statement creates the two result sets, it joins them based on the columns in each table that contain the state and the sum of invoice totals. The final result set contains the state, name, and sum of invoice totals for the vendor in each state with the largest sum of invoice totals. This result set is sorted by state.

At this point, you might be wondering if there is an easier solution to this problem. For example, you might think that you could solve the problem by joining the Vendors and Invoices tables, grouping by vendor state, and calculating the sum of invoices for each vendor. However, if you group by vendor state, you can't include the name of the vendor in the result set. And if you group by vendor state and vendor name, the result set includes all vendors, not just the top vendor from each state. As a result, the query presented here is a fairly straightforward way of solving the problem.

A complex query that uses three subqueries

```
SELECT t1.vendor_state, vendor_name, t1.sum_of_invoices
FROM
    (
        -- sum of invoice totals by vendor
        SELECT vendor_state, vendor_name,
               SUM(invoice_total) AS sum_of_invoices
        FROM vendors v JOIN invoices i
            ON v.vendor_id = i.vendor_id
        GROUP BY vendor_state, vendor_name
    ) t1
    JOIN
        (
            -- largest sum of invoice totals by state
            SELECT vendor_state,
                   MAX(sum_of_invoices) AS sum_of_invoices
            FROM
            (
                -- sum of invoice totals by vendor
                SELECT vendor_state, vendor_name,
                       SUM(invoice_total) AS sum_of_invoices
                FROM vendors v JOIN invoices i
                    ON v.vendor_id = i.vendor_id
                GROUP BY vendor_state, vendor_name
            ) t2
            GROUP BY vendor_state
        ) t3
    ON t1.vendor_state = t3.vendor_state AND
       t1.sum_of_invoices = t3.sum_of_invoices
ORDER BY vendor_state
```

The result set

	VENDOR_STATE	VENDOR_NAME	SUM_OF_INVOICES
1	AZ	Wells Fargo Bank	662
2	CA	Digital Dreamworks	7125.34
3	DC	Reiter's Scientific & Pro Books	600
4	MA	Dean Witter Reynolds	1367.5

(10 rows)

How the query works

- This query retrieves the vendor from each state that has the largest sum of invoice totals.
- This query uses comments to clearly identify its three subqueries.
- The subqueries named t1 and t2 return the same result set. This result set contains each vendor's state, name, and sum of invoice totals.
- The subquery named t3 returns the vendor state and the largest sum of invoices for any vendor in that state. To do that, this subquery uses the nested subquery named t2.
- The subqueries named t1 and t3 are joined on both the vendor_state and sum_of_invoices columns.

Figure 6-11 A complex query that uses subqueries

A procedure for building complex queries

To build a complex query like the one in the previous figure, you can use a procedure like the one in figure 6-12. To start, you should state the question in plain language so you're clear about what you want the query to answer. In this case, the question is, "Which vendor in each state has the largest sum of invoice totals?"

Once you're clear about the problem, you should outline the query using *pseudocode*. Pseudocode represents the intent of the query, but doesn't necessarily use SQL code. The pseudocode shown in this figure, for example, uses part SQL code and part English. This pseudocode identifies the three columns returned by the main query, two subqueries, and the join condition for the two subqueries.

The next step in the procedure is to code and test the subqueries to be sure they work the way you want them to. For example, this figure shows the code for the first subquery along with its result set. This returns all of the data that you want, but it also includes extra rows that you don't want. To remove the extra rows from the first query, you can code the second subquery shown in this figure. This subquery uses the first subquery as a nested subquery. Although this removes the extra rows, it also removes the vendor_name column.

Once you're sure that both subqueries work the way you want them to, you can use them in the main query. The pseudocode in this figure shows that you should join the result sets returned by the subqueries on the vendor_state and sum_of_invoices columns. In addition, the code in the previous figure shows how these two subqueries are used in the main query. This allows you to get all of the columns you want in the final result set without any of the extra rows that you don't want.

Writing complex queries can be difficult, but following a procedure like the one shown in this figure can make it a little easier. Once you get thee subqueries working correctly, you can begin coding your main query, and you can cut and paste the subqueries into the main query.

A procedure for building complex queries

1. State the problem to be solved by the query in plain language.
2. Use pseudocode to outline the query.
3. Code the subqueries and test them to be sure that they return the correct data.
4. Code and test the final query.

The problem to be solved by the query in figure 6-11

Which vendor in each state has the largest sum of invoice totals?

Pseudocode for the complete query

```
SELECT vendor_state, vendor_name, sum_of_invoices
FROM (subquery1 returning vendor_state, vendor_name & sum_of_invoices)
JOIN (subquery2 returning vendor_state & largest_sum_of_invoices)
    ON vendor_state AND sum_of_invoices
ORDER BY vendor_state
```

The code for the first subquery

```
SELECT vendor_state, vendor_name, SUM(invoice_total) AS sum_of_invoices
FROM vendors v JOIN invoices i
    ON v.vendor_id = i.vendor_id
GROUP BY vendor_state, vendor_name
```

The result for the first subquery

	VENDOR_STATE	VENDOR_NAME	SUM_OF_INVOICES
1	CA	IBM	1200.12
2	CA	Blue Cross	564
3	CA	Fresno County Tax Collector	856.92
4	MI	Data Reproductions Corp	21927.31

(34 rows)

The code for the second subquery

```
SELECT vendor_state, MAX(sum_of_invoices) AS sum_of_invoices
FROM
(
   SELECT vendor_state, vendor_name, SUM(invoice_total) AS sum_of_invoices
   FROM vendors v JOIN invoices i
       ON v.vendor_id = i.vendor_id
   GROUP BY vendor_state, vendor_name
) t
GROUP BY vendor_state
```

The result of the second subquery

	VENDOR_STATE	SUM_OF_INVOICES
1	CA	7125.34
2	MI	119892.41
3	PA	265.36
4	DC	600

(10 rows)

Figure 6-12 A procedure for building complex queries

How to use subquery factoring

As you just saw, subqueries can quickly grow complex. To make your code easier to write, read, and maintain, you can use subquery factoring. In addition to reducing the amount of code you need in a complex statement, subquery factoring allows you to loop through tables recursively in a way you'd otherwise need to use PL/SQL for.

How to use the WITH clause

The WITH clause allows you to define and name blocks of code for use in the rest of the SQL statement. It is frequently used with the *subquery factoring clause*, which allows you to define and name a subquery. Once named, you can then reference the subquery without having to retype the full subquery each time. This is called *subquery factoring*, and a subquery defined this way is also called a *common table expression (CTE)*.

Figure 6-13 uses subquery factoring to simplify the complex query from figure 6-11. To start, the statement begins with the WITH keyword to start the WITH clause. Then, it starts a subquery factoring clause by coding a name for the subquery ("summary") followed by the AS keyword, an opening parenthesis, a SELECT statement that defines the subquery, and a closing parenthesis. This is the same subquery that was named t1 and t2 in figure 6-11.

Next, a second subquery factoring clause defines a subquery named top_in_state. This subquery references the first subquery in its FROM clause. This works because the summary subquery has already been defined in the statement.

When coding multiple subqueries within a subquery factoring clause, a subquery can refer to any subquery coded before it, but it can't refer to subqueries coded after it. For example, this statement wouldn't work if the two subqueries were coded in the opposite order.

Finally, the SELECT statement that's coded immediately after the two named subqueries uses both just as if they were tables. To avoid ambiguous references, each column is qualified by the name for the corresponding subquery.

If you compare this code with figure 6-11, you'll see this version is easier to read and understand. In addition, this version is easier to maintain because the summary query is coded in one place, not in two. So, if any changes need to be made later, you only need to make them in one place.

The syntax of the WITH clause with a subquery factoring clause

```
WITH subquery_name1 AS (subquery_definition1)
[, subquery_name2 AS (subquery_definition2)]
[...]
sql_statement
```

Two named subqueries and a query that uses them

```
WITH summary AS
(
    SELECT vendor_state, vendor_name, SUM(invoice_total) AS sum_of_invoices
    FROM invoices
        JOIN vendors ON invoices.vendor_id = vendors.vendor_id
    GROUP BY vendor_state, vendor_name
),
top_in_state AS
(
    SELECT vendor_state, MAX(sum_of_invoices) AS sum_of_invoices
    FROM summary
    GROUP BY vendor_state
)
SELECT summary.vendor_state, summary.vendor_name,
    top_in_state.sum_of_invoices
FROM summary JOIN top_in_state
    ON summary.vendor_state = top_in_state.vendor_state AND
        summary.sum_of_invoices = top_in_state.sum_of_invoices
ORDER BY summary.vendor_state
```

The result set

	VENDOR_STATE	VENDOR_NAME	SUM_OF_INVOICES
1	AZ	Wells Fargo Bank	662
2	CA	Digital Dreamworks	7125.34
3	DC	Reiter's Scientific & Pro Books	600
4	MA	Dean Witter Reynolds	1367.5
5	MI	Malloy Lithographing Inc	119892.41
6	NV	United Parcel Service	23177.96
7	OH	Edward Data Services	207.78

(10 rows)

Description

- The WITH clause is used to define and name blocks of code.
- The *subquery factoring clause* is a subclause of the WITH clause. It defines and names a subquery, allowing you to code just the subquery name in the rest of the SQL statement.
- To define multiple subqueries, separate them with commas.
- Each subclause in a WITH clause can refer to itself and any previously defined subqueries in the same WITH clause.
- The *subquery factoring clause* can also be called a *common table expression (CTE).*

Figure 6-13 How to use the WITH clause

How to code a recursive query

Another use of the subquery factoring clause is to create a *recursive query*. A recursive query is a query that refers to itself iteratively. Because the subquery factoring clause allows you to name and reuse subqueries without retyping them, you can use it to easily create recursive subquery factoring.

In a relational database, recursive queries are commonly used to return hierarchical data. For example, an organizational chart might have a parent element (a manager) with one or more child elements (direct reports), and each child element may have one or more child elements of their own, and so on. Let's say you wanted to calculate the rank of each employee with the CEO as rank 1, her direct reports as rank 2, their direct reports as rank 3, and so on. How would you do that?

The top of this figure shows an Employees table from the EX database where the manager_id column is used to identify the manager for each employee. Here, Cindy Smith is the CEO, and as such doesn't have a manager. Elmer Jones and Paulo Locario report to Cindy, and so on.

The code in this figure begins by coding a WITH clause and a name (org_chart) for the subquery, just like the subqueries in the last figure. However, it then codes a list of the column names that the subquery will return in parentheses. This list is optional for most subquery factoring clauses, but is required when you code a recursive query. Furthermore, the order in which you code the column names must match the order of the column names in both subqueries, which must also match each other.

Then, the subquery defines two select statements joined by the UNION ALL operator. The first statement is non-recursive and is called the *anchor member*. In this example, it uses the WHERE clause to select all rows in the Employees table for which the manager_id value is null. In this case, that's just Cindy Smith. It also defines a column named ranking and assigns a literal numeric value of 1 to it. You can think of the result set returned by this query as the first version of the org_chart table.

The second statement is the *recursive member*. It uses an inner join to get all rows in the Employees table that have a manager_id that matches an employee_id in the current version of the org_chart table. For each of these rows, it sets the value of the ranking column to the rank of the manager plus 1. The first time this recursive member executes, it returns a result set with two rows, for Jones and Locario. Then, it is combined with the first version of the org_chart table to get a table that has three rows (Smith, Jones, and Locario).

The recursive member then executes again, looking for all employees that have a manager_id matching an employee_id in the now larger version of org_chart and assigning a value of 1 + their manager's rank to the ranking column.

This continues until there are no new rows found by the recursive member that aren't already in the current version of the org_chart table. At this point, you can refer to the result set created by the recursive query as you would any other table.

Finally, note that when you code a recursive query you cannot use the DISTINCT operator or the GROUP BY clause in the recursive member.

The Employees table

	EMPLOYEE_ID	LAST_NAME	FIRST_NAME	DEPARTMENT_NUMBER	MANAGER_ID
1	1	Smith	Cindy	2	(null)
2	2	Jones	Elmer	4	1
3	3	Simonian	Ralph	2	2
4	4	Hernandez	Olivia	1	9
5	5	Aaronsen	Robert	2	4
6	6	Watson	Denise	6	8
7	7	Hardy	Thomas	5	2
8	8	O'Leary	Rhea	4	9
9	9	Locario	Paulo	6	1

A recursive subquery factoring clause

```
WITH org_chart (employee_id, last_name, manager_id, ranking) AS
(
        -- anchor member
        -- get all employees without managers (null value for manager_id)
        -- set the value of ranking column for these employees to 1
        SELECT employee_id, last_name, manager_id, 1 AS ranking
        FROM employees
        WHERE manager_id IS NULL
    UNION ALL
        -- recursive member
        -- get the direct reports of all employees in the table so far
        -- and set ranking for these employees to ranking+1
        SELECT employees.employee_id, employees.last_name,
               employees.manager_id, org_chart.ranking + 1
        FROM employees
        JOIN org_chart
             ON employees.manager_id = org_chart.employee_id
)
SELECT * FROM org_chart
ORDER BY ranking, employee_id
```

The result set

	EMPLOYEE_ID	LAST_NAME	MANAGER_ID	RANKING
1	1	Smith	(null)	1
2	2	Jones	1	2
3	9	Locario	1	2
4	3	Simonian	2	3
5	4	Hernandez	9	3

(9 rows)

Description

- A *recursive query* is a query that refers to itself iteratively.
- A recursive subquery factoring clause combines the result set of a non-recursive SELECT statement (the *anchor member*) with the result of a recursive SELECT statement (the *recursive member*).
- A recursive subquery factoring clause must use the UNION ALL operator.
- The recursive member cannot contain the DISTINCT keyword or a GROUP BY clause.

Figure 6-14 How to code a recursive query

How to use the hierarchical query clause

Because the most common use of recursive queries in SQL is to return hierarchal data, Oracle provides a clause called the *hierarchical query clause*. When you use this clause, you don't need to use a subquery to create a recursive loop. Because of that, it's convenient, concise, and easy to read and understand.

The example in figure 6-15 shows how this works. It returns the same hierarchical data as in the previous figure, but its code is clearly much shorter.

First, notice that the list of columns in the SELECT clause includes one named LEVEL, even though LEVEL is not a column in the Employees table or defined anywhere in the SELECT statement. That's because LEVEL is a *pseudocolumn*.

A pseudocolumn is not actually a column, but a function that doesn't take any parameters. Pseudocolumns get their name from the fact that you can use one as if it were a column in the table you are selecting from, but not perform any other actions that you normally would with a column. For example, you can't insert, update, or delete the values stored in a pseudocolumn as there aren't any stored values to change.

Not all SQL dialects provide pseudocolumns, and of the ones that do, most uses for pseudocolumns have been superceded over the years by newer functionality. However, they can still be quite useful in the right circumstances, such as this one.

When you use the hierarchical query clause, the LEVEL pseudocolumn automatically becomes available to your statement. It returns the level of the row based on the relationship you specify in the CONNECT BY clause.

The CONNECT BY clause is coded after a FROM or WHERE clause and identifies the relationship between parent rows and child rows. To indicate which row is the parent row, code the PRIOR keyword before it. In this figure, for example, the PRIOR keyword specifies that a row is a child row if the row's employee_id column is equal to the manager_id column of another row.

After the CONNECT BY clause, this query uses the START WITH clause to identify the row to be used as the root of the hierarchy. In this example, that's the row with an employee_id equal to 1, or Cindy Smith. However, you could get the same result by coding the clause as follows:

```
START WITH manager_id IS NULL
```

Finally, note the order of the result set. The hierarchical query clause orders its results according to a depth-first search of the hierarchy based on the original order of the table. If you want, you can use the ORDER SIBLINGS BY clause to order siblings. For example, to list Hardy before Simonian, you would code the following:

```
ORDER SIBLINGS BY last_name
```

Or, to return the result set in the same order as in the previous figure, you can use the ORDER BY clause and sort by the LEVEL pseudocolumn.

The syntax of the hierarchical query clause

```
CONNECT BY [NOCYCLE] relationship
[START WITH row_specification]
[ORDER SIBLINGS BY order_by_list]
```

The hierarchy of the Employees table

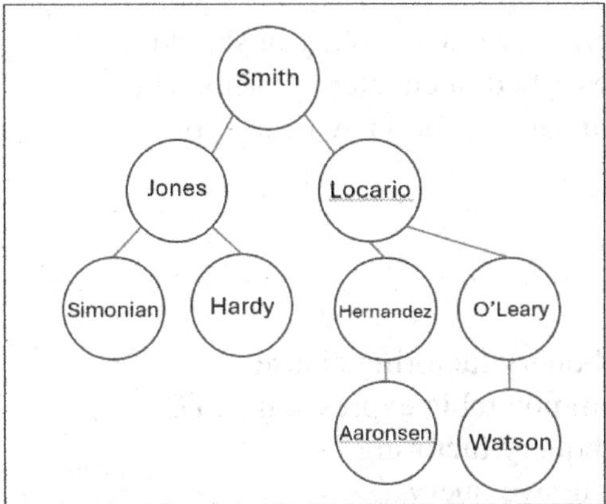

How to use the hierarchical query clause

```
-- LEVEL is a pseudocolumn available when you use CONNECT BY
SELECT employee_id, last_name, manager_id, LEVEL
FROM employees
CONNECT BY manager_id = PRIOR employee_id
START WITH employee_id = 1
```

The result set

	EMPLOYEE_ID	LAST_NAME	MANAGER_ID	LEVEL
1	1	Smith	(null)	1
2	2	Jones	1	2
3	3	Simonian	2	3
4	7	Hardy	2	3
5	9	Locario	1	2
6	4	Hernandez	9	3
7	5	Aaronsen	4	4
8	8	O'Leary	9	3
9	6	Watson	8	4

Description

- A *hierarchical query* is a query that returns rows in a hierarchical order.
- You can use the LEVEL *pseudocolumn* to identify the level for each row.
- Use the START WITH clause to identify the row or rows to be used as the root of the hierarchical query.
- The connecting relationship must use the PRIOR operator to indicate the parent row.
- To order rows of siblings of the same parent, use the ORDER SIBLINGS BY clause.

Figure 6-15 How to use the hierarchical query clause

Perspective

Subqueries provide a powerful tool for solving difficult problems. Before you use a subquery, however, remember that a subquery can often be restated more clearly by using a join. If so, you'll typically want to use a join instead of a subquery.

If you find yourself coding the same subqueries over and over, you should consider creating a view for that subquery, as described in chapter 11. This will help you develop queries more quickly since you can use the view instead of coding the subquery again.

Terms

subquery
introduce a subquery
nested subquery
correlated subquery
noncorrelated subquery
uncorrelated subquery
inline view
pseudocode
subquery factoring clause
common table expression (CTE)
subquery factoring
recursive query
anchor member
recursive member
hierarchical query
pseudocolumn

Exercises

1. Write a SELECT statement that returns the same result set as this SELECT statement but don't use a join. Instead, use a subquery in a WHERE clause that uses the IN keyword.

```
SELECT DISTINCT vendor_name
FROM vendors JOIN invoices
    ON vendors.vendor_id = invoices.vendor_id
ORDER BY vendor_name
```

2. Write a SELECT statement with a subquery in the WHERE clause that answers this question: Which invoices have a payment_total that's greater than the average payment_total for all invoices that have at least some amount paid? Return the invoice_number and invoice_total for each of these invoices.
3. Write a SELECT statement that returns two columns from the General_Ledger_Accounts table: account_number and account_description. The result set should have one row for each account number that has never been used in the Invoice_Line_Items table. Use a subquery in the WHERE clause with the NOT EXISTS operator, and sort the final result set by account_number.
4. Write a SELECT statement that returns four columns: vendor_name, invoice_id, invoice_sequence, and line_item_amt for each invoice that has more than one line item in the Invoice_Line_Items table. Use joins in the FROM clause, and a subquery in the WHERE clause that tests whether invoice_sequence is greater than 1.

5. Write a SELECT statement that returns a single value that represents the sum of the largest unpaid invoices for each vendor (just one for each vendor). Use a subquery in the FROM clause that returns MAX(invoice_total) grouped by vendor_id, filtering for invoices with a balance due.
6. Rewrite exercise 5 so it uses a subquery factoring clause.
7. Write a SELECT statement that returns the name, city, and state of each vendor that's located in a unique city and state. In other words, don't include vendors that have a city and state in common with another vendor.
8. Use a correlated subquery in the WHERE clause to return one row per vendor, representing the vendor's minimum invoice (the one with the earliest date). Each row should include these four columns: vendor_name, invoice_number, invoice_date, and invoice_total.
9. Rewrite exercise 8 so it gets the same result but uses a noncorrelated subquery in the FROM clause instead of a correlated subquery in the WHERE clause.

7

How to insert, update, and delete data

In the last four chapters, you learned how to code the SELECT statement to retrieve and summarize data. Now, you'll learn how to code INSERT, UPDATE, and DELETE statements to modify the data in a table. When you're done with this chapter, you'll know how to code the four statements that are used every day by professional application developers.

How to create test tables **194**
How to re-create the tables for this book 194
How to create a table from a SELECT statement 194
How to insert new rows **196**
How to insert a single row 196
How to insert default values and null values 198
How to use a subquery to insert multiple rows 200
How to update existing rows **202**
How to update rows 202
How to use a subquery in an UPDATE statement 204
How to delete existing rows **206**
How to delete rows 206
How to use a subquery in a DELETE statement 206
How to commit and roll back changes **208**
Perspective **210**

How to create test tables

Before you begin experimenting with INSERT, UPDATE, and DELETE statements, you need to make sure that your experimentation won't affect "live" data that's used by other people at your business or school. The easiest way to do that is to use test tables.

How to re-create the tables for this book

If you're only working with the tables for this book, you can use the procedure shown in appendix A (Windows) or B (macOS) to re-create the tables for this book and restore them to their original state at any time. Then, you can experiment all you want without worrying about how much you change these tables because you can always re-create them.

How to create a table from a SELECT statement

If you're working with tables that are running on a server that's available from your business or school, it's usually a good idea to create a copy of some or all of a table before you do any testing. To do that, you can use the CREATE TABLE AS statement with a SELECT statement as shown in figure 5-1. When you use this technique, the result set that's defined by the SELECT statement is copied into a new table. Then, you can experiment all you want with the test table and delete it when you're done.

When you use this technique to create tables, Oracle only copies the column definitions and data. In other words, Oracle doesn't retain other parts of the column definitions such as primary keys, foreign keys, and indexes. As a result, when you experiment with copied tables, you may get different results than you would get with the original tables. Still, this is usually preferable to experimenting with live data.

The examples in this figure show how to use the CREATE TABLE AS statement. The first example copies all of the columns and all of the rows in the Invoices table into a new table named Invoices_Copy. The second example copies all of the columns in the Invoices table into a new table named Old_Invoices, but only for rows where the balance due is zero. And the third example creates a table that contains summary data from the Invoices table.

When you're done experimenting with test tables, you can use the DROP TABLE statement to delete any tables you don't need anymore. In this figure, for instance, the fourth example shows how to drop the Old_Invoices table.

Finally, note that the CREATE TABLE AS statement is different from the CREATE TABLE statement. Although they start the same way, their syntax and purpose are quite different.

The simplified syntax of the CREATE TABLE AS statement

```
CREATE TABLE table_name AS select_statement
```

Create a copy of the Invoices table

```
CREATE TABLE invoices_copy AS
SELECT *
FROM invoices
```

Create a partial copy of the Invoices table

```
CREATE TABLE old_invoices AS
SELECT *
FROM invoices
WHERE invoice_total - payment_total - credit_total = 0
```

Create a table with summary rows from the Invoices table

```
CREATE TABLE vendor_balances AS
SELECT vendor_id, SUM(invoice_total) AS sum_of_invoices
FROM invoices
WHERE (invoice_total - payment_total - credit_total) <> 0
GROUP BY vendor_id
```

Delete a table

```
DROP TABLE old_invoices
```

Description

- You can use the CREATE TABLE AS statement to create a new table based on the result set defined by a SELECT statement.
- Each column name in the SELECT clause must be unique. If you use a calculated value in the select list, you must name the column.
- You can code the other clauses of the SELECT statement just as you would for any other SELECT statement, including grouping, aggregates, joins, and subqueries.
- The table you name must not already exist.
- You can delete a table by using the DROP TABLE statement.

Warning

- When you use the SELECT statement to create a table, only the column definitions and data are copied. Definitions of primary keys, foreign keys, indexes, default values, and so on are not included in the new table.

Figure 7-1 How to create a table from a SELECT statement

How to insert new rows

To add new rows to a table, you use the INSERT statement. This statement lets you insert a single row with the values you specify or selected rows from another table. However, the row is not permanently saved to the database until you commit the change as described later in this chapter.

How to insert a single row

Figure 7-2 shows how to code an INSERT statement to insert a single row. Because the examples in this figure insert rows into the Invoices table, this figure reviews the column definitions for this table. This shows the sequence of the columns in the table and which columns have default values or allow null values.

When you code an INSERT statement, you name the table in the INSERT clause, followed by an optional list of columns. Then, you list the values to be inserted in the VALUES clause.

The two examples in this figure illustrate how this works. The first example doesn't include a column list. Because of that, the VALUES clause must include a value for every column in the table, and those values must be listed in exact the same sequence that the columns appear in the table.

The second INSERT statement includes a column list. However, this list doesn't include three columns. It doesn't include the payment_total and credit_total columns since these columns provide a default value of 0. And it doesn't include the payment_date column since this column allows a null value.

In addition, the columns in this example aren't listed in the same sequence as the columns in the table. When you include a column list, you can code the columns in any sequence you like. Then, you just need to be sure you code the values in the VALUES clause in the same sequence.

In the second example, all of the same values are used as in the first example with the exception of the invoice ID. That's because the ID number must be unique for each row in the Invoices table. Otherwise, if you ran both examples you would get an error after the second indicating that the new row has violated the unique constraint on the table.

Finally, note that the values given must be of compatible data types with the columns they are given for. Otherwise, Oracle will return an error and not insert the row.

The syntax of the INSERT statement for inserting a single row

```
INSERT INTO table_name [(column_list)] VALUES (value_list)
```

The columns of the Invoices table

```
invoice_id          NUMBER             NOT NULL
vendor_id           NUMBER             NOT NULL
invoice_number      VARCHAR2(50 BYTE)  NOT NULL
invoice_date        DATE               NOT NULL
invoice_total       NUMBER(9,2)        NOT NULL
payment_total       NUMBER(9,2)                       DEFAULT 0
credit_total        NUMBER(9,2)                       DEFAULT 0
terms_id            NUMBER             NOT NULL
invoice_due_date    DATE               NOT NULL
payment_date        DATE
```

Add a row without using a column list

```
-- values must be in the same order as the columns
INSERT INTO invoices
VALUES (115, 97, '456789', '01-AUG-24', 8344.50, 0, 0, 1, '31-AUG-24', NULL)
```

Add a row using a column list

```
-- columns may be listed in any order
INSERT INTO invoices
    (invoice_id, vendor_id, invoice_number, invoice_total, terms_id,
     invoice_date, invoice_due_date)
VALUES (116, 97, '456789', 8344.50, 1, '01-AUG-24', '31-AUG-24')
```

Description

- You use the INSERT statement to add a new row to a table.
- In the INSERT clause, you specify the name of the table to add a row to, along with an optional column list, followed by a VALUES clause with a list of values.
- If you don't include a column list, you must list the values for the new row in the same order as they appear in the table, and you must code a value for each column.
- If you include a column list, you must list the values for the new row in the same order as they appear in the column list. You can omit columns with that have default values or accept null values.

Figure 7-2 How to insert a single row

How to insert default values and null values

If a column allows null values, you can use the INSERT statement to insert a null value into that column. Similarly, if a column is defined with a default value, you can use the INSERT statement to insert that default.

Figure 7-3 begins by listing the columns for the Color_Sample table in the EX schema. The column color_id represents the internal ID for the column. The column color_number is defined with a default value of 0. And the column color_name allows null values.

The first two statements illustrate how you assign a default or a null value using a column list. To do that, you simply omit the column from the list. In the first statement, for example, the column list names only the color_id and color_number, so color_name is assigned a null value. Similarly, the column list in the second statement names only color_id and color_name, so color_number is assigned its default value of 0.

The next three statements show how you assign a default or null value to a column without including a column list by using the DEFAULT and NULL keywords. For example, the third statement specifies a value for color_name, but uses the DEFAULT keyword for color_number. Because of that, Oracle assigns a value of 0 to this column. The fourth statement assigns a value of 808 to color_number and uses the NULL keyword to assign a null value to color_name.

Finally, the fifth statement uses both the DEFAULT and NULL keywords to assign a value of zero to the color_number column and a null value to the color_name column.

If you want, you can use the DEFAULT and NULL keywords with a column list even though it's not necessary. Generally, you can choose whatever is easiest to both code and read.

The definition of the Color_Sample table

```
color_id          NUMBER             NOT NULL
color_number      NUMBER             NOT NULL      DEFAULT 0
color_name        VARCHAR2(10 BYTE)
```

Add five rows to the table

```
-- color_name allows nulls, so can be omitted from column list
INSERT INTO color_sample (color_id, color_number)
VALUES (1, 606);

-- color_number has a default value, so can be omitted from column list
INSERT INTO color_sample (color_id, color_name)
VALUES (2, 'Yellow');

-- use DEFAULT to insert a default value
INSERT INTO color_sample
VALUES (3, DEFAULT, 'Orange');

-- use NULL to insert a null value
INSERT INTO color_sample
VALUES (4, 808, NULL);

INSERT INTO color_sample
VALUES (5, DEFAULT, NULL);
```

The table after the rows are inserted

	COLOR_ID	COLOR_NUMBER	COLOR_NAME
1	1	606	(null)
2	2	0	Yellow
3	3	0	Orange
4	4	808	(null)
5	5	0	(null)

Description

- If a column allows null values, you can use the NULL keyword to insert a null value into that column.
- If a column has a default value, you can use the DEFAULT keyword to insert the default value for that column.
- If you include a column list, you can omit columns with default values and null values. Then, the default value or null value is assigned automatically.

Figure 7-3 How to insert default values and null values

How to use a subquery to insert multiple rows

Instead of using the VALUES clause of the INSERT statement to specify the values for a single row, you can use a subquery to select the rows you want to insert from another table.

The first two examples in figure 7-4 retrieve rows from the Invoices table and insert them into a table named Invoice_Archive. This table is defined with the same columns as the Invoices table. However, the payment_total and credit_total columns aren't defined with default values. Because of that, you must include values for these columns.

The first example uses a subquery in an INSERT statement without coding a column list. In addition, the SELECT clause of the subquery is coded with an asterisk so that all the columns in the Invoices table will be retrieved. Then, after the search condition in the WHERE clause is applied, all the rows in the result set are inserted into the Invoice_Archive table. This works because the columns in both tables are the same and in the same order.

The second example shows how you can use a column list in the INSERT clause when you use a subquery to retrieve rows. Just as when you use the VALUES clause, you can list the columns in any sequence. However, the columns must be listed in the same sequence in the SELECT clause of the subquery. In addition, you can omit columns that are defined with default values or that allow null values.

Notice that the subqueries in these statements aren't coded within parentheses as a subquery in a SELECT statement is. That's because they're not coded within a clause of the INSERT statement. Instead, they're coded in place of the VALUES clause.

The third example shows how you can use a single INSERT statement to insert three new rows of literal values into a table. This statement uses UNION ALL to combine three SELECT statements that use the Dual table into a single result set that can be inserted into the Color_Sample table from the previous figure. When you code literal values like this, you can use NULL but not DEFAULT. However, if you include a column list, you can omit columns that allow default or null values, just like you would for any INSERT statement.

Before you execute INSERT statements like these, you'll want to be sure that the rows and columns retrieved by the subquery are the ones you want to insert. To do that, you can execute the SELECT statement by itself. Then, when you're sure it retrieves the correct data, you can add the INSERT clause to insert the rows into another table.

The syntax for inserting rows selected from another table

```
INSERT INTO table_name [(column_list)]
SELECT column_list
FROM table_source
[WHERE search_condition]
```

Insert paid invoices into the Invoice_Archive table

```
INSERT INTO invoice_archive
SELECT *
FROM invoices
WHERE invoice_total - payment_total - credit_total = 0
(74 rows inserted.)
```

The same statement with a column list

```
INSERT INTO invoice_archive
    (invoice_id, vendor_id, invoice_number, invoice_total, credit_total,
    payment_total, terms_id, invoice_date, invoice_due_date)
SELECT
    invoice_id, vendor_id, invoice_number, invoice_total, credit_total,
    payment_total, terms_id, invoice_date, invoice_due_date
FROM invoices
WHERE invoice_total - payment_total - credit_total = 0
(74 rows inserted.)
```

Insert multiple rows of literal values into the Color_Sample table

```
-- NULL can be used, but not DEFAULT
INSERT INTO color_sample (color_id, color_number, color_name)
SELECT 6, 0,    'Green' FROM dual UNION ALL
SELECT 7, 340,  NULL  FROM dual UNION ALL
SELECT 8, 76,   'Blue'  FROM dual;
(3 rows inserted.)
```

Description

- To insert rows selected from one or more tables into another table, you can code a subquery in place of the VALUES clause.
- If you don't code a column list in the INSERT clause, the subquery must return values for all the columns in the table where the rows will be inserted, and the columns must be returned in the same order as they appear in that table.
- If you include a column list in the INSERT clause, the subquery must return values for those columns in the same order as they appear in the column list.
- To insert multiple rows of literal values with one INSERT statement, you can use UNION ALL to combine a series of SELECT statements that return literal values from the Dual table.
- With Oracle 23ai or later, you can omit the FROM clause when coding a SELECT statement that uses the Dual table.

Figure 7-4 How to use a subquery to insert multiple rows

How to update existing rows

To modify the data in one or more rows of a table, you use the UPDATE statement. Although most of the UPDATE statements you code will perform simple updates, you can also code more complex UPDATE statements that include subqueries. Like INSERT, changes made with UPDATE will not be saved to the database until you commit them.

How to update rows

Figure 7-5 presents the simplified syntax of the UPDATE statement. The UPDATE clause names the table to be updated, the SET clause names the columns to be updated and the values to be assigned to those columns, and the WHERE clause specifies the condition a row must meet to be updated.

Although the WHERE clause is optional, you'll almost always want to include it. If you don't, all of the rows in the table will be updated, which usually isn't what you want.

The first UPDATE statement in this figure modifies the values of two columns in the Invoices table: payment_date and payment_total. Because the WHERE clause in this statement identifies a specific invoice number, only the columns for that invoice will be updated.

Notice in this example that the value to be assigned to payment_date is coded as a literal. However, you can assign any valid expression to a column as long as it evaluates to a value that's compatible with the data type of the column. You can also use the NULL keyword to assign a null value to a column that allows nulls, and you can use the DEFAULT keyword to assign the default value to a column that's defined with one.

The second UPDATE statement modifies the terms_id column. This time, however, the WHERE clause specifies that all the rows for vendor 95 should be updated. Because this vendor has six rows in the Invoices table, all six rows are updated.

The third UPDATE statement illustrates how you can use an expression to assign a value to a column. In this case, the expression increases the value of the credit_total column by 100. Like the first UPDATE statement, this statement updates a single row.

Before you execute an UPDATE statement, you'll want to be sure that you've selected the correct rows. To do that, you can execute a SELECT statement with the same search condition. Then, if the SELECT statement returns the correct rows, you can change it to an UPDATE statement.

The simplified syntax of the UPDATE statement

```
UPDATE table_name
SET column_name_1 = expression_1 [, column_name_2 = expression_2]...
[WHERE search_condition]
```

Assign new values to two columns of a single row

```
UPDATE invoices
SET payment_date = '21-SEP-24',
    payment_total = 19351.18
WHERE invoice_number = '97/522'
(1 row updated.)
```

Assign a new value to one column for all invoices for a vendor

```
UPDATE invoices
SET terms_id = 1
WHERE vendor_id = 95
(6 rows updated.)
```

Use an arithmetic expression to assign a new value to a column

```
UPDATE invoices
SET credit_total = credit_total + 100
WHERE invoice_number = '97/522'
(1 row updated.)
```

Description

- You use the UPDATE statement to modify one or more rows in a table.
- In the SET clause, you name each column and its new value. You can specify the value for a column as a literal or an expression.
- In the WHERE clause, you can specify the conditions that must be met for a row to be updated.
- You can use the DEFAULT and NULL keywords to specify default and null values.

Warning

- If you omit the WHERE clause, all rows in the table will be updated.

Figure 7-5 How to update rows

How to use a subquery in an UPDATE statement

Figure 7-6 presents three UPDATE statements that illustrate how you can use a subquery in an update operation. In the first statement, a subquery is used in the SET clause to retrieve the latest (maximum) invoice due date from the Invoices table. Then, that value is assigned to the invoice_due_date column for invoice number 97/522.

In the second statement, a subquery is used in the WHERE clause to identify the invoices to be updated. This subquery returns the vendor_id value for the vendor in the Vendors table with the name "Pacific Bell." Then, all the invoices with that vendor_id value are updated.

The third UPDATE statement also uses a subquery in the WHERE clause. This subquery returns a list of the vendor_id values for all vendors in California, Arizona, and Nevada. Then, the IN operator is used to update all the invoices with vendor_id values in that list. Note that although the subquery returns 80 vendors, many of those vendors don't have invoices. As a result, the UPDATE statement only affects 51 invoices.

Assign the latest due date in the Invoices table to a specific invoice

```
UPDATE invoices
SET credit_total = credit_total + 100,
    invoice_due_date = (SELECT MAX(invoice_due_date) FROM invoices)
WHERE invoice_number = '97/522'
(1 row updated.)
```

Update all invoices for a vendor based on the vendor's name

```
UPDATE invoices
SET terms_id = 1
WHERE vendor_id =
   (SELECT vendor_id
    FROM vendors
    WHERE vendor_name = 'Pacific Bell')
(6 rows updated.)
```

Change the terms of all invoices for vendors in specified states

```
UPDATE invoices
SET terms_id = 1
WHERE vendor_id IN
   (SELECT vendor_id
    FROM vendors
    WHERE vendor_state IN ('CA', 'AZ', 'NV'))
(51 rows updated.)
```

Description

- You can use a subquery in the SET clause to return the value that's assigned to a column.
- You can use a subquery in the WHERE clause to provide one or more values used in a search condition.

Figure 7-6 How to use subqueries in an UPDATE statement

How to delete existing rows

To delete one or more rows from a table, you use the DELETE statement. As with the UPDATE statement, you can use subqueries in a DELETE statement to help identify the rows to be deleted. In addition, deleted rows are not permanently deleted until you commit the change.

How to delete rows

Figure 7-7 presents the simplified syntax of the DELETE statement along with four examples that show how this statement works. To start, you specify the name of the table in the DELETE clause. In this clause, the FROM keyword is optional, but you can include it for clarity.

To identify specific rows to be deleted, you code a WHERE clause. Although this clause is optional, you'll almost always include it. If you don't, all of the rows in the table are deleted. This is a common coding mistake, and it can be disastrous if you're working with live data.

If you want to make sure that you've selected the correct rows before you run the DELETE statement, you can use a SELECT statement to test the WHERE clause. Then, if the correct rows are retrieved, you can use the same WHERE clause for the DELETE statement.

The first DELETE statement in this figure deletes a single row from the General_Ledger_Accounts table. To do that, it specifies the account_number value of the row to be deleted in the WHERE clause.

The second DELETE statement deletes a single row from the Invoice_Line_Items table. To do that, it specifies the invoice_id value and the invoice_sequence value of the row to be deleted in the WHERE clause.

The third DELETE statement deletes four rows from the Invoice_Line_Items table. To do that, it specifies 100 as the invoice_id value of the row to be deleted. Since the invoice for this ID has four line items, this deletes all four line items.

How to use a subquery in a DELETE statement

If you try to delete a row that's related to other rows with a foreign-key constraint, Oracle will return an error message and won't delete the row. For example, if you try to delete a row from the Vendors table that has related rows in the Invoices and Invoice_Line_Items tables, Oracle will return an error message, and it won't delete the vendor. Usually, that's what you want.

However, if you want to delete a vendor that has one or more child rows, you start by deleting all of the invoices and line items for that vendor. To do that, you can start by using a statement like the one in the fourth example to delete the line items for the vendor from the Invoice_Line_Items table. In this case, the statement uses a subquery to delete all line items for the vendor with the ID of 115. Then, you can use a similar statement to delete all of the invoices for that vendor from the Invoices table. Finally, you can use a simple DELETE statement to delete the row for the vendor from the Vendors table.

The syntax of the DELETE statement

```
DELETE [FROM] table_name
[WHERE search_condition]
```

Delete one row

```
DELETE FROM general_ledger_accounts
WHERE account_number = 306;
(1 row deleted.)
```

Delete one row using a compound condition

```
DELETE FROM invoice_line_items
WHERE invoice_id = 19 AND invoice_sequence = 1
(1 row deleted.)
```

Delete four rows

```
DELETE FROM invoice_line_items
WHERE invoice_id = 100
(4 rows deleted)
```

Delete all line items for a vendor

```
DELETE FROM invoice_line_items
WHERE invoice_id IN
     (SELECT invoice_id
      FROM invoices
      WHERE vendor_id = 115)
(4 rows deleted.)
```

Description

- You can use the DELETE statement to delete one or more rows from a table.
- You specify the conditions that must be met for a row to be deleted in the WHERE clause.
- You can use a subquery within the WHERE clause.
- A foreign-key constraint may prevent you from deleting a row. In that case, you can only delete the row if you delete all child rows for that row first.

Warning

- If you omit the WHERE clause from a DELETE statement, all the rows in the table will be deleted.

Figure 7-7 How to delete rows

How to commit and roll back changes

When you use SQL Developer to execute INSERT, UPDATE, and DELETE statements, Oracle Database automatically adds those statements to the current *transaction*. A transaction is a group of SQL statements that must all be executed successfully before they are saved to the database. To make the changes to the database permanent, you must explicitly *commit* the changes to the database. Or, you can undo, or *roll back*, the changes.

For example, the INSERT statement in figure 7-8 adds a single row to the Invoices table. If you use SQL Developer to execute this statement, the row will be added to the Invoices table. It will appear in the table's data and in a result set if you select it in a query. However, any other users using the database won't see this new row unless and until you commit the change to the database.

When you're practicing with INSERT, UPDATE, and DELETE statements, you may want them always rolled back. If you want to make permanent changes to a production system, though, you'll need to make sure to commit the changes.

You can make the changes to the database permanent by executing the COMMIT statement shown in this figure. When you do, SQL Developer will display a message that indicates whether the commit succeeded. Or, if you're using SQL Developer, you can commit the changes by clicking on the Commit button or pressing F11. However, you won't see a message indicating if the commit succeeded. Similarly, you can roll back a transaction by using the ROLLBACK statement, clicking the Rollback button, or pressing F12.

Most of the time, you'll want to use one of these techniques to manually commit your changes. However, if you want SQL Developer to automatically commit changes immediately after they are made, you can enable Autocommit in the Preferences.

When you exit SQL Developer without committing your transactions, Oracle Database will ask if you want to commit or roll back your changes. This includes all INSERT, UPDATE, and DELETE statements that haven't explicitly been committed. At this time, you can choose to commit, roll back, or not close SQL Developer. However, it's not a good idea to rely on this method to commit changes in case SQL Developer closes unexpectedly.

Add a new row to a table

```
INSERT INTO invoices
VALUES (115, 97, '456789', '01-AUG-24', 8344.50, 0, 0, 1, '31-AUG-24', NULL)
```

The system response

```
1 row inserted.
```

Commit the changes

```
COMMIT
```

The system response

```
Commit complete.
```

Roll back the changes

```
ROLLBACK
```

The system response

```
Rollback complete.
```

Closing SQL Developer without committing changes

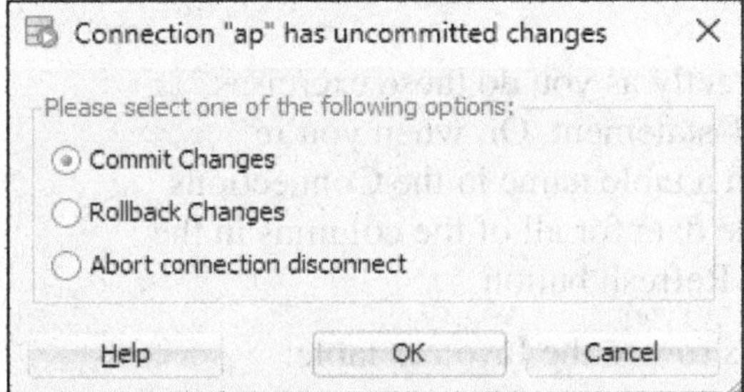

Description

- A *transaction* is a group of SQL statements that must all be executed together.
- By default, Oracle adds all INSERT, UPDATE, and DELETE statements to the current transaction.
- To *commit* a transaction, you can code a COMMIT statement, click the Commit button, or press F11.
- To reverse, or *roll back*, a transaction, you can code a ROLLBACK statement, click the Rollback button, or press F12.
- When you use the buttons or shortcut keys to commit or roll back changes, the system does not print a response.
- If you don't commit your changes, Oracle will ask if you want to roll them back when you exit SQL Developer.
- To have SQL Developer automatically commit changes after each INSERT, UPDATE, or DELETE statement is executed, select Tools→Preferences, Then, expand the Database node, click Advanced, and finally check "Autocommit".

Figure 7-8 How to commit and roll back changes

Perspective

In this chapter, you learned how to use the INSERT, UPDATE, and DELETE statements to modify the data in a database. In chapters 9 and 10, you'll learn more about the table definitions that can affect the way these statements work. And in chapter 15, you'll learn more about executing groups of INSERT, UPDATE, and DELETE statements as a single transaction.

If you followed along with the examples in this chapter and committed any of your changes, now is a good time to re-create the databases and restore them to their original state before continuing.

Terms

transaction
commit
roll back

Exercises

To test whether a table has been modified correctly as you do these exercises, you can write and run an appropriate SELECT statement. Or, when you're using Oracle SQL Developer, you can click on a table name in the Connections window and then on the Data tab to display the data for all of the columns in the table. To refresh the data on this tab, click the Refresh button.

1. Write an INSERT statement that adds this row to the Invoices table:

```
invoice_id:          115
vendor_id:           32
invoice_number:      AX-024-019
invoice_date:        8/15/2024
invoice_total:       $512.95
payment_total:       $0.00
credit_total:        $0.00
terms_id:            2
invoice_due_date:    9/15/2024
payment_date:        null
```

2. Write an UPDATE statement that modifies the Vendors table. Change the default account number for each vendor that has a default account number of 400 to 403 instead.
3. Write an UPDATE statement that modifies the Invoices table. For each invoice for a vendor with a default_terms_id of 2, set the terms_id column to 2.
4. Write a DELETE statement that deletes the row that you added to the Invoices table in exercise 1.
5. After you have verified that all of the modifications for the first four exercises have been successful, roll back the changes. Then, verify that they have been rolled back.

8

How to work with data types and functions

So far, you have been using SQL statements to work with the three most common types of data: strings, numbers, and dates. Now, this chapter takes a more in-depth look at the data types that are available with Oracle Database and shows some basic functions for working with them.

Data type overview **212**

How to work with character data **214**

The character data types 214

How to use character functions 216

How to parse a string 220

How to work with numeric data **222**

The numeric data types 222

Common number format elements 224

How to use numeric functions 226

How to search for floating-point numbers 228

How to work with temporal data **230**

The temporal data types 230

Common datetime format elements 232

How to use datetime functions 234

How to perform a date search 236

Common timestamp and interval formats 238

How to use timestamp functions 240

How to use interval functions 242

How to use the EXTRACT function 244

How to convert data from one type to another **246**

How to convert characters, numbers, and dates 246

How to sort strings in numerical sequence 248

How to perform a time search 250

How to convert characters to and from their numeric codes 252

Other functions you should know about **254**

How to use the CASE expression 254

How to use the COALESCE, NVL, and NVL2 functions 256

How to use the GROUPING function 258

Perspective **260**

Data type overview

A column's *data type* specifies the type of information that the column is intended to store. A column's data type also determines the types of operations that can be performed on the data.

The built-in Oracle data types can be divided into the categories shown in the first table in figure 8-1. There are a few more categories for specialized use, but these are the data types you'll see most often.

The *character data types* are used to store a string of one or more characters. They can include letters, numbers, special characters, or symbols. The terms character, string, and text are used interchangeably to describe this type of data.

The *numeric data types* store numbers that can be used for mathematical calculations. These numbers can be positive or negative and can include a decimal point. They can also be floating-point numbers that are used to store approximate values for very large and very small numbers.

The *temporal data types* are used primarily to store dates and times. They can also be called datetime or just date types. In addition to dates and times, this group includes time intervals and timestamps.

The *Boolean data type* stores one of just three values: true, false, or unknown (null). It is available with Oracle 23ai or later.

The *large object (LOB) data types* are used to store large amounts of binary data such as image, audio, and video files or very large amounts of text. The files can be stored either internally in the database or as paths to files stored elsewhere.

Spatial data types store data for working with geographical coordinates. To use them, the free Oracle Spatial and Graph component must be installed on your system.

The *JSON data type*, naturally enough, is used to store *JavaScript Object Notation (JSON)* documents. Similarly, the *XML data type* is used to store XML (Extensible Markup Language) documents.

Built-in data type categories

Category	Description
Character	Strings of characters
Number	Integer, decimal, and floating-point numbers
Temporal	Dates, times, timestamps, and intervals
Boolean	True or false
Large object (LOB)	Text, images, sound, and video
Spatial	Geographical values
JSON	JSON documents
XML	XML documents

Description

- Oracle provides built-in *data types* for storing many types of data.
- *Character data* can include numbers, letters, symbols, and emojis.
- The *Boolean data type* is available as of Oracle 23ai, and can store only true, false, or unknown (null).
- The *large object (LOB) data types* are useful for storing images, audio, video, and large amounts of text.
- The *spatial data types* are useful for storing geographical values such as global positioning system (GPS) data.
- The *JSON data type* is used for storing JavaScript Object Notation (JSON) documents.
- The *XML data type* is used for storing Extensible Markup Language (XML) documents.
- There are additional data types beyond the ones listed here, but these are the types most commonly used by database programmers.

Figure 8-1 Data type overview

How to work with character data

Character data consists of strings of letters, numbers, symbols, and even emojis depending on your database's character set. You've already worked with character data in previous chapters, but now you'll learn more about different types of character data and some common functions for working with strings.

The character data types

Figure 8-2 presents the four built-in character data types. The CHAR and NCHAR data types store *fixed-length strings*. Data stored using these data types always occupies the same number of bytes, regardless of the actual number of characters in the string. These data types are typically used to define columns that have a fixed number of characters. For example, the vendor_state column in the Vendors table is defined as CHAR(2) because it always contains two characters.

The VARCHAR2 and NVARCHAR2 data types, on the other hand, store *variable-length strings*. Data stored using these data types uses only the number of bytes needed to store the characters in the string. These types are typically used to define columns whose lengths vary from one row to the next.

The CHAR and VARCHAR2 data types use the database's *character set* to store strings. A character set is a collection of letters, numbers, and symbols indexed by a code. ASCII and Unicode are both examples of character sets.

By default, the database character set is AL32UTF8. Generally, this set uses 1 byte of storage for ASCII characters, 2 bytes for Western European characters, 3 bytes for Asian characters, and 4 bytes for some additional characters.

NCHAR and NVARCHAR2 use the database's *national character set*. The N prefix refers to when Unicode characters were also called national characters. By default, the national character set is AL16UTF16, which uses 2 bytes for most characters and 4 bytes for some additional characters.

You can view the character sets for your database by executing the query shown in this figure. It returns a list of all the NLS (National Language Support) parameters and their current values.

Although you typically store numeric values using numeric data types, the character data types may be a better choice for some numeric values. For example, you typically store zip codes, telephone numbers, and social security numbers in character columns even though they contain only numbers. That's because their values aren't used in arithmetic operations. In addition, if you store these numbers in numeric columns, leading zeroes are stripped, which usually isn't what you want. You also can't include formatting like parentheses and dashes.

The character data types

Type	Description
`CHAR[(size [BYTE|CHAR])]`	Fixed-length strings of character data where size is the number of characters or bytes. The default is 1 character, and the max size is 2000 bytes.
`NCHAR[(size)]`	Fixed-length strings of character data where size is the number of characters. Maximum size is determined by the national character set definition, with an upper limit of 2000 bytes.
`VARCHAR2(size [BYTE|CHAR])`	Variable-length strings of character data where size is the maximum number of characters or bytes, with an upper limit of 4000 bytes. The size argument is required.
`NVARCHAR2(size)`	Variable-length strings of character data where size is the maximum number of characters. Maximum size is determined by the national character set definition, with an upper limit of 4000 bytes. The size argument is required.

How to view the character sets for your database

```
SELECT * FROM nls_database_parameters
```

The result set

	PARAMETER	VALUE
15	NLS_NCHAR_CHARACTERSET	AL16UTF16
16	NLS_CHARACTERSET	AL32UTF8

```
(20 rows)
```

Description

- The CHAR and NCHAR data types are used for *fixed-length strings*. These data types use the same amount of storage regardless of the actual length of the string.
- If you insert a string for a CHAR or NCHAR column that's smaller than the specified size, the end of the string will be padded with spaces. If you insert a string that's too big, Oracle returns an error.
- The VARCHAR2 and NVARCHAR2 types are used for *variable-length strings*. These data types use only the amount of storage needed for a given string.
- By default, Oracle Database uses the AL32UTF8 character set for CHAR and VARCHAR and the AL16UTF16 character set for NCHAR and NVARCHAR.
- The AL32UTF8 character set uses 1 byte of storage for ASCII characters, 2 bytes for Western European characters, 3 bytes for Asian characters, and 4 bytes for some additional characters.
- The AL16UTF16 character set uses 2 bytes for most characters and 4 bytes for some additional characters.

Figure 8-2 The character data types

How to use character functions

Figure 8-3 shows how to use some of the common character functions available from Oracle. To start, you can use the LTRIM and RTRIM functions to remove leading or trailing characters from the left or right side of a string. Or, you can use the TRIM function to remove leading and trailing characters from both sides of a string. Most of the time, you'll use these functions to remove leading and trailing spaces. However, you can also use these functions to remove other characters such as zeros, periods, and so on.

Conversely, you can use the LPAD and RPAD functions to add leading or trailing characters to the left or right side of a string. Again, most of the time, you'll use these functions to add leading and trailing spaces. However, you can also use these functions to add other characters.

You can use the LOWER and UPPER functions to convert the characters in a string to lower or uppercase. In addition, you can use the INITCAP function to convert the characters in a string so the initial letter in each word is capitalized and the rest of the word is lowercase.

You can use the NVL and NVL2 functions to provide a substitute string if the string value is a NULL value. For example, you can use these functions to display a value of "Unknown" instead of displaying the default value of null.

You can use the last four functions presented in this figure to modify a string. To start, you can use the SUBSTR function to return the specified number of characters from anywhere in a string. When you use this function, you can use the second argument to specify the starting point, where 1 is the first character in the string. In addition, you can use the third argument to specify the number of characters that you want to return. If you don't specify this argument, this function will return all characters from the starting point to the end of the string.

To specify the starting point and length arguments for the SUBSTR function, it's common to use the INSTR and LENGTH functions to return integer values. For example, if you want to find the position of the first space in a string, you can use an INSTR function to return an integer value for its position. Then, if necessary, you can perform arithmetic operations on this integer, and you can nest the resulting expression within a SUBSTR function. Similarly, you can use the LENGTH function to return the number of characters in a string, you can include this function in an arithmetic expression, and you can nest the expression within a SUBSTR function.

Finally, you can use the REPLACE function to replace a substring within a string with another substring. For example, you might want to use this function to replace hyphens with periods.

Common character functions

Function	Description
`LTRIM(string[, trim_chars])`	Removes all characters specified in trim_chars from the left of the string. If trim_chars is not given, it defaults to a blank space.
`RTRIM(string[, trim_chars])`	Same as LTRIM but removes characters from the right side of the string instead of the left.
`TRIM([trim_char FROM ]string)`	Removes the specified trim character from both sides of the string. If trim_char is not specified, defaults to a blank space. Trim_char must be a single character, not a set.
`LPAD(string, length[, pad_string])`	Pads the left side of the string to the specified length with spaces or with the characters specified by pad_string.
`RPAD(string, length[, pad_string])`	Same as LPAD, but pads the right side of the string.
`LOWER(string)`	Converts the string to lowercase letters.
`UPPER(string)`	Converts the string to uppercase letters.
`INITCAP(string)`	Converts the initial letter in each word to uppercase and all other letters to lowercase.
`NVL(string, value)`	If the string is null, returns value. Otherwise, returns the string.
`NVL2(string, value1, value2)`	If the string is not null, returns value1. If the string is null, returns value2.
`SUBSTR(string, start[, length])`	Returns the specified number of characters (length) from the string starting from the specified position (start).
`LENGTH(string)`	Returns the number of characters in the string.
`INSTR(string, substr[, start])`	Returns the position of the first occurrence of the substring in the string starting from the beginning or the specified position (start). If not found, returns 0.
`REPLACE(string, substr[, replace])`	Returns the string with all occurrences of the substring removed or replaced with the specified replace string.

Description

- Most of these functions work with the CHAR, VARCHAR2, NCHAR, NVARCHAR2, CLOB (character large object), and NCLOB (national character large object) data types.
- Some functions, like LOWER and UPPER, have linguistic-sensitive versions prefixed with NLS_ for use with languages other than English.
- Some of these functions have additional arguments or variations for more functionality. Check the documentation for more information.

Figure 8-3 How to use character functions (part 1 of 2)

Part 2 of figure 8-3 shows examples of the string functions described in part 1. To start, the LTRIM and RTRIM examples remove spaces from the left and right sides of a string. Then, the TRIM example removes spaces from both sides of a string. Here, the result column uses single quotes to identify strings. This distinguishes string values from number values, and it shows the leading and trailing spaces for the result.

The second LTRIM example shows how to remove other characters besides spaces from a string. Here, the second argument specifies that all dollar signs and zeros should be removed from the left side of the string. This shows that the second argument of the LTRIM and RTRIM functions allows you to trim multiple characters.

Conversely, the second TRIM example uses a different syntax that only allows you to specify a single character. Here, you must specify that character, followed by the FROM keyword, followed by the string.

The LPAD and RPAD examples show how to add spaces or other characters to the left or right side of a string. If you want, you can use these functions to align the columns of a result set. In particular, note how the LPAD function can be used to align numbers with the right side of a column.

The LOWER, UPPER, and INITCAP examples show how to change the case of a string. Note how the INITCAP example works the same regardless of whether the input string is uppercase or lowercase.

The NVL and NVL2 examples show how to substitute a string for a null value. In these examples, string literals are used to specify a string value, and the NULL keyword is used to specify a null value.

The SUBSTR examples show how to return part of a string. To start, the first example returns the first five characters of a string. To do that, the second argument specifies 1 as the position for first character in the string, and the second argument specifies 5 as the length of the string. The second SUBSTR example works similarly, but it returns a string that begins at the seventh character and is three characters long. Unlike the first two examples, the third SUBSTR example doesn't include a third argument. As a result, it returns all characters in the string from the seventh character to the end of the string.

The INSTR examples show how to search for a string within a string and to return an integer value for its starting position. To start, the first example shows how to return an integer for the first hyphen in the string. Next, the second example shows how to return an integer for the second hyphen in the string. To do that, this example uses the third argument of the INSTR function to start the search at the fifth character in the string. Finally, the third example shows how to return the position for a string literal of "1212". This shows that you can use this function to search for a single character or a string of multiple characters.

The LENGTH examples show how to return an integer for the length of a string. Note that this includes any leading or trailing spaces.

The REPLACE examples show how to replace part of a string with another string. Here, the first example replaces all of the hyphens within the string with periods. Then, the second example replaces all of the hyphens within the string with nothing.

Character function examples

Example	Result
`LTRIM('  John Smith  ')`	`'John Smith  '`
`RTRIM('  John Smith  ')`	`'  John Smith'`
`TRIM('  John Smith  ')`	`'John Smith'`
`LTRIM('$0019.99', '$0')`	`'19.99'`
`TRIM('$' FROM '$0019.99')`	`'0019.99'`
`LPAD('$19.99', 15)`	`'         $19.99'`
`LPAD('$2150.78', 15)`	`'       $2150.78'`
`LPAD('$2150.78', 15, '.')`	`'.......$2150.78'`
`RPAD('John', 15)`	`'John           '`
`RPAD('John', 15, '.-')`	`'John.-.-.-.-.-.'`
`LOWER('CA')`	`'ca'`
`UPPER('ca')`	`'CA'`
`INITCAP('john SMITH')`	`'John Smith'`
`NVL('Fresh Corn Records', 'Unknown Company Name')`	`'Fresh Corn Records'`
`NVL(NULL, 'Unknown Company Name')`	`'Unknown Company Name'`
`NVL2('Fresh Corn Records', 'Known', 'Unknown')`	`'Known'`
`NVL2(NULL, 'Known', 'Unknown')`	`'Unknown'`
`SUBSTR('(559) 555-1212', 1, 5)`	`'(559)'`
`SUBSTR('(559) 555-1212', 7, 3)`	`'555'`
`SUBSTR('(559) 555-1212', 7)`	`'555-1212'`
`INSTR('559-555-1212', '-')`	`4`
`INSTR('559-555-1212', '-', 5)`	`8`
`INSTR('559-555-1212', '1212')`	`9`
`LENGTH('(559) 555-1212')`	`14`
`LENGTH('  (559) 555-1212  ')`	`18`
`REPLACE('559-555-1212', '-')`	`'5595551212'`
`REPLACE('559-555-1212', '-', '.')`	`'559.555.1212'`

Figure 8-3 How to use character functions (part 2 of 2)

How to parse a string

Figure 8-4 shows how to parse a string. To start, the first SELECT statement shows how you can use the SUBSTR function to format columns in a result set. Here, the second column begins by getting the first name of the vendor contact and appending a space to the end of this name. Then, it uses the SUBSTR function to return the first initial of the last name for the vendor contract and appends this initial to the string. Finally, this column adds a period after the initial for the last name.

The third column displays the vendor's phone number without an area code. To accomplish that, this column specification uses the SUBSTR function to return all characters of the vendor_phone column starting at the seventh character. Of course, this only works correctly because all of the phone numbers are stored in the same format with the area code in parentheses.

This example also shows how you can use a function in the search condition of a WHERE clause. This condition uses the SUBSTRING function to select only those rows with an area code of 559. To do that, it retrieves three characters from the vendor_phone column starting with the second character. Again, this assumes that the phone numbers are all in the same format and that the area code is enclosed in parentheses.

The second SELECT statement shows how to extract multiple values from a single column. In this example, both first and last names are stored in the Name column of the String_Sample table. As a result, if you want to work with the first and last names independently, you have to parse the string using the string functions. In this example, the first name is considered to be every character up to the first space, and the last name is considered to be every character after the first space.

To extract the first name, this statement uses the SUBSTR and INSTR functions. First, it uses the INSTR function to locate the first space in the Name column. Then, it uses the SUBSTR function to extract all of the characters up to that space. Note that a value of one is subtracted from the value that's returned by the INSTR function, so the space itself isn't included in the first name.

To extract the last name, this statement uses the same functions. First, it uses the INSTR function to locate the first space in the Name column. Then, it uses the SUBSTR function to extract all of the characters from that space to the end of the string. Note that a value of one is added to the value that's returned by the INSTR function, so the space itself isn't included in the last name.

As you review this example, note that this code won't work for all names. If, for example, a first name contains a space, such as in the name Jean Paul, this code won't work properly. That illustrates the importance of designing a database so that this type of problem doesn't occur.

Use SUBSTR to format columns

```
SELECT vendor_name,
       vendor_contact_first_name || ' ' ||
           SUBSTR(vendor_contact_last_name, 1, 1) || '.'
           AS contact_name,
       SUBSTR(vendor_phone, 7) AS phone
FROM vendors
WHERE SUBSTR(vendor_phone, 2, 3) = '559'
ORDER BY vendor_name
```

The result set

	VENDOR_NAME	CONTACT_NAME	PHONE
1	Abbey Office Furnishings	Kyra F.	555-8300
2	BFI Industries	Erick K.	555-1551
3	Bill Marvin Electric Inc	Kaitlin H.	555-5106
4	Cal State Termite	Demetrius H.	555-1534

(34 rows)

The String_Sample table in the EX schema

	ID	NAME
1	1	Lizbeth Darien
2	2	Darnell O'Sullivan
3	17	Lance Pinos-Potter
4	20	Jean Paul Renard
5	3	Alisha von Strump

A INSTR function nested in a SUBSTR function

```
SELECT SUBSTR(name, 1, (INSTR(name, ' ') - 1)) AS first_name,
       SUBSTR(name, (INSTR(name, ' ') + 1)) AS last_name
FROM string_sample
```

The result set

	FIRST_NAME	LAST_NAME
1	Lizbeth	Darien
2	Darnell	O'Sullivan
3	Lance	Pinos-Potter
4	Jean	Paul Renard
5	Alisha	von Strump

Description

- Functions can be used to format the values in columns.
- It's common to nest one function within another.

Figure 8-4 How to parse a string

How to work with numeric data

Numeric data consists of integers, decimals, and floating-point values. They can also store NaN (not a number), inf (infinity), and -inf (negative infinity).

The numeric data types

Figure 8-5 presents Oracle's built-in numeric data types. The NUMBER data type is used to store positive and negative numbers with a fixed number of digits. For business applications, this will usually be the only numeric data type that you need.

When you specify a NUMBER, you give the precision and scale in parentheses. The *precision* is the maximum number of *significant digits* that can be stored in the column. A significant digit is the number of digits that are certain, or meaningful, in a number. The *scale* is the number of digits from decimal point to the least significant digit, with positive values for a least-significant digit to the right of the decimal point and negative values for a least-significant digit to the left. For example, the number 123.45 has a precision of 5 and a scale of 2. On the other hand, 12000 has a precision of 2 and a scale of -3.

In contrast to the fixed numbers stored by the NUMBER data type, the FLOAT data type is used to store *floating-point numbers*. These numbers are useful for storing the values for very large or small numbers, but with a limited number of significant digits. Be warned that the FLOAT data type doesn't always store exact values.

To express the value of a floating-point number, Oracle allows you to use *scientific notation*. To use this notation, you type the significant digits, then the letter E followed by a power of 10. For instance, 3.65E+9 is equal to 3.65×10^9, or 3,650,000,000.

The BINARY_FLOAT and BINARY_DOUBLE data types conform to the IEEE (Institute of Electrical and Electronic Engineers) standard for floating-point arithmetic. BINARY_FLOAT is used to represent *single-precision*, 32-bit, floating-point numbers, while BINARY_DOUBLE is used to represent *double-precision*, 64-bit, floating-point numbers.

The BOOLEAN data type is available with Oracle 23ai and later. It stores just one of three possible values: TRUE, FALSE, or NULL (unknown). When working with a column defined using this data type, "no", "n", "off", and "f" as well as the number 0 are all considered FALSE by Oracle. Conversely, "yes", "y", "on", and "t" as well as all numbers other than 0 are considered TRUE. If you are using an earlier version of Oracle, you can store a Boolean value by using the NUMBER type to store either 1 for true or 0 for false.

The numeric data types

Type	Bytes	Description
`NUMBER [(p[, s])]`	1 to 22	Stores numbers with absolute values from 1.0 x 10^{-130} to, but not including, 1.0 x 10^{126}. The precision (p) can range from 1 to 38. The scale (s) can range from -84 to 127.
`FLOAT [(p)]`	1 to 22	Stores floating-point numbers. The precision (p) can range from 1 to 126 bits. A FLOAT value is represented internally as a NUMBER value.
`BINARY_FLOAT`	5	Stores 32-bit (single-precision) floating-point values.
`BINARY_DOUBLE`	9	Stores 64-bit (double-precision) floating-point values.

The Boolean data type

Type	Description
`BOOLEAN`	Stores either TRUE, FALSE, or NULL (unknown).

Description

- The numeric data types are used to store numbers. They may also store NaN (not a number), inf (infinity), and -inf (negative infinity).
- The *precision* for a NUMBER indicates the maximum number of *significant digits*.
- The *scale* for a NUMBER indicates the number of digits from the decimal point to the least significant digit.
- A positive scale indicates the number of significant digits to the right of the decimal point. A negative scale indicates the number of significant digits to the left.
- Scale can be greater than precision when using scientific notation.
- *Floating-point numbers* have a limited number of significant digits. They can store very large or very small values, but may not represent a value exactly.
- The BINARY_FLOAT and BINARY_DOUBLE types use binary precision instead of decimal precision.
- A *single-precision* binary floating-point number provides for up to 7 significant digits. A *double-precision* binary floating-point number provides for up to 16 significant digits.
- The BOOLEAN data type stores either TRUE, FALSE, or NULL (unknown). It is available as of Oracle 23ai.
- The values “no”, “n”, “off”, and “f” as well as the number 0 map to FALSE when used with a BOOLEAN column.
- The values “yes”, “y”, “on”, and “t” as well as all numbers other than 0 map to TRUE when used with a BOOLEAN column.

Figure 8-5 The numeric data types

Common number format elements

When you are working with numeric data, it can be helpful to format it for ease of reading. Figure 8-6 summarizes the most common number format elements and provides some examples. These examples show how to specify the currency symbols, group separators, and signs that are used to format numbers.

Since these format elements often provide several ways to get the same result, you can use the format element that makes the most sense for your situation. For example, if you always want the currency symbol to be a dollar sign ($), you can use this sign to specify the currency symbol. However, if you want to use the default currency symbol for the database, you can use one of the other elements to specify the currency symbol. Then, the symbol that's used will vary depending on the NLS parameters that are stored within the database.

When coding some format elements, you can code the element before or after the format for the number. For example, you can code the MI element (a minus sign) before or after the format for the number. In other cases, you must code the format element before the format for the number. For example, you must code the FM element (remove spaces) before the format for the number.

Common number format elements

Element	Description
9	Digit
.	Decimal point
D	Decimal point
,	Comma
G	Group separator
0	Leading or trailing zeros
$	Dollar sign
L	Local currency symbol
U	Dual currency symbol
C	Currency symbol
S	Minus sign for negative numbers, plus sign for positive numbers
MI	Minus sign for negative numbers
PR	Negative numbers in brackets, one space before and after positive numbers
FM	Removes leading or trailing spaces or zeros
EEEE	Scientific notation

Number format examples

Value	Format	Output
1975.5	9999	1976
1975.5	9,999.9	1,975.5
1975.5	9G999D9	1,975.5
1975.5	99,999.99	1,975.50
1975.5	09,999.990	01,975.500
1975.5	$99,999.99	$1,975.50
1975.5	L9,999.99	$1,975.50
1975.5	U9,999.99	$1,975.50
1975.5	C9,999.99	USD1,975.50
1975.5	S9,999.99	+1,975.50
-1975.5	9,999.99S	1,975.50-
-1975.5	9,999.99MI	1,975.50-
1975.5	9,999.99MI	1,975.50
-1975.5	9,999.99PR	<1,975.50>
1975.5	9,999.99PR	1,975.50
01975.50	FM9,999.99	1,975.5
1975.5	9.99EEEE	1.98E+03

Description

- Only one decimal point may be specified.

Figure 8-6 Common number format elements

How to use numeric functions

Figure 8-7 summarizes some of the common numeric functions that Oracle provides. The function you'll probably use most often is the ROUND function. This function rounds a number to the specified precision. Note that you can round the digits to the left of the decimal point by coding a negative value for this argument. However, you're more likely to code a positive number to round the digits to the right of the decimal point.

Another function that you might use regularly is the TRUNC function. This function works like the ROUND function, but it truncates the number instead of rounding to the nearest number. In other words, this function chops off the end of the number without doing any rounding. For example, if you round 19.99 to the nearest integer, you get a value of 20. However, if you truncate 19.99, you get a value of 19.

The other functions in this figure perform the mathematical operations you would expect them to. Furthermore, in addition to the functions shown here, Oracle provides a variety of other functions for performing calculations. In particular, it provides functions for performing trigonometric calculations, such as SIN for calculating the sine of an angle in radians, and ASIN for calculating the arc sine. If you need a function that isn't shown here but you suspect it exists, consult the Oracle Database documentation.

Finally, note that some functions with the same names as those in this table are available for different data types. You'll want to run a test and be sure that the function works the way you expect it to before you use any of these functions to modify data.

Common numeric functions

Function	Description
`ROUND(number[, p])`	Returns the number rounded to the specified precision (p). If p is positive, the digits to the right of the decimal point are rounded. If negative, the digits to the left are rounded.
`TRUNC(number[, p])`	Returns the number truncated to the specified precision (p). If p is negative, this function truncates the significant digits to the left of the decimal point.
`SIGN(number)`	Returns -1 if the number is less than 0, 0 if equal to 0, and 1 if greater than 0.
`CEIL(number)`	Returns the number rounded up to the nearest integer.
`FLOOR(number)`	Returns the number rounded down to the nearest integer.
`ABS(number)`	Returns the absolute value of the number.
`MOD(number, divisor)`	Returns the remainder of the number divided by the divisor.
`POWER(number, exponent)`	Returns the number raised to the power of the exponent.
`SQRT(number)`	Returns the square root of the number.

Examples that use the numeric functions

Example	Result	Example	Result
`ROUND(12.5)`	13	`ABS(1.25)`	1.25
`ROUND(12.4999, 0)`	12	`ABS(-1.25)`	1.25
`ROUND(12.4999, 1)`	12.5		
`ROUND(12.4944, 2)`	12.49	`SIGN(1.25)`	1
`ROUND(1264.99, -2)`	1300	`SIGN(0)`	0
		`SIGN(-1.25)`	-1
`TRUNC(12.5)`	12		
`TRUNC(12.4999, 1)`	12.4	`MOD(10, 10)`	0
`TRUNC(12.4944, 2)`	12.49	`MOD(10, 9)`	1
`TRUNC(1264.99, -2)`	1200		
		`POWER(2, 2)`	4
`CEIL(1.25)`	2	`POWER(2, 2.5)`	5.65685...
`CEIL(-1.25)`	-1		
`FLOOR(1.25)`	1	`SQRT(4)`	2
`FLOOR(-1.25)`	-2	`SQRT(5)`	2.23606...

Note

- You can use these functions to work with any numeric data type.
- There are also functions for mathematical operations like LOG, SIN, COS, TAN, etc. When in doubt, consult the documentation.
- Some functions with the same names but different parameters are available for different data types, such as dates and intervals.

Figure 8-7 How to use numeric functions

How to search for floating-point numbers

Floating-point numbers don't always contain exact values. From a practical point of view, that means that you don't want to search for an exact value when you're working with floating-point numbers. If you do, you'll miss values that are essentially equal to the value you want.

To illustrate, consider the Float_Sample table shown in figure 8-8. This table is in the EX schema and includes a column named float_value that's defined with the BINARY_DOUBLE data type. Here, the first example shows what happens when a SELECT statement retrieves all rows where the value stored in the float_value column is equal to 1. As you can see, the result set includes only the second row, even though the first and third rows also contain values approximately equal to 1.

To solve this problem, you can search for an approximate value when you perform a search on a column with a floating-point data type. To do that, you can search for a range of values as shown by the second example in this figure. This SELECT statement searches for values between .99 and 1.01.

Another option is to round the value that you're searching for. This is illustrated by the third example. In both the second and third examples, the statement returns the three rows in the Float_Sample table that have a float value that's approximately equal to 1.

The Float_Sample table in the EX schema

	FLOAT_ID	FLOAT_VALUE
1	1	0.999999999999999
2	2	1.0
3	3	1.000000000000001
4	4	1234.56789012345
5	5	999.04440209348
6	6	24.04849

Search for an exact value

```
SELECT * FROM float_sample
WHERE float_value = 1
```

The result set

	FLOAT_ID	FLOAT_VALUE
1	2	1.0

Search for a range of values

```
SELECT * FROM float_sample
WHERE float_value BETWEEN 0.99 AND 1.01
```

Search using rounded values

```
SELECT * FROM float_sample
WHERE ROUND(float_value, 2) = 1
```

The result set for both queries

	FLOAT_ID	FLOAT_VALUE
1	1	0.999999999999999
2	2	1.0
3	3	1.000000000000001

Description

- Because the values of floating-point numbers aren't always exact, you'll want to search for approximate values when you retrieve floating-point numbers. To do that, you can specify a range of values, or you can use the ROUND function to search for rounded values.

Figure 8-8 How to search for floating-point numbers

How to work with temporal data

Temporal data includes dates, times, and intervals. Oracle provides a number of convenient functions for working with temporal data of all kinds.

The temporal data types

Figure 8-9 presents the temporal data types that are used to store dates, times, and intervals. The first table summarizes the datetime types. Each of these data types stores a set of fields rather than a single value. For example, the DATE type stores a year, month, day, hour, minute and second. The three TIMESTAMP types are similar to DATE, but also store fractions of a second and, depending on the type, a time zone.

If you don't enter a value for one of the fields stored by a DATE or TIMESTAMP, a default is used instead. The defaults are the current year for the year, the current month for the month, 1 for the day, and 0 for the hour, minute, and second.

The second table presents the INTERVAL types. These data types make it easier to work with time intervals like 2 days, 2 hours, and 12 minutes.

The INTERVAL YEAR TO MONTH type stores a time interval in years and months. For example, you can store an interval of 1 year and 3 months. When you use this type, you can specify a value from 0 to 9 for year precision, which is the number of digits that Oracle uses to store the years. By default, this value is set to 2, which allows you to store a maximum of 99 years.

The INTERVAL DAY TO SECOND type stores a time interval in days, hours, minutes and seconds. When you use this type, you can specify a value from 0 to 9 for the day precision, which is the maximum number of digits that Oracle uses to store the days. By default, this value is set to 2, which allows you to store a maximum of 99 days. In addition, you can specify fractional second precision. This works the same as fractional seconds for the TIMESTAMP types.

If you aren't sure what your local time zone is, you can get the database time zone (DBTIMEZONE) and session time zone (SESSIONTIMEZONE) as shown in the first example.

You can get more information about how your database stores temporal data by viewing the NLS_Database_Parameters table as shown in the second example. This includes default formats and the calendar in use by your database.

The datetime data types

Type	Bytes	Description
`DATE`	7	Stores dates and times divided into fields for year, month, day, minute, hour, and second. Valid dates range from January 1, 4712 BC to December 31, 9999 AD.
`TIMESTAMP[(fsp)]`	7 to 11	An extension of DATE that optionally stores fractional parts of seconds. The fractional second precision (fsp) is the number of decimal places used to store the second. fsp can range from 0 to 9, with a default of 6.
`TIMESTAMP [(fsp)] WITH TIME ZONE`	13	Like TIMESTAMP, but includes either a time zone region name or a time zone offset from UTC.
`TIMESTAMP [(fsp)] WITH LOCAL TIME ZONE`	7 to 11	Like TIMESTAMP WITH TIME ZONE, but uses the database time zone instead of storing a zone.

The interval data types

Type	Bytes	Description
`INTERVAL YEAR [(yp)] TO MONTH`	5	Stores a time interval in years and months. Year precision (yp) is the digits in the year, with a default of 2.
`INTERVAL DAY [(dp)] TO SECOND [(fsp)]`	11	Stores a time interval in days, hours, minutes and seconds. Day precision (dp) is the max digits in the day, with a default of 2 and max of 9. Fractional second precision (fsp) is the digits for fractional seconds, with a default of 6 and max of 9.

View your database and session time zones

```
SELECT DBTIMEZONE, SESSIONTIMEZONE FROM dual
```

The result set

	DBTIMEZONE	SESSIONTIMEZONE
1	+00:00	America/Los_Angeles

View your default datetime format

```
SELECT * FROM nls_database_parameters
```

The result set

	PARAMETER	VALUE
6	NLS_TIMESTAMP_TZ_FORMAT	DD-MON-RR HH.MI.SSXFF AM TZR

(20 rows)

Description

- If you do not enter a field for a DATE, the defaults are the current year for the year, the current month for the month, 1 for the day, and 0 for the hour, minute, and second.
- The default datetime formats are determined by the database's NLS parameters.

Figure 8-9 The temporal data types

Common datetime format elements

Figures 8-10 shows how to use the most common datetime format elements. Most of them are self-explanatory. However, the RR format element requires some explanation. It interprets years from 00 to 49 as 2000 to 2049, and years 50 through 99 are interpreted as 1950 through 1999. As a result, 98 is interpreted as 1998, not 2098. Although RR is the default format for dates, it's generally considered a good practice to use YYYY to avoid ambiguity.

You will also see that the MONTH and DAY elements pad the end of the value that's returned with spaces to accommodate the longest value that can be returned. Since this makes it easy to align the month and day elements of a date in columns, this is often what you want. If it isn't what you want, you can use the TRIM function to trim the spaces from the ends of these elements.

Last, if a format element returns letters, the capitalization that you use when you specify the format element corresponds with the capitalization for the returned value. If, for example, you specify "MONTH", the month will be returned in all caps (AUGUST). If you specify "Month", the month will be returned with an initial cap (August). And if you specify "month", the month will be returned in lowercase (august). Note that this also applies to the date format elements for both the full and abbreviated names of months and days.

Common datetime format elements

Element	Description	Element	Description
AD	Anno Domini	DAY	Name of day padded with spaces
BC	Before Christ	DY	Abbreviated name of day
CC	Century	DDD	Day of year (1-366)
YEAR	Year spelled out	DD	Day of month (1-31)
YYYY	Four-digit year	D	Day of week (1-7)
YY	Two-digit year	HH	Hour of day (1-12)
RR	Two-digit round year	HH24	Hour of day (0-23)
Q	Quarter of year (1-4)	MI	Minute (0-59)
MONTH	Name of month padded with spaces	SS	Second (0-59)
MON	Abbreviated name of month	SSSSS	Seconds past midnight (0-86399)
MM	Month (1-12)	AM	Ante Meridian
WW	Week of year (1-53)	PM	Post Meridian
W	Week of month (1-5)	-/,.;:	Allowable punctuation

Datetime format examples

Format	Example
DD-MON-YY	19-AUG-24
DD-Mon-RR	19-Aug-24
MM/DD/YY	08/19/24
YYYY-MM-DD	2024-08-19
Dy Mon DD, YY	Mon Aug 19, 24
MONTH DD, YYYY BC	AUGUST 19, 2024 AD
Month DD, YYYY B.C.	August 19, 2024 A.D.
HH:MI	04:20
HH24:MI:SS	16:20:36
HH:MI AM	04:20 PM
HH:MI A.M.	04:20 P.M.
HH:MI:SS	04:20:36
YYYY-MM-DD HH:MI:SS AM	2024-08-19 04:20:36 PM

Description

- Some elements, like MON and DAY, depend on the country and language of your database.
- To change the capitalization of an element, capitalize the corresponding format element accordingly.
- If you use the RR format, years from 00 to 49 are interpreted as 2000 to 2049, and years 50 through 99 are interpreted as 1950 through 1999.

Figure 8-10 Common datetime format elements

How to use datetime functions

Figure 8-11 shows how common datetime functions work and how the plus and minus signs can be used to work with dates. Then, it shows some examples of using these functions with date literals.

To get the current date and time, you can use the SYSDATE or CURRENT_DATE function, although the SYSDATE function is more commonly used. In most cases, both of these functions return the same value. However, if a session time zone has been set, the value returned by the CURRENT_DATE function will be adjusted to accommodate that time zone.

When you use the ROUND and TRUNC functions without a format, the default is to round or truncate a date to 00:00:00 on a 24-hour clock. But when you specify a date format like MI for minutes, the datetime value is rounded or truncated to that time component.

When you use the MONTHS_BETWEEN function, it returns a whole number of months when the day component is the same for both date arguments. However, if the day component isn't the same for both date arguments, this function returns a decimal number.

When you use the ADD_MONTHS function, you can specify a positive value to add months to the specified date or a negative number to subtract months from the specified date. If necessary, the function will increase or decrease the year component.

LAST_DAY returns the last day of the month for the given date. And when you use the NEXT_DAY function, you specify a day of the week as the second argument. Then, the function returns the next day of the week that comes after the specified date. For instance, the first example of this function returns the first Friday after August 15, 2024, which is August 16, and the second example returns the first Thursday after August 15, 2024, which is August 22. Note that you can use full or abbreviated names for the second argument.

When you use the addition (+) and subtraction (-) operators, you can add a specific number of days to a date or subtract a specific number of days from a date. You can also use the subtraction operator to subtract one date from another. If the first date comes after the second date, a positive number is returned. Otherwise, a negative number is returned.

Common datetime functions

Function	Description
`SYSDATE`	Returns the current local date and time based on the operating system's clock.
`CURRENT_DATE`	Returns the local date and time adjusted for the current session time zone.
`ROUND(date[, format])`	Returns the date rounded to the field specified by the format. If the format is omitted, rounds to the nearest day.
`TRUNC(date[, format])`	Like ROUND, but truncates the date.
`MONTHS_BETWEEN(date1, date2)`	Returns the number of months between date1 and date2.
`ADD_MONTHS(date, months)`	Adds the specified number of months to the specified date and returns the resulting date.
`LAST_DAY(date)`	Returns the date for last day of the month for the specified date.
`NEXT_DAY(date, day_of_week)`	Returns the date for the next day of the week that comes after the specified date.

Two operators for working with dates

Operator	Description
+	Adds the specified number of days to a date. Two dates can't be added together.
-	Subtracts a specified number of days from a date. Or, subtracts one date from another and returns the number of days between the two dates.

Examples that use the datetime functions

Example	Result
`SYSDATE`	`19-AUG-24 04:20:36 PM`
`ROUND(SYSDATE)`	`20-AUG-24 12:00:00 AM`
`TRUNC(SYSDATE, 'MI')`	`19-AUG-24 04:20:00 PM`
`MONTHS_BETWEEN('01-JUL-24','01-SEP-24')`	`-2`
`MONTHS_BETWEEN('15-SEP-24','01-AUG-24')`	`1.4516129...`
`ADD_MONTHS('19-AUG-24', -1)`	`19-JUL-24`
`ADD_MONTHS('19-AUG-24', 11)`	`19-JUL-25`
`LAST_DAY('15-FEB-24')`	`29-FEB-24`
`NEXT_DAY('15-AUG-24', 'FRIDAY')`	`16-AUG-24`
`NEXT_DAY('15-AUG-24', 'THURS')`	`22-AUG-24`
`SYSDATE - 1`	`18-AUG-24`
`SYSDATE + 7`	`26-AUG-24`
`SYSDATE - TO_DATE('01-JAN-24')`	`231.60368...`
`TO_DATE('01-JAN-24') - SYSDATE`	`-231.60368...`

Figure 8-11 How to use datetime functions

How to perform a date search

In Oracle, datetime values always contain both a date and a time component. However, the times are hidden by default when viewing DATE values in SQL Developer. To view the times, you can use an ALTER SESSION statement as shown at the beginning of figure 8-12 to change the NLS_DATE_FORMAT parameter to a format that includes the time component. Then, when you view DATE values like the ones in the Date_Sample table, you'll see not just the dates but also the hours, minutes, and seconds for each value.

In this table, note that the time components in the first three rows each have a value of 00:00:00. In contrast, the time components in the next three rows have non-zero time components. Non-zero times like these can complicate a search for rows with a specific date if you don't account for them in your search condition.

The first SELECT statement shows a typical problem that you can encounter when searching for dates. Here, the statement searches for rows in the Date_Sample table with a date value of "28-FEB-24". Since a time component isn't specified by the date literal in the WHERE clause, a default time component of 00:00:00 is added to the date when it is converted to a DATE value. However, because the only row with a date of February 28 has a time that isn't 00:00:00, no rows are returned by this statement.

To solve this problem, you can search for a range of dates that includes only the dates you're looking for. This is shown by the second SELECT statement in this figure. In this example, the search is for any date greater than or equal to February 28, 2024 and less than February 29, 2024. As a result, this search finds any date on February 28, 2024, no matter what the time component is.

Another method is to use the TRUNC function to truncate the time component from the datetime values as shown in the third SELECT statement. In this example, the time components of the start dates are all set to 00:00:00, which is the same as the time component for the searched date literal.

A third approach to this problem is to extract the month, day, and year components of each date in the WHERE clause using EXTRACT. Then, the condition in the WHERE clause can search for the right month, day, and year values. Usually, though, the techniques in this figure are easier to use.

Set the session date format to display times

```
ALTER SESSION SET NLS_DATE_FORMAT = 'DD-MON-RR HH24:MI:SS';
```

The Date_Sample table in the EX schema

	DATE_ID	START_DATE
1	1	01-MAR-79 00:00:00
2	2	28-FEB-99 00:00:00
3	3	31-OCT-06 00:00:00
4	4	28-FEB-23 10:00:00
5	5	28-FEB-24 13:58:32
6	6	01-MAR-24 09:02:25

A search that returns nothing

```
SELECT * FROM date_sample
WHERE start_date = '28-FEB-2024'
```

The result set

DATE_ID	START_D...

Search for a range of dates

```
SELECT * FROM date_sample
WHERE start_date >= '28-FEB-2024' AND start_date < '29-FEB-2024'
```

Search using TRUNC

```
SELECT * FROM date_sample
WHERE TRUNC(start_date) = '28-FEB-2024'
```

The result set for both searches

	DATE_ID	START_DATE
1	5	28-FEB-24 13:58:32

Description

- To view the time components of DATE values, you can use ALTER SESSION to change the NLS_DATE_FORMAT parameter to a date format that includes a time component.
- If you perform a search using a date that doesn't include a time component, Oracle sets the searched date's time to 00:00:00. This can result in searches that don't return expected values.
- You can accommodate non-zero time components by searching for a range of dates or using the TRUNC function.

Figure 8-12 How to perform a date search

Common timestamp and interval formats

Figure 8-13 summarizes the most common format elements for timestamps and intervals and provides some examples that show how to use format elements to specify fractional seconds and time zones.

It's important to note that a TIMESTAMP may store a time zone as an offset from UTC, a region name, or an abbreviation. None of these are guaranteed to be unique. For that reason, it is recommended that you store time zones using offsets, which can always be used to calculate the local time relative to Universal Time. If, however, you prefer to use region names or abbreviations, you can see a list of more than 2,000 acceptable names by querying v$timezone_names.

Common timestamp and interval format elements

Element	Description
FF[1-9]	Fractional seconds
TZH	Time zone hour offset
TZM	Time zone minutes offset
TZR	Time zone region (UTC, PST, US/Pacific, America/Los_Angeles, etc)
TZD	Time zone with daylight savings

Time formats that use fractional seconds and time zones

Format	Example
HH:MI:SS.FF AM	**04:20:36.123456 PM**
HH:MI:SS.FF3 AM	**04:20:36.123 PM**
HH:MI:SS.FF9 AM TZH:TZM	**04:20:36.123456000 PM -07:00**
HH:MI:SS.FF9 am Tzr	**04:20:36.123456789 pm America/Los_Angeles**
HH:MI:SS.FF9 A.M. TZD	**04:20:36.123456789 P.M. PDT**

View time zone region names and abbreviations

```
SELECT * FROM v$timezone_names
```

The result set

	TZNAME	TZABBREV	CON_ID
1	Africa/Abidjan	LMT	0
2	Africa/Abidjan	GMT	0
3	Africa/Accra	LMT	0
4	Africa/Accra	GMT	0
5	Africa/Accra	+0020	0
6	Africa/Addis_Ababa	LMT	0

(2382 rows)

Description

- A time zone can be stored as an offset from UTC, a region name, or an abbreviation. Names and abbreviations are not necessarily unique.

Figure 8-13 Common timestamp and interval formats

How to use timestamp functions

Figure 8-14 shows how to use some of the common functions for working with timestamps. To start, you can use the first three functions to return a TIMESTAMP or TIMESTAMP WITH TIME ZONE value. When you use these functions, you can use the fractional second precision parameter to control the maximum number of fractional seconds that are returned by the function. Or, if you omit this parameter, the default is 6. However, the maximum number of fractional seconds may be limited by the system clock on the computer that's hosting the database. On many systems, these functions only return 3 fractional seconds

The first four examples in this figure show how to use the functions that return a timestamp for the current time. Here, the SYSTIMESTAMP function returns a timestamp with a time zone offset. The CURRENT_TIMESTAMP function returns a timestamp with a time zone name. The LOCALTIMESTAMP function returns a timestamp that doesn't include the time zone portion of the timestamp. And the second SYSTIMESTAMP function specifies a fractional second precision of 2 so the fractional seconds for this timestamp are rounded to 2 decimal places.

The last three examples show how to use the FROM_TZ and SYS_EXTRACT_UTC functions. Here, the FROM_TZ function converts the current timestamp from the PST time zone to the EST time zone. Note that this function returns the newly converted timestamp with the UTC time zone. Then, the two SYS_EXTRACT_UTC functions convert two different timestamps to the UTC time zone.

Common functions for working with timestamps

Function	Description
`SYSTIMESTAMP([fsp])`	Returns a TIMESTAMP WITH TIME ZONE value for the current date and time in the database time zone.
`CURRENT_TIMESTAMP([fsp])`	Returns a TIMESTAMP WITH TIME ZONE value for the current date and time in the session time zone.
`LOCALTIMESTAMP([fsp])`	Returns a TIMESTAMP value for the current date and time in the session time zone.
`FROM_TZ(timestamp, time_zone)`	Converts a TIMESTAMP value and a time zone to a TIMESTAMP WITH TIME ZONE value.
`SYS_EXTRACT_UTC(timestamp)`	Extracts the UTC time from a TIMESTAMP value with a time zone offset or time zone region name.
`TO_TIMESTAMP(expr[, fmt])`	Converts the result of an expression to a value of the TIMESTAMP type.
`TO_TIMESTAMP_TZ(expr[, fmt])`	Converts the result of an expression to a value of the TIMESTAMP WITH TIME ZONE type.

Examples that use functions to work with timestamps

Example	Result
`SYSTIMESTAMP`	`19-AUG-24 04.20.16.906000000 PM -07:00`
`CURRENT_TIMESTAMP`	`19-AUG-24 04.20.16.906000000 PM PST`
`LOCALTIMESTAMP`	`19-AUG-24 04.20.16.906000000 PM`
`SYSTIMESTAMP(2)`	`19-AUG-24 04.20.16.910000000 PM -07:00`
`FROM_TZ(LOCALTIMESTAMP, 'EST')`	`19-AUG-24 08.20.16.906000000 PM UTC`
`SYS_EXTRACT_UTC(SYSTIMESTAMP)`	`19-AUG-24 11.20.16.906000000 PM`
`SYS_EXTRACT_UTC(CURRENT_TIMESTAMP)`	`19-AUG-24 11.20.16.906000000 PM`

Description

- If you omit the fractional second precision (fsp) parameter, the default is 6.
- The maximum number of fractional seconds that's returned by the TIMESTAMP functions is determined by the system on which the database resides. On many systems, the TIMESTAMP function returns only 3 fractional seconds.
- These examples assume that current date and time is August 19 2024 at 04:20:16.906 PM Pacific Standard Time with daylight savings.

Figure 8-14 How to use timestamp functions

How to use interval functions

Figure 8-15 shows some of the common functions for working with intervals. You can use the first two functions to convert numbers to INTERVAL values. And you can use the last two functions to convert strings to INTERVAL values.

The first four examples in the table show how you can use the NUMTOYMINTERVAL function to convert a number to an INTERVAL YEAR TO MONTH value. Within these statements, you can use the YEAR or MONTH keywords to specify the conversion unit. For example, the first two functions convert numbers to year values, and the next two functions convert numbers to month values. Because the INTERVAL YEAR TO MONTH type doesn't have a component that can store the decimal portion of a month, the decimal part of the number in the fourth function is truncated.

The next six examples show how to use the NUMTODSINTERVAL function to convert a number to an INTERVAL DAY TO SECOND value. As you can see, this function works much like the NUMTOYMINTERVAL function.

The last four examples in the table show how to use the TO_YMINTERVAL and TO_DSINTERVAL functions. Here, the TO_YMINTERVAL functions use two valid string literal formats for the same YEAR TO MONTH interval, but the first format specifies some leading zeros that aren't necessary. Then, the two TO_DSINTERVAL functions show two valid string literal formats for the same DAY TO SECOND interval, but the first format specifies some leading zeros that aren't necessary. Either way, the results of the functions are the same.

The final example after the table shows how you can use these functions within a SELECT statement. Here, the only column that's returned by the SELECT statement uses the NUMTODSINTERVAL function to return an INTERVAL DAY TO SECOND value for the number of days that's calculated by subtracting the invoice date from the payment date in a table of invoices. This should give you some how idea of how the INTERVAL data types and functions can make it easier for you to work with intervals in your own applications.

Common functions for working with intervals

Function	Description
`NUMTOYMINTERVAL(n, 'ym_unit')`	Converts a number to an INTERVAL YEAR TO MONTH value. The second argument must be YEAR or MONTH.
`NUMTODSINTERVAL(n, 'ds_unit')`	Converts a number to an INTERVAL DAY TO SECOND value. The second argument must be DAY, HOUR, MINUTE, or SECOND.
`TO_YMINTERVAL('ym_literal')`	Converts a string that contains a valid INTERVAL YEAR TO MONTH literal to the data type.
`TO_DSINTERVAL('ds_literal')`	Converts a string that contains a valid INTERVAL DAY TO SECOND literal to the data type.

Examples that use functions to work with intervals

Example	Result
`NUMTOYMINTERVAL(2,'YEAR')`	+02-00
`NUMTOYMINTERVAL(1.25,'YEAR')`	+01-03
`NUMTOYMINTERVAL(2,'MONTH')`	+00-02
`NUMTOYMINTERVAL(1.25,'MONTH')`	+00-01
`NUMTODSINTERVAL(30,'DAY')`	+30 00:00:00.000000
`NUMTODSINTERVAL(30,'HOUR')`	+01 06:00:00.000000
`NUMTODSINTERVAL(30,'MINUTE')`	+00 00:30:00.000000
`NUMTODSINTERVAL(30,'SECOND')`	+00 00:00:30.000000
`NUMTODSINTERVAL(90,'SECOND')`	+00 00:01:30.000000
`NUMTODSINTERVAL(30.123,'SECOND')`	+00 00:00:30.123000
`TO_YMINTERVAL('01-03')`	+01-03
`TO_YMINTERVAL('1-3')`	+01-03
`TO_DSINTERVAL('1 06:00:00.00')`	+01 06:00:00.000000
`TO_DSINTERVAL('1 6:0:0')`	+01 06:00:00.000000

Retrieve an interval

```
SELECT NUMTODSINTERVAL(payment_date - invoice_date, 'DAY')
       AS payment_interval
FROM invoices
WHERE payment_date IS NOT NULL
```

Result set

	PAYMENT_INTERVAL
1	+46 00:00:00.000000
2	+61 00:00:00.000000
3	+26 00:00:00.000000

Figure 8-15 How to use interval functions

How to use the EXTRACT function

Figure 8-16 shows you how to use the EXTRACT function to return the various parts of a temporal value as strings. To do that, you specify the appropriate element for the part of the value that you want to return. For example, MONTH returns a number representing the month of the specified date.

Not all fields can be used with all temporal data types. Most of these restrictions are intuitive. For example, you can't use the YEAR or MONTH fields with an INTERVAL DAY TO SECOND type because that data type doesn't contain a year or a month part. However, you can't use the HOUR, MINUTE, and SECOND fields with the DATE type even though this type contains hour and minute parts. If you want to extract these parts from a DATE, you can first convert the date to a timestamp or string as shown later in this chapter.

The EXTRACT function

Function	Description
`EXTRACT(field FROM expr)`	Returns the specified field from the expression. Expr must have a temporal data type.

Fields you can use with EXTRACT

Field	DATE	TIMESTAMP	INTERVAL (YtM)	INTERVAL (DtS)
YEAR	Yes	Yes	Yes	No
MONTH	Yes	Yes	Yes	No
DAY	Yes	Yes	No	Yes
HOUR	No	Yes	No	Yes
MINUTE	No	Yes	No	Yes
SECOND	No	Yes	No	Yes
TIMEZONE_HOUR	Timestamps with time zones only			
TIMEZONE_MINUTE	Timestamps with time zones only			
TIMEZONE_REGION	Timestamps with time zones only			
TIMEZONE_ABBR	Timestamps with time zones only			

Examples that use EXTRACT

Example	Result
`SYSDATE`	`19-AUG-24 04:20:36 PM`
`SYSTIMESTAMP`	`19-AUG-24 04.20.16.906000000 PM -07:00`
`EXTRACT(YEAR FROM SYSDATE)`	`2024`
`EXTRACT(MONTH FROM SYSDATE)`	`8`
`EXTRACT(MINUTE FROM SYSDATE)`	`(error)`
`EXTRACT(MINUTE FROM SYSTIMESTAMP)`	`20`

Description

- If there is ambiguity in a field, such as a timezone region, EXTRACT returns UNKNOWN or UNK.

Figure 8-16 How to use the EXTRACT function

How to convert data from one type to another

As you work with the various data types, you'll find that you frequently need to convert a value from one data type to another. To do that, you can use the functions that are described next.

How to convert characters, numbers, and dates

Figure 8-17 shows how to use three of the Oracle TO_ functions to convert data from one type to another. As you might expect, the TO_CHAR function converts an expression to the VARCHAR2 data type. The TO_NUMBER function converts an expression to the NUMBER type. And the TO_DATE function converts an expression to the DATE type. You can code these functions with or without format specifications.

Alternately, you can use the CAST function. This is an ANSI-standard function that lets you convert, or *cast*, an expression to the data type you specify. This is useful if you want to specify a precision and scale for a number or a maximum size for a string. However, although CAST appears to allow you to convert any type of data to any other type, it does have a variety of restrictions specific to the type of conversion being attempted. It's best to consult the Oracle documentation before using CAST for any but the most common data types.

To show how these functions work, these examples use literal values to specify numbers and dates. As a result, date literals are enclosed in single quotation marks, and numeric literals aren't enclosed. In practice, though, you're more likely to use these functions on columns or expressions.

Besides the three TO_ functions in this figure, Oracle provides TO_ functions for all of its built-in types. For example, you can use the TO_NCHAR function to convert a number or datetime value to an NVARCHAR2 type, and you can use the TO_TIMESTAMP function to convert a character type to a TIMESTAMP type. For more information about TO_ functions, consult the Oracle Database documentation.

Some functions for converting data

Function	Description
`TO_CHAR(expr[, format])`	Converts an expression to a value of the VARCHAR2 type.
`TO_NUMBER(expr[, format])`	Converts an expression to a value of the NUMBER type.
`TO_DATE(expr[, format])`	Converts an expression to a value of the DATE type.
`CAST(expr AS type[,format])`	Converts an expression to a value of the specified type.

Examples of converting data

Example	Resulting Value
`TO_CHAR(1975.5)`	`'1975.5'`
`TO_CHAR(1975.5, '$99,999.99')`	`'$1,975.50'`
`TO_CHAR(SYSDATE)`	`'19-AUG-24'`
`TO_CHAR(SYSDATE, 'DD-MON-YYYY HH24:MI:SS')`	`'19-AUG-2024 10:53:56'`
`TO_NUMBER('1975.5')`	`1975.5`
`TO_NUMBER('$1,975.5')`	`(error)`
`TO_NUMBER('$1,975.5', '$99,999.99')`	`1975.5`
`TO_DATE('15-APR-24')`	`15-APR-2024 00:00:00`
`TO_CHAR(TO_DATE('15-APR-24'), 'DD-MON-YYYY HH24:MI:SS')`	`'15-APR-2014 00:00:00'`
`CAST('Hello' AS CHAR(2))`	`'He'`
`CAST('1528.786' AS NUMBER(6,2))`	`1528.79`
`CAST('15-APR-24' AS DATE, 'DD-MON-YYYY' )`	`15-APR-2024`

Description

- Oracle provides TO functions for most of the built-in data types.
- TO_CHAR always returns a VARCHAR2 value, but functions slightly differently depending on if the expression passed to it is a number, date, string, or Boolean value.
- If no format element is specified, Oracle uses the default format for the data type.
- CAST can be used for conversions if you want to specify data type values like precision and scale.

Figure 8-17 How to convert characters, numbers, and dates

How to sort strings in numerical sequence

Figure 8-18 addresses a common problem that occurs when you store numeric data in a character column and then want to sort the column in numeric sequence. In this figure, the columns in the String_Sample table in the EX schema are defined with character data types. As a result, the first example sorts the values in the ID column in alphabetical sequence instead of numeric sequence, which isn't what you want.

One way to solve this problem is to convert the values in the ID column to integers for sorting purposes as shown in the second example. To do that, this example uses the TO_NUMBER function to convert the ID column to a NUMBER value. As a result, this example sorts the rows in numeric sequence.

Another way to solve this problem is to pad the numbers with leading zeros or spaces as shown in the third example. To do that, this example uses the LPAD function to pad the left side of the ID column with zeros. When it does, it specifies an alias for this column of lpad_id. Then, it sorts the result set by this alias, which causes the rows to be returned in numeric sequence.

Sort by a character column

```
SELECT * FROM string_sample
ORDER BY id
```

The result set

	ID	NAME
1	1	Lizbeth Darien
2	17	Lance Pinos-Potter
3	2	Darnell O'Sullivan
4	20	Jean Paul Renard
5	3	Alisha von Strump

Sort by a character column as if it were a numeric column

```
SELECT * FROM string_sample
ORDER BY TO_NUMBER(id)
```

The result set

	ID	NAME
1	1	Lizbeth Darien
2	2	Darnell O'Sullivan
3	3	Alisha von Strump
4	17	Lance Pinos-Potter
5	20	Jean Paul Renard

Sorted by a character column padded with leading zeros

```
SELECT LPAD(id, 2, '0') AS lpad_id, name
FROM string_sample
ORDER BY lpad_id
```

The result set

	LPAD_ID	NAME
1	01	Lizbeth Darien
2	02	Darnell O'Sullivan
3	03	Alisha von Strump
4	17	Lance Pinos-Potter
5	20	Jean Paul Renard

Description

- If you sort by a character column that contains numbers, you may receive unexpected results.
- You can sort a character column numerically by converting it to a numeric data type or by padding the left side of the values with leading zeros or spaces.

Figure 8-18 How to sort strings in numerical sequence

How to perform a time search

When you search for a time value without specifying a date component, Oracle uses the default values for the date parts. That is, it sets the year to the current year, the month to the current month, and the day to 1. This can cause unexpected results when searching for a value with a specific time component.

The first SELECT statement in figure 8-19 searches for values with a time component of 10:00:00. However, the statement doesn't return any rows because even though one row has the correct time value, that row doesn't have the correct date value.

The second SELECT statement in this figure shows one way you can solve this problem. Here, the search condition uses the TO_CHAR function to convert the datetime values in the start_date column to string values that don't contain a date component. To do that, it uses a date format element of "HH24:MI:SS". That way, the string that's returned will match the string literal for the time that's specified in the search condition.

Like the first SELECT statement, the third SELECT statement doesn't return any rows. Again, the problem is the same as in the first statement. To solve this problem, you can use the TO_CHAR function to remove the date component as shown in the fourth SELECT statement. Once you do that, this SELECT statement will return all rows that have a time within the range that's specified in the search condition.

Alternately, you can use the EXTRACT function to compare and search for specific hours, minutes, and seconds. Which method is easier to use will depend on the specific nature of your search.

Set the session date format to display times

```
ALTER SESSION SET NLS_DATE_FORMAT = 'DD-MON-RR HH24:MI:SS';
```

The Date_Sample table in the EX schema

	DATE_ID	START_DATE
1	1	01-MAR-79 00:00:00
2	2	28-FEB-99 00:00:00
3	3	31-OCT-06 00:00:00
4	4	28-FEB-23 10:00:00
5	5	28-FEB-24 13:58:32
6	6	01-MAR-24 09:02:25

A search that fails to return a row

```
SELECT * FROM date_sample
WHERE start_date = TO_DATE('10:00:00', 'HH24:MI:SS')
```

A search that ignores the date component

```
SELECT * FROM date_sample
WHERE TO_CHAR(start_date, 'HH24:MI:SS') = '10:00:00'
```

The result set

	DATE_ID	START_DATE
1	4	28-FEB-23 10:00:00

Another search that fails to return a row

```
SELECT * FROM date_sample
WHERE start_date >= TO_DATE('09:00:00', 'HH24:MI:SS')
  AND start_date  < TO_DATE('12:59:59', 'HH24:MI:SS')
```

Another search that ignores the date component

```
SELECT * FROM date_sample
WHERE TO_CHAR(start_date, 'HH24:MI:SS') >= '09:00:00'
  AND TO_CHAR(start_date, 'HH24:MI:SS')  < '12:59:59'
```

The result set

	DATE_ID	START_DATE
1	4	28-FEB-23 10:00:00
2	6	01-MAR-24 09:02:25

Description

- If you use the TO_DATE function to return a date that only contains a time component, the date component is set to the default (the current year, the current month, and day 1).
- To ignore the date component of a datetime value, you can convert it to a CHAR value and only return the part of the time value that you want to use.

Figure 8-19 How to perform a time search

How to convert characters to and from their numeric codes

Figure 8-20 shows three functions that are used to convert characters to and from their equivalent numeric codes according to the database character set.

For instance, the CHR function in this figure converts the number 97 to its equivalent ASCII code, the letter “a”. Conversely, the ASCII function converts the letter “a” to its numeric equivalent of 97. Although the string passed to the ASCII function can include more than one character, only the first character is converted.

The NCHR function works like the CHR function. However, it uses the national character set instead.

The CHR function is frequently used to output ASCII control characters that can’t be typed on your keyboard. The three most common control characters are presented in this figure. These characters can be used to format output so it’s easier to read. The SELECT statement in this figure, for example, uses the CHR(13) control character to start a new line after the vendor name and vendor address in the output. Note, however, that the formatting only appears correctly when you view the script output tab, not query results.

Functions for converting characters to and from their numeric codes

Function	Description
`ASCII(character)`	Returns a NUMBER value that corresponds to the character according to the database's character set.
`CHR(number)`	Returns a VARCHAR2 value that corresponds to the specified NUMBER value according to the database's character set.
`NCHR(number)`	Returns a VARCHAR2 value that corresponds to the specified NUMBER value in the database's national character set.

Examples that use the ASCII, CHR, and NCHR functions

Example	Result
`ASCII('a')`	97
`ASCII('abc')`	97
`CHR(97)`	a
`NCHR(97)`	a
`NCHR(332)`	Ō

ASCII codes for common control characters

Control character	Value
Tab	`CHR(9)`
Line feed	`CHR(10)`
Carriage return	`CHR(13)`

Format output for a script with CHR

```
SELECT vendor_name || CHR(13)
       || vendor_address1 || CHR(13)
       || vendor_city || ', ' || vendor_state || ' ' || vendor_zip_code
       AS vendor_address
FROM vendors
WHERE vendor_id = 1
```

The script output

```
VENDOR_ADDRESS
--------------------------------------------------------------------------------
US Postal Service
Attn:  Supt. Window Services
Madison, WI 53707
```

Figure 8-20 How to convert characters to and from their numeric codes

Other functions you should know about

In addition to the functions presented so far in this chapter, Oracle provides some other general-purpose functions that you should know about.

How to use the CASE expression

The CASE function allows you to specify different values depending on an expression or condition. Figure 8-21 presents the two formats of the CASE function and two examples showing how this function works.

The first example uses the simple CASE expression. When you use this format, Oracle compares the input expression in the CASE clause with the expressions in the WHEN clauses. It returns the value specified by the first WHEN clause that matches the input expression.

In this example, the input expression is the terms_id column of the Invoices table, and the WHEN expressions are the five valid values for this column. If the value of the terms_id column is 3, for example, this function returns the value "Net due 30 days."

Although it's not shown in this example for space reasons, you can also code an ELSE clause at the end of the CASE expression. Then, if none of the when expressions are equal to the input expression, the function returns the value specified in the ELSE clause.

The simple CASE expression is typically used with columns that can contain a limited number of values, such as the terms_id column in this example. In contrast, the searched CASE expression can be used for a wide variety of purposes. For example, you can test for conditions other than equal with this function. In addition, each condition can be based on a different column or expression. The second example in this figure shows how this function works.

This example determines the status of the invoices in the Invoices table. To do that, the searched CASE function determines the number of days between the July 18, 2024 and the invoice due date. If the difference is greater than 30, the CASE function returns the value "Over 30 days past due." Similarly, if the difference is greater than 0, the function returns the value "1 to 30 days past due."

Note that if an invoice is 45 days old, both of these conditions are true. In that case, the function returns the expression associated with the first condition since this condition is evaluated first. In other words, the sequence of the conditions is critical to getting logical results. If neither of the conditions is true, the function uses the ELSE clause to return a value "Current."

The syntax of the simple CASE expression

```
CASE input_expression
    WHEN when_expression_1 THEN result_expression_1
   [WHEN when_expression_2 THEN result_expression_2]...
   [ELSE else_result_expression]
END
```

The syntax of the searched CASE expression

```
CASE
    WHEN conditional_expression_1 THEN result_expression_1
   [WHEN conditional_expression_2 THEN result_expression_2]...
   [ELSE else_result_expression]
END
```

A simple CASE expression

```
SELECT invoice_number, terms_id,
    CASE terms_id
        WHEN 1 THEN 'Net due 10 days'
        WHEN 2 THEN 'Net due 20 days'
        WHEN 3 THEN 'Net due 30 days'
        WHEN 4 THEN 'Net due 60 days'
        WHEN 5 THEN 'Net due 90 days'
    END AS terms
FROM invoices
```

The result set

	INVOICE_NUMBER	TERMS_ID	TERMS
1	QP58872	4	Net due 60 days
2	Q545443	4	Net due 60 days
3	P-0608	5	Net due 90 days

A searched CASE expression

```
SELECT invoice_number, invoice_total, invoice_date, invoice_due_date,
    CASE
        WHEN (TO_DATE('18-JUL-24') - invoice_due_date) > 30
            THEN 'Over 30 days past due'
        WHEN (TO_DATE('18-JUL-24') - invoice_due_date) > 0
            THEN '1 to 30 days past due'
        ELSE 'Current'
    END AS status
FROM invoices
WHERE invoice_total - payment_total - credit_total > 0
```

The result set

	INVOICE_NUMBER	INVOICE_TOTAL	INVOICE_DATE	INVOICE_DUE_DATE	STATUS
37	547480102	224	19-MAY-24	24-JUN-24	1 to 30 days past due
38	547481328	224	20-MAY-24	25-JUN-24	1 to 30 days past due
39	40318	21842	18-JUL-24	20-JUL-24	Current
40	31361833	579.42	23-MAY-24	09-JUN-24	Over 30 days past due

Figure 8-21 How to use the CASE expression

How to use the COALESCE, NVL, and NVL2 functions

Figure 8-22 presents three functions that you can use to substitute non-null values for null values: COALESCE, NVL, and NVL2. Although these functions are similar, COALESCE is the most flexible because it lets you specify a list of values. Then, it returns the first non-null value in the list. In contrast, the NVL and NVL2 functions aren't as flexible since they only let you substitute a single non-null value for a null value. Still, these functions are useful in some situations.

The first example uses the COALESCE function to return the value of the payment_date column if that column doesn't contain a null value. Otherwise, it returns the value of the invoice_due_date column if that column doesn't contain a null value. Otherwise, it returns a date of January 1, 1900. However, as none of the invoices have a null value for the due date, this value is unused in the result.

Note that all of the arguments for this function must be of the same data type. In this case, all three arguments are of the DATE type.

The second example performs a similar task using the NVL function. In this example, the NVL function substitutes a string value of "Unpaid" for a null payment date. Since both arguments must be of the same data type, this example uses the TO_CHAR function to convert the payment_date column to a string.

The third example performs a similar task using the NVL2 function. In this example, the NVL2 function substitutes a string value of "Unpaid" for a null payment date, and it substitutes a string value of "Paid" for a non-null payment date. With the NVL2 function, only the second and third arguments need to be of the same data type. As a result, it isn't necessary to convert the first argument to a string.

The syntax of the COALESCE, NVL, and NVL2 functions

```
COALESCE(expression1 [, expression2][, expression3]...)
NVL(expression, null_replacement)
NVL2(expression, not_null_replacement, null_replacement)
```

Use the COALESCE function

```
SELECT payment_date, invoice_due_date,
       COALESCE(payment_date, invoice_due_date, TO_DATE('01-JAN-1900'))
       AS payment_date_2
FROM invoices
```

	PAYMENT_DATE	INVOICE_DUE_DATE	PAYMENT_DATE_2
1	11-APR-24	22-APR-24	11-APR-24
2	14-MAY-24	23-MAY-24	14-MAY-24
3	(null)	30-JUN-24	30-JUN-24
4	12-MAY-24	16-MAY-24	12-MAY-24
5	13-MAY-24	16-MAY-24	13-MAY-24
6	(null)	26-JUN-24	26-JUN-24

Use the NVL function

```
SELECT payment_date,
       NVL(TO_CHAR(payment_date), 'Unpaid') AS payment_date_2
FROM invoices
```

	PAYMENT_DATE	PAYMENT_DATE_2
1	11-APR-24	11-APR-24
2	14-MAY-24	14-MAY-24
3	(null)	Unpaid
4	12-MAY-24	12-MAY-24

Use the NVL2 function

```
SELECT payment_date,
       NVL2(payment_date, 'Paid', 'Unpaid') AS payment_date_2
FROM invoices
```

	PAYMENT_DATE	PAYMENT_D...
1	11-APR-24	Paid
2	14-MAY-24	Paid
3	(null)	Unpaid
4	12-MAY-24	Paid

Description

- COALESCE, NVL, and NVL2 let you substitute non-null values for null values.
- The COALESCE function returns the first expression in a list of expressions that isn't null. All of the expressions in the list must have the same data type. If all of the expressions are null, this function returns a null value.

Figure 8-22 How to use the COALESCE, NVL, and NVL2 functions

How to use the GROUPING function

In chapter 5, you learned how to use the ROLLUP CUBE clause to add summary rows to a summary query. When you do that, a null value is assigned to any column in a summary row that isn't being summarized.

If you want to assign a value other than null to these columns, you can do that by using the GROUPING function. This function accepts the name of a column as its argument and returns 1 if the row is a summary row for that column. Otherwise, it returns 0. The column you specify must be one of the columns named in a GROUP BY clause that includes the ROLLUP CUBE clause.

The example in figure 8-23 shows how you can use the GROUPING function within a CASE expression to determine the values that are assigned to the grouped columns. If a row summarizes the vendor_state column, for example, the value of the GROUPING function for that column is 1. Then, the CASE expression in this example assigns a literal string value of "==========" to that column. Otherwise, it retrieves the value of the column from the Vendors table.

Similarly, if a row summarizes the vendor_city column, a literal string value of "==========" is assigned to that column. As you can see in the result set shown here, this makes it more obvious which columns are being summarized.

This technique is particularly useful if the columns you're summarizing can contain null values. In that case, it would be difficult to determine which rows are summary rows and which rows simply contain null values. Then, you may not only want to use the GROUPING function to replace the null values in summary rows, but you may want to use the COALESCE or NVL function to replace the null values retrieved from the base table.

The syntax of the GROUPING function

```
GROUPING(column_name)
```

Use the GROUPING function with CASE

```
SELECT
    CASE
        WHEN GROUPING(vendor_state) = 1 THEN '=========='
        ELSE vendor_state
    END AS vendor_state,
    CASE
        WHEN GROUPING(vendor_city) = 1 THEN '=========='
        ELSE vendor_city
    END AS vendor_city,
    COUNT(*) AS qty_vendors
FROM vendors
WHERE vendor_state IN ('IA', 'NJ')
GROUP BY ROLLUP(vendor_state, vendor_city)
ORDER BY vendor_state DESC, vendor_city DESC
```

The result set

	VENDOR_STATE	VENDOR_CITY	QTY_VENDORS
1	NJ	Washington	1
2	NJ	Fairfield	1
3	NJ	East Brunswick	2
4	NJ	==========	4
5	IA	Washington	1
6	IA	Fairfield	1
7	IA	==========	2
8	==========	==========	6

Description

- You can use the GROUPING function to determine when a null value is assigned to a column as the result of the ROLLUP or CUBE operator. The column you name in this function must be one of the columns named in the GROUP BY clause.
- If a null value is assigned to the specified column as the result of the ROLLUP or CUBE operator, the GROUPING function returns a value of 1. Otherwise, it returns a value of 0.
- You typically use the GROUPING function with the CASE expression. Then, if the GROUPING function returns a value of 1, you can assign a value other than null to the column.

Figure 8-23 How to use the GROUPING function

Perspective

In this chapter, you learned about the most common Oracle data types. You also learned how to use the common functions for working with strings, numbers, and dates. At this point, you have the essential skills you need to write SQL statements at a professional level.

Terms

data type
character data type
Boolean data type
large object (LOB) data type
spatial data type
JSON data type
XML data type
numeric data type
datetime data type
fixed-length string
variable-length string
character set
national character set
precision
significant digit
scale
floating-point number
single precision
double precision
scientific notation
casting

Exercises

1. Write a SELECT statement that returns these columns from the Invoices table:

   The invoice_total column

   The invoice_total column with a comma after the thousands place and with 2 digits to the right of the decimal point (use the TO_CHAR function)

   The invoice_total column with no commas and no digits to the right of the decimal point and no decimal point (use the TO_CHAR function)

   The invoice_total column as an integer with up to seven digits (use the CAST function)

2. Write a SELECT statement that returns these columns from the Invoices table:

   The invoice_date column

   The invoice_date column with its full date and time including a four-digit year on a 24-hour clock (use the TO_CHAR function)

   The invoice_date column with its full date and time including a four-digit year on a 12-hour clock with an am/pm indicator (use the TO_CHAR function)

   The invoice_date column as VARCHAR2(10) (use the CAST function)

3. Write a SELECT statement that returns these columns from the Vendors table:

   The vendor_name column

   The vendor_name column in all capital letters

   The vendor_phone column

   The last four digits of each phone number

   The second word in each vendor name if there is one or "N\A" if there isn't one (use the INSTR, SUBSTR, and NVL functions)

   The vendor_phone column with the parts of the number separated by dots (as in 555.555.5555)

4. Write a SELECT statement that returns these columns from the Invoices table:

   The invoice_date column

   The invoice_date column plus 30 days

   The payment_date column

   A column named days_to_pay that shows the number of days between the invoice date and the payment date

   The number of days between the invoice date and the payment date converted to an INTERVAL DAY TO SECOND value (use 'DAY' for the second argument)

   The number of the invoice_date's month (use the EXTRACT function)

   The last day of the invoice date's month

5. Write a SELECT statement that returns these columns from the Vendors table:

   The vendor_city column

   The vendor_state column

   A column that uses a searched CASE expression and the IN operator to display "West Coast", "East Coast", or "Other" based on the following specifications:

Category	**State codes**
West Coast	CA, OR, WA
East Coast	ME, NH, MA, RI, CT, NY, NJ, DE, MD, VA, DC
Other	All other state codes

Section 3

Database design and implementation

In large organizations, database administrators are usually responsible for designing the databases that are used by production applications, and they may also be responsible for the databases that are used for testing those applications. Often, though, programmers are asked to design, create, or maintain small databases that are used for testing. And in a small organization, programmers may also be responsible for the production databases.

So whether you're a database administrator or a SQL programmer, you need the skills and knowledge presented in this section. That's true even if you aren't ever called upon to design or maintain a database. By understanding what's going on behind the scenes, you'll be able to use SQL more effectively.

In chapter 9, you'll learn how to design a database. In chapter 10, you'll learn how to use the Data Definition Language (DDL) statements to create and maintain the tables, indexes, and sequences of a database. In chapter 11, you'll learn how to create and maintain views, which are database objects that provide another way to look at tables. Finally, in chapter 12, you'll learn how to design the security for your database by creating users that have restricted access to your database.

9

How to design a database

In this chapter, you'll learn how to design a new database. This is useful information whether or not you ever design a database on your own. To illustrate this process, this chapter uses the accounts payable (AP) database that you've seen throughout this book.

How to design a data structure 266

The basic steps for designing a data structure 266

How to identify the data elements 268

How to subdivide the data elements 270

How to identify the tables and assign columns 272

How to identify the primary and foreign keys 274

How to enforce the relationships between tables 276

How normalization works 278

How to identify the columns to be indexed 280

How to normalize a data structure 282

The seven normal forms 282

How to apply the first normal form 284

How to apply the second normal form 286

How to apply the third normal form 288

When and how to denormalize a data structure 290

Perspective 292

How to design a data structure

Databases are often designed by database administrators (DBAs) or design specialists. This is especially true for large, multiuser databases. How well this is done can directly affect your job as a SQL programmer. In general, a well-designed database is easy to understand and query, while a poorly designed database is difficult to work with. In fact, when you work with a poorly designed database, you will often need to figure out how it is designed before you can code your queries appropriately.

The topics that follow will teach you a basic approach for designing a *data structure*. We use that term to refer to a model of the database rather than the database itself. Once you design the data structure, you can use the techniques presented in the next two chapters to create a database with that design. By understanding the right way to design a database, you'll work more effectively as a SQL programmer.

The basic steps for designing a data structure

In many cases, you can design a data structure based on an existing real-world system. The illustration at the top of figure 9-1 presents a conceptual view of how this works. Here, you can see that all of the information about the people, documents, and facilities within a real-world system is mapped to the tables, columns, and rows of a database system.

As you design a data structure, each table represents one object, or *entity*, in the real-world system. Then, within each table, each column stores one item of information, or *attribute*, for the entity, and each row stores one occurrence, or *instance*, of the entity.

This figure also presents the six steps you can follow to design a data structure. You'll learn more about each of these steps in the figures that follow. In general, though, step 1 is to identify all the data elements that need to be stored in the database. Step 2 is to break complex elements down into smaller components whenever that makes sense. Step 3 is to identify the tables that will make up the system and to determine which data elements are assigned as columns in each table. Step 4 is to define the relationships between the tables by identifying the primary and foreign keys. Step 5 is to normalize the database to reduce data redundancy. And step 6 is to identify the indexes that are needed for each table.

To model a database system after a real-world system, you can use a technique called *entity-relationship (ER) modeling*. Although this book doesn't present ER modeling in its entirety, this chapter applies some of the basic elements of this technique to its design diagrams. As a result, this chapter presents some of the basics of this modeling technique.

A database system is modeled after a real-world system

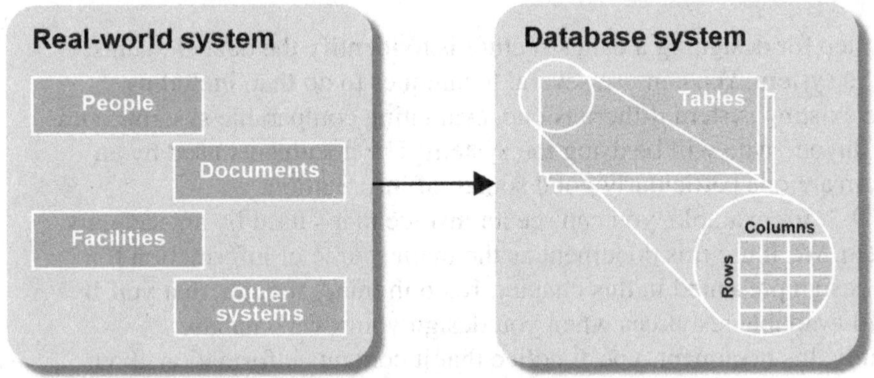

The six basic steps for designing a data structure

Step 1: Identify the data elements

Step 2: Subdivide each element into its smallest useful components

Step 3: Identify the tables and assign columns

Step 4: Identify the primary and foreign keys

Step 5: Review whether the data structure is normalized

Step 6: Identify the indexes

Description

- A relational database system should model the real-world environment where it's used. The job of the designer is to analyze the real-world system and then map it onto a relational database system.
- A table in a relational database typically represents an object, or *entity*, in the real world. Each column of a table is used to store an *attribute* associated with the entity, and each row represents one *instance* of the entity.
- To model a database and the relationships between its tables after a real-world system, you can use a technique called *entity-relationship (ER) modeling*. Some of the diagrams you'll see in this chapter apply the basic elements of ER modeling.

Figure 9-1 The basic steps for designing a data structure

How to identify the data elements

The first step for designing a data structure is to identify the data elements required by the system. You can use several techniques to do that, including analyzing the existing system if there is one, evaluating comparable systems, and interviewing anyone who will be using the system. The documents used by an existing system are one particularly good source of information.

In figure 9-2, for example, you can see an invoice that's used by an accounts payable system. We'll use this document as the main source of information for the database design presented in this chapter. Keep in mind, though, that you'll want to use all available resources when you design your own database.

If you study this document, you'll notice that it contains information about three different entities: vendors, invoices, and line items. First, the form itself has preprinted information about the vendor who issued the invoice, such as the vendor's name and address. If this vendor were to issue another invoice, this information wouldn't change.

This document also contains specific information about the invoice. Some of this information, such as the invoice number, invoice date, and invoice total, is general in nature. Although the actual information will vary from one invoice to the next, each invoice will include this information. In addition to this general information, each invoice includes information about the items that were purchased. Although each line item contains similar information, each invoice can contain a different number of line items.

One of the things you need to consider as you review a document like this is how much information your system needs to track. For an accounts payable system, for example, you may not need to store detailed data such as the information about each line item. Instead, you may just need to store summary data like the invoice total. As you think about what data elements to include in the database, then, you should have an idea of what information you'll need to get back out of the system.

An invoice that can be used to identify data elements

Acme Fabrication, Inc.	
Custom Contraptions, Contrivances and Confabulations	Invoice Number: I01-1088
1234 West Industrial Way East Los Angeles California 90022	Invoice Date: 10/05/24
800.555.1212 fax 562.555.1213 www.acmefabrication.com	Terms: Net 30

Part No.	Qty.	Description	Unit Price	Extension
CUST345	12	Design service, hr	100.00	1200.00
457332	7	Baling wire, 25x3ft roll	79.90	559.30
50173	4375	Duct tape, black, yd	1.09	4768.75
328771	2	Rubber tubing, 100ft roll	4.79	9.58
CUST281	7	Assembly, hr	75.00	525.00
CUST917	2	Testing, hr	125.00	250.00
		Sales Tax		245.20

Your salesperson:	Ruben Goldberg, ext 4512
Accounts receivable:	Inigo Jones, ext 4901

$7,557.83
PLEASE PAY THIS AMOUNT

Thanks for your business!

The data elements identified on the invoice document

Vendor name	Invoice date	Item extension
Vendor address	Invoice terms	Vendor sales contact name
Vendor phone number	Item part number	Vendor sales contact extension
Vendor fax number	Item quantity	Vendor AR contact name
Vendor web address	Item description	Vendor AR contact extension
Invoice number	Item unit price	Invoice total

Description

- Depending on the nature of the system, you can identify data elements in a variety of ways, including interviewing users, analyzing existing systems, and evaluating comparable systems.
- The documents used by a real-world system, such as the invoice shown above, can often help you identify the data elements of the system.
- As you identify the data elements of a system, you should begin thinking about the entities that those elements are associated with. That will help you identify the tables of the database later on.

Figure 9-2 How to identify the data elements

How to subdivide the data elements

Some of the data elements you identify in step 1 of the design procedure will consist of multiple components. The next step, then, is to divide these elements into their smallest useful values. Figure 9-3 shows how you can do that.

The first example in this figure shows how you can divide the name of the sales contact for a vendor. Here, the name is divided into two elements: a first name and a last name. When you divide a name like this, you can easily perform operations like sorting by last name and using the first name in a salutation, such as "Dear Ruben." In contrast, if the full name is stored in a single column, you have to use the string functions to extract the component you need. But as you learned in the last chapter, that can lead to inefficient and complicated code.

In general, then, you should separate a name like this whenever you'll need to use the name components separately. Later, when you need to use the full name, you can combine the first and last names using concatenation. However, you should also be aware that some cultures use different naming styles. Does your database need to distinguish between the presence of a middle name and a person with two last names? Deciding early on how your system will handle a name like Antoine Marie Jean-Baptiste Roger can save you a lot of headaches in the future.

The second example shows how you typically divide an address. Notice in this example that the street number and street name are stored in a single column. Although you could store these components in separate columns, that usually doesn't make sense since these values are typically used together. That's what I mean when I say the data elements should be divided into their smallest *useful* values.

With that guideline in mind, you might even need to divide a single string into two or more components. A bulk mail system, for example, might require a separate column for the first three digits of the zip code. And a telephone number could require as many as four columns: one for the area code, one for the three-digit prefix, one for the four-digit number, and one for the extension.

As in the previous step, knowledge of the real-world system and of the information that will be extracted from the database is critical. In some circumstances, it may be okay to store data elements with multiple components in a single column. That can simplify your design and reduce the overall number of columns. In general, though, most designers divide data elements as much as possible. That way, it's easy to accommodate almost any query, and you don't have to change the database design later on when you realize that you need to use just part of a column value.

A name that's divided into first and last names

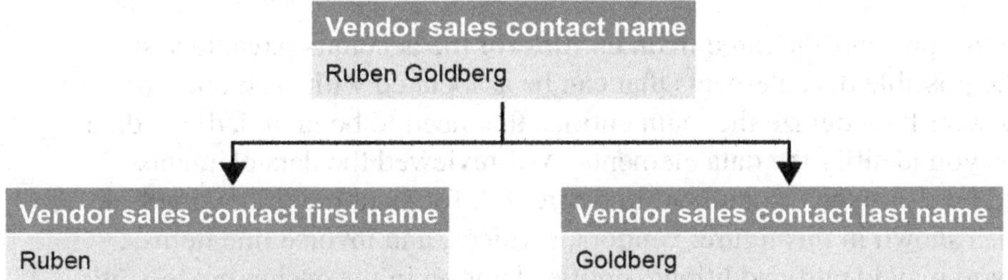

An address that's divided into street address, city, state, and zip code

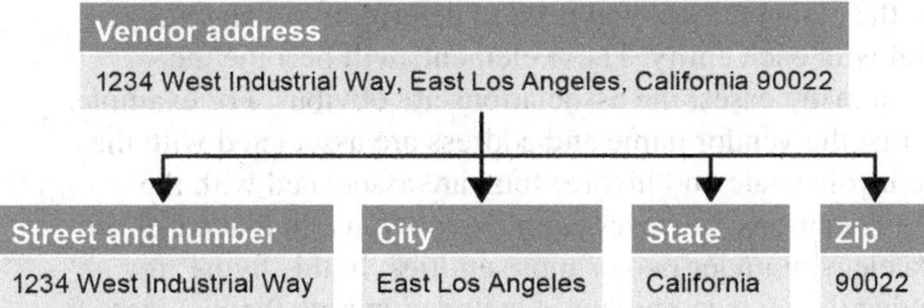

Description

- If a data element contains two or more components, you should consider subdividing the element into those components. That way, you won't need to parse the element each time you use it.
- The extent to which you subdivide a data element depends on how it will be used. Because it's difficult to predict all future uses for the data, most designers subdivide data elements as much as possible.
- When you subdivide a data element, you can easily rebuild it when necessary by concatenating the individual components.

Figure 9-3 How to subdivide the data elements

How to identify the tables and assign columns

Figure 9-4 presents the three main entities for the accounts payable system and lists the possible data elements that can be associated with each one. In most cases, you'll recognize the main entities that need to be included in a data structure as you identify the data elements. As I reviewed the data elements represented on the invoice document in figure 9-2, for example, I identified the three entities shown in this figure: vendors, invoices, and invoice line items. Although you may identify additional entities later on in the design process, it's sufficient to identify the main entities at this point. These entities will become the tables of the database.

After you identify the main entities, you need to determine which data elements are associated with each entity. These elements will become the columns of the tables. In many cases, the associations are obvious. For example, it's easy to determine that the vendor name and address are associated with the vendors entity and the invoice date and invoice total are associated with the invoices entity. Some associations, however, aren't so obvious. In that case, you may need to list a data element under two or more entities. In this figure, for example, you can see that the invoice number is included in both the invoices and invoice line items entities and the account number is included in all three entities. Later, when you normalize the data structure, you may be able to remove these repeated elements. For now, though, it's okay to include them.

In this figure, any data elements that weren't identified in previous steps are shown in italics. Although you should be able to identify most of the data elements in the first two steps of the design process, you'll occasionally think of additional elements during the third step.

Similarly, you may decide during this step that you don't need some of the data elements you've identified. For example, I decided that I didn't need the fax number or web address of each vendor. So I used strikethrough to indicate that these data elements should not be included.

Finally, I identified the data elements that are included in two or more tables by coding an asterisk after them. Although you can use any notation you like for this step of the design process, you'll want to be sure that you document your design decisions.

By the way, a couple of the new data elements I added may not be clear to you if you haven't worked with a corporate accounts payable system before. "Terms" refers to the payment terms that the vendor offers. For example, the terms might be net 30 (the invoice must be paid in 30 days) or might include a discount for early payment. "Account number" refers to the general ledger accounts that a company uses to track its expenses. For example, one account number might be assigned for advertising expenses, while another might be for office supplies. Each invoice that's paid is assigned to an account, and in some cases, different line items on an invoice are assigned to different accounts.

Possible tables and columns for an accounts payable system

Vendors	Invoices	Invoice line items
Vendor name	Invoice number*	Invoice number*
Vendor address	Invoice date	~~Item part number~~
Vendor city	Terms*	Item quantity
Vendor state	Invoice total	Item description
Vendor zip code	*Payment date*	Item unit price
Vendor phone number	*Payment total*	Item extension
~~Vendor fax number~~	*Invoice due date*	*Account number**
~~Vendor web address~~	*Credit total*	*Sequence number*
Vendor contact first name	*Account number**	
Vendor contact last name		
~~Vendor contact phone~~		
~~Vendor AR first name~~		
~~Vendor AR last name~~		
~~Vendor AR phone~~		
*Terms**		
*Account number**		

Description

- After you identify and subdivide all of the data elements for a database, you should group them by the entities with which they're associated. These entities will later become the tables of the database, and the elements will become the columns.
- If a data element relates to more than one entity, you can include it under all of the entities it relates to. Then, when you normalize the database, you may be able to remove the duplicate elements.
- As you assign the elements to entities, you should omit elements that aren't needed, and you should add any additional elements that are needed.

The notation used in this figure

- Data elements that were previously identified but aren't needed are crossed out.
- Data elements that were added are displayed in italics.
- Data elements that are related to two or more entities are followed by an asterisk.

How to identify the primary and foreign keys

Once you identify the entities and data elements of a system, the next step is to identify the relationships between the tables. To do that, you need to identify the primary and foreign keys as shown in figure 9-5.

As you know, a primary key is used to uniquely identify each row in a table. In some cases, you can use an existing column as the primary key. For example, you might consider using the vendor_name column as the primary key of the Vendors table. Because the values for this column can be long, however, and because it would be easy to enter a value like that incorrectly, that's not a good candidate for a primary key. Instead, you should use an ID column like vendor_id that's incremented by one for each new record.

Similarly, you might consider using the invoice_number column as the primary key of the Invoices table. However, it's possible for different vendors to use the same invoice number, so this value isn't necessarily unique. Because of that, another ID column like invoice_id can be used as the primary key.

To uniquely identify the rows in the Invoice_Line_Items table, this design uses a composite key. This composite key uses two columns to identify each row. The first column is the invoice_id column from the Invoices table, and the second column is the invoice_sequence column. This is necessary because this table may contain more than one row (line item) for each invoice. And that means that the invoice_id value by itself may not be unique.

After you identify the primary key of each table, you need to identify the relationships between the tables and add foreign key columns as necessary. In most cases, two tables will have a one-to-many relationship with each other. For example, each vendor can have many invoices, and each invoice can have many line items. To identify the vendor that each invoice is associated with, a vendor_id column is included in the Invoices table. Because the Invoice_Line_Items table already contains an invoice_id column, it's not necessary to add another column to this table.

The diagram at the top of this figure illustrates the relationships I identified between the tables in the accounts payable system. As you can see, the primary keys are displayed in bold. Then, the lines between the tables indicate how the primary key in one table is related to the foreign key in another table. Here, a small, round connector indicates the one side of the relationship, and the connector with a triangle indicates the many side of the relationship.

In addition to the one-to-many relationships shown in this diagram, you can also use many-to-many relationships and one-to-one relationships. The second diagram in this figure, for example, shows a many-to-many relationship between an Employees table and a Committees table. As you can see, this type of relationship can be implemented by creating a *linking table*, also called a *connecting table* or an *associate table*. This table contains the primary key columns from the two tables. Then, each table has a one-to-many relationship with the linking table. Notice that the linking table doesn't have its own primary key. Because this table doesn't correspond to an entity and because it's used only in conjunction with the Employees and Committees tables, a primary key isn't needed.

The relationships between the tables in the accounts payable system

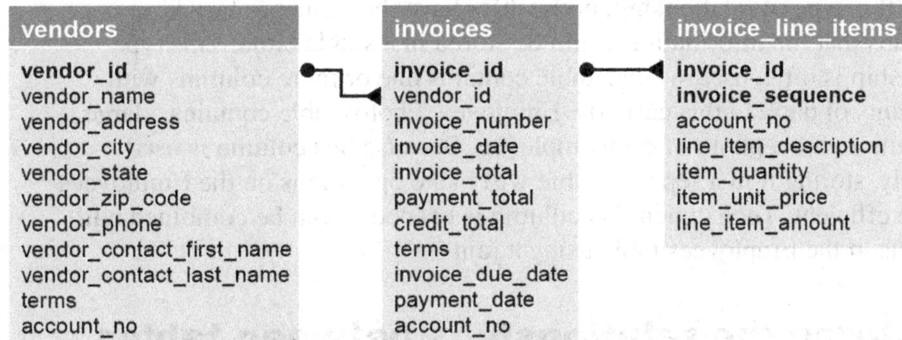

Two tables with a many-to-many relationship

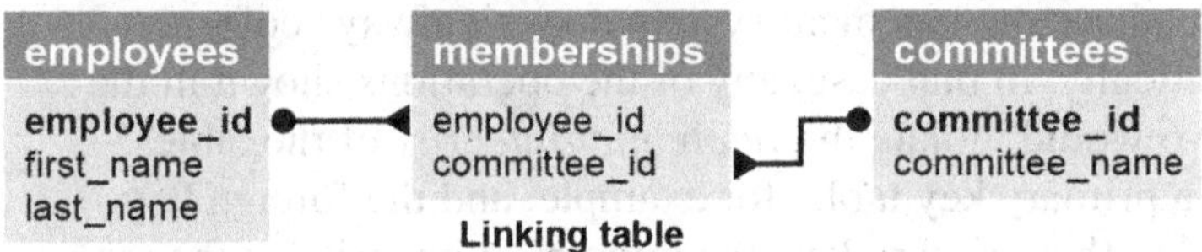

Two tables with a one-to-one relationship

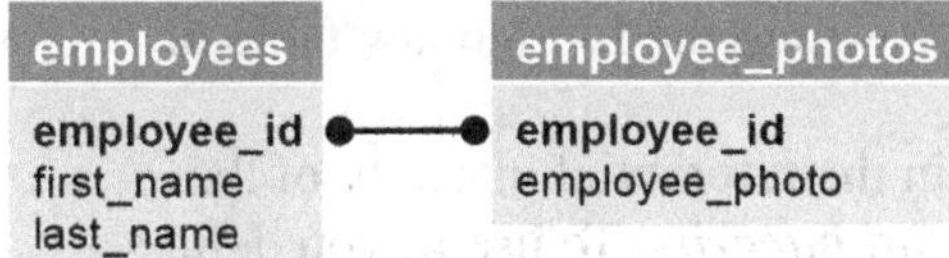

Description

- Each table should have a primary key that uniquely identifies each row. If possible, you should use an existing column for the primary key.
- The values of the primary keys should seldom, if ever, change. The values should also be short and easy to enter correctly.
- If a suitable column doesn't exist for a primary key, you can create an ID column that is incremented by one for each new row as the primary key.
- If two tables have a one-to-many relationship, you may need to add a foreign key column to the table on the "many" side. The foreign key column must have the same data type as the primary key column it's related to.
- If two tables have a many-to-many relationship, you'll need to define a *linking table* to relate them. Then, each of the tables in the many-to-many relationship will have a one-to-many relationship with the linking table. The linking table doesn't usually have a primary key.
- If two tables have a one-to-one relationship, they should be related by their primary keys. This type of relationship is typically used to improve performance. Then, columns with large amounts of data can be stored in a separate table.

Figure 9-5 How to identify the primary and foreign keys

The third example illustrates two tables that have a one-to-one relationship. With this type of relationship, both tables have the same primary key, which means that the information could be stored in a single table. This type of relationship is often used when a table contains one or more columns with large amounts of data. In this case, the Employee_Photos table contains a large binary column with a photo of each employee. Because this column is used infrequently, storing it in a separate table will make operations on the Employees table more efficient. Then, when this column is needed, it can be combined with the columns in the Employees table using a join.

How to enforce the relationships between tables

Although the primary keys and foreign keys indicate how the tables in a database are related, the database management system doesn't always enforce those relationships automatically. In that case, any of the operations shown in the table at the top of figure 9-6 would violate the *referential integrity* of the tables. If you deleted a row from a primary key table, for example, and the foreign key table included rows related to that primary key, the referential integrity of the two tables would be destroyed. In that case, the rows in the foreign key table that no longer have a related row in the primary key table would be *orphaned.* Similar problems can occur when you insert a row into the foreign key table or update a primary key or foreign value.

To enforce those relationships and maintain the referential integrity of the tables, Oracle provides for *declarative referential integrity*. To use it, you define *foreign key constraints* that indicate how the referential integrity between the tables is enforced. You'll learn more about defining foreign key constraints in the next chapter. For now, just realize that these constraints can prevent all of the operations listed in this figure that violate referential integrity.

Operations that can violate referential integrity

This operation...	Violates referential integrity if...
Delete a row from the primary key table	The foreign key table contains one or more rows related to the deleted row
Insert a row in the foreign key table	The foreign key value doesn't have a matching primary key value in the related table
Update the value of a foreign key	The new foreign key value doesn't have a matching primary key value in the related table
Update the value of a primary key	The foreign key table contains one or more rows related to the row that's changed

Description

- *Referential integrity* means that the relationships between tables are maintained correctly. That means that a table with a foreign key doesn't have rows with foreign key values that don't have matching primary key values in the related table.
- In Oracle, you can enforce referential integrity by using declarative referential integrity or by defining triggers.
- To use *declarative referential integrity (DRI)*, you define *foreign key constraints*. You'll learn how to do that in the next chapter.
- When you define foreign key constraints, you can specify how referential integrity is enforced when a row is deleted from the primary key table. The options are to return an error, to delete the related rows in the foreign key table, or to set the foreign key values in the related rows to null.
- If referential integrity isn't enforced and a row is deleted from the primary key table that has related rows in the foreign key table, the rows in the foreign key table are said to be *orphaned*.

Figure 9-6 How to enforce the relationships between tables

How normalization works

The next step in the design process is to review whether the data structure is *normalized*. To do that, you look at how the data is separated into related tables. If you follow the first four steps for designing a database presented in this chapter, your database will already be partially normalized when you get to this step. However, almost every design can be normalized further.

Figure 9-7 illustrates how *normalization* works. The first two tables in this figure show some of the problems caused by an *unnormalized* data structure. In the first table, you can see that each row represents an invoice. Because an invoice can have one or more line items, however, the item_description column must be repeated to provide for the maximum number of line items. But since most invoices have fewer line items than the maximum, this can waste storage space.

In the second table, each line item is stored in a separate row. That eliminates the problem caused by repeating the item_description column, but it introduces a new problem: the invoice number must be repeated in each row. This, too, can cause storage problems, particularly if the repeated column is large. In addition, it can cause maintenance problems if the column contains a value that's likely to change. Then, when the value changes, each row that contains the value must be updated. And if a repeated value must be reentered for each new row, it would be easy for the value to vary from one row to another.

To eliminate the problems caused by *data redundancy*, you can normalize the data structure. To do that, you apply the *normal forms* you'll learn about later in this chapter. As you'll see, there are a total of seven normal forms. However, it's common to apply only the first three. The diagram in this figure, for example, shows the accounts payable system in third normal form. Although it may not be obvious at this point how this reduces data redundancy, that will become clearer as you learn about the different normal forms.

A table that contains repeating columns

	INVOICE_NUMBER	ITEM_DESCRIPTION_1	ITEM_DESCRIPTION_2	ITEM_DESCRIPTION_3
1	112897	VB ad	SQL ad	Library directory
2	97/552	Catalogs	SQL flyer	(null)
3	97/553B	Card revision	(null)	(null)

A table that contains redundant data

	VENDOR_NAME	INVOICE_NUMBER	ITEM_DESCRIPTION
1	Cahners Publishing	112897	VB ad
2	Cahners Publishing	112897	SQL ad
3	Cahners Publishing	112897	Library directory
4	Zylka Design	97/522	Catalogs
5	Zylka Design	97/522	SQL flyer
6	Zylka Design	97/533B	Card revision

The accounts payable system in third normal form

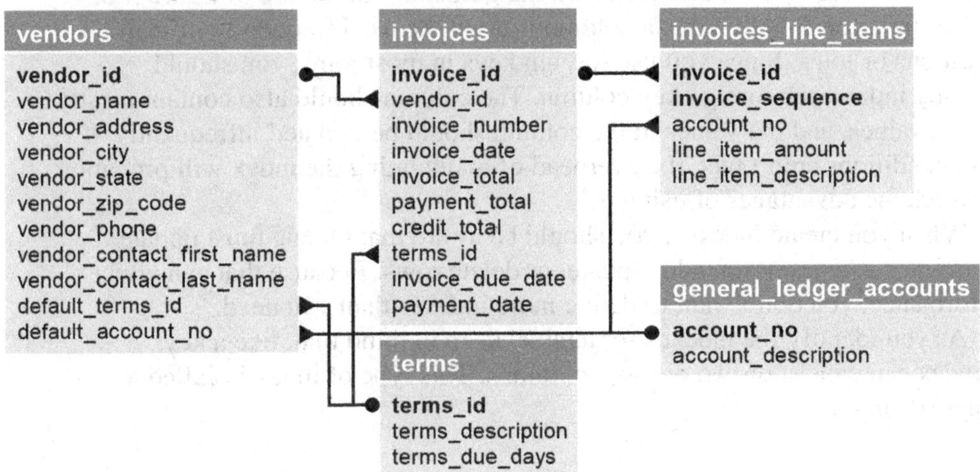

Description

- *Normalization* is a formal process you can use to separate the data in a data structure into related tables. Normalization reduces *data redundancy*, which can cause storage and maintenance problems.
- In an *unnormalized data structure*, a table can contain information about two or more entities. It can also contain repeating columns, columns that contain repeating values, and data that's repeated in two or more rows.
- In a *normalized data structure*, each table contains information about a single entity, and each piece of information is stored in exactly one place.
- To normalize a data structure, you apply the *normal forms* in sequence. Although there are a total of seven normal forms, a data structure is typically considered normalized if the first three normal forms are applied.

Figure 9-7 How normalization works

How to identify the columns to be indexed

The last step in the design process is to identify the columns that should be indexed. An *index* is a structure that provides for locating one or more rows directly. Without an index, a database management system has to perform a *table scan*, which involves searching through the entire table.

Just as the index of a book has page numbers that direct you to a specific subject, a database index has pointers that direct the system to a specific row. This can speed performance not only when you're searching for rows based on a search condition, but also when you're joining data from tables. If a join is done based on a primary key to foreign key relationship, for example, and an index is defined for the foreign key column, the database management system can use that index to locate the rows for each primary key value.

When you use Oracle, an index is automatically created for the primary key in each table that you create. But you should consider creating indexes for other columns in some of the tables based on the guidelines at the top of figure 9-8.

To start, you should index a column if it will be used frequently in search conditions or joins. Since you use foreign keys in most joins, you should typically index each foreign key column. The column should also contain mostly distinct values, and the values in the column should be updated infrequently. If these conditions aren't met, the overhead of maintaining the index will probably outweigh the advantages of using it.

When you create indexes, you should be aware that Oracle must update the indexes whenever you add, update, or delete rows. Because that can affect performance, you don't want to define more indexes than you need.

As you identify the indexes for a table, keep in mind that, like a key, an index can consist of two or more columns. This type of index is called a *composite index.*

When to create an index

- When the column is a foreign key
- When the column is used frequently in search conditions or joins
- When the column contains a large number of distinct values
- When the column is updated infrequently

Description

- Oracle automatically creates an index for a primary key.
- An *index* provides a way for a database management system to locate information more quickly. When it uses an index, the database management system can go directly to a specific row rather than having to search through all the rows until it finds it.
- Indexes speed performance when searching and joining tables.
- You can create *composite indexes* that include two or more columns. You should use this type of index when the columns in the index are updated infrequently or when the index will cover almost every search condition on the table.
- Because indexes must be updated each time you add, update, or delete a row, you shouldn't create more indexes than you need.

Figure 9-8 How to identify the columns to be indexed

How to normalize a data structure

The topics that follow describe the seven normal forms and teach you how to apply the first three. Then, the last topic explains when and how to denormalize a data structure. When you finish these topics, you'll have the basic skills for designing databases that are efficient and easy to use.

The seven normal forms

Figure 9-9 summarizes the seven normal forms. Each normal form assumes that the previous forms have already been applied. Before you can apply the third normal form, for example, the design must already be in the second normal form.

Strictly speaking, a data structure isn't normalized until it's in the fifth or sixth normal form. However, the normal forms past the third normal form are applied infrequently. Because of that, I won't present those forms in detail here. Instead, I'll just describe them briefly so you'll have an idea of how to apply them if you need to.

The *Boyce-Codd normal form* is a slightly stronger version of the third normal form that can be used to eliminate *transitive dependencies*. With this type of dependency, one column depends on another column, which depends on a third column. Most tables that are in the third normal form are also in the Boyce-Codd normal form.

The fourth normal form can be used to eliminate multiple *multivalued dependencies* from a table. A multivalued dependency is one where a primary key column has a one-to-many relationship with a non-key column. This normal form gets rid of misleading many-to-many relationships.

To apply the fifth normal form, you continue to divide the tables of the data structure into smaller tables until all redundancy has been removed. When further splitting would result in tables that couldn't be used to reconstruct the original table, the data structure is in fifth normal form. In this form, most tables consist of little more than key columns with one or two data elements.

The *domain-key normal form*, sometimes called the sixth normal form, further reduces all tables to just one key column and no more than one additional column. Although it isn't commonly used for production databases, the sixth normal form is sometimes used in data warehousing systems.

This figure also lists the benefits of normalizing a data structure. To summarize, normalization produces smaller, more efficient tables. In addition, it reduces data redundancy, which makes the data easier to maintain and reduces the amount of storage needed for the database. Because of these benefits, you should always consider normalizing your data structures.

In the academic study of computer science, normalization is considered a form of design perfection that should always be strived for. In practice, though, database designers and DBAs tend to use normalization as a flexible design guideline.

The seven normal forms

Normal form	Description
First (1NF)	The value stored at the intersection of each row and column must be a scalar value, and a table must not contain any repeating columns.
Second (2NF)	Every non-key column must depend on the entire primary key.
Third (3NF)	Every non-key column must depend only on the primary key.
Boyce-Codd (BCNF)	A non-key column can't be dependent on another non-key column. This prevents *transitive dependencies*, where column A depends on column C and column B depends on column C. Since both A and B depend on C, A and B should be moved into another table with C as the key.
Fourth (4NF)	A table must not have more than one *multivalued dependency*, where the primary key has a one-to-many relationship to non-key columns. This form gets rid of misleading many-to-many relationships.
Fifth (5NF)	The data structure is split into smaller and smaller tables until all redundancy has been eliminated. If further splitting would result in tables that couldn't be joined to recreate the original table, the structure is in fifth normal form.
Domain-key (DKNF) or Sixth (6NF)	Every constraint on the relationship is dependent only on key constraints and domain constraints, where a *domain* is the set of allowable values for a column. This form prevents the insertion of any unacceptable data by enforcing constraints at the level of a relationship, rather than at the table or column level.

The benefits of normalization

- Since a normalized database has more tables than an unnormalized database, and since each table has an index on its primary key, the database has more indexes. That makes data retrieval more efficient.
- Since each table contains information about a single entity, each index has fewer columns (usually one) and fewer rows. That makes data retrieval and insert, update, and delete operations more efficient.
- Each table has fewer indexes, which makes insert, update, and delete operations more efficient.
- Data redundancy is minimized, which simplifies maintenance and reduces storage.

Description

- Each normal form assumes that the design is already in the previous normal form.
- A database is typically considered to be normalized if it is in third normal form. The other four forms are not commonly used and are not covered in detail in this book.

Figure 9-9 The seven normal forms

How to apply the first normal form

Figure 9-10 illustrates how you apply the first normal form to an unnormalized invoice data structure consisting of the data elements that are shown in figure 9-2. The first two tables in this figure illustrate structures that aren't in first normal form. Both of these tables contain a single row for each invoice. Because each invoice can contain one or more line items, however, the first table allows for repeating values in the item_description column. The second table is similar, except it includes a separate column for each line item description. Neither of these structures is acceptable in first normal form.

The third table in this figure has eliminated the repeating values and columns. To do that, it includes one row for each line item. Notice, however, that this has increased the data redundancy. Specifically, the vendor name and invoice number are now repeated for each line item. This problem can be solved by applying the second normal form.

Before I describe the second normal form, I want you to realize that I intentionally omitted many of the columns in the invoice data structure from the examples in this figure and the next figure. In addition to the columns shown here, for example, each of these tables would also contain the vendor address, invoice date, invoice total, etc. By eliminating these columns, it will be easier for you to focus on the columns that are affected by applying the normal forms.

The invoice data with a column that contains repeating values

	VENDOR_NAME	INVOICE_NUMBER	ITEM_DESCRIPTION
1	Cahners Publishing	112897	VB ad, SQL ad, Library directory
2	Zylka Design	97/522	Catalogs, SQL Flyer
3	Zylka Design	97/533B	Card revision

The invoice data with repeating columns

	VENDOR_NAME	INVOICE_NUMBER	ITEM_DESCRIPTION_1	ITEM_DESCRIPTION_2	ITEM_DESCRIPTION_3
1	Cahners Publishing	112897	VB ad	SQL ad	Library directory
2	Zylka Design	97/552	Catalogs	SQL flyer	(null)
3	Zylka Design	97/553B	Card revision	(null)	(null)

The invoice data in first normal form

	VENDOR_NAME	INVOICE_NUMBER	ITEM_DESCRIPTION
1	Cahners Publishing	112897	VB ad
2	Cahners Publishing	112897	SQL ad
3	Cahners Publishing	112897	Library directory
4	Zylka Design	97/522	Catalogs
5	Zylka Design	97/522	SQL flyer
6	Zylka Design	97/533B	Card revision

Description

- For a table to be in first normal form, its columns must not contain repeating values. Instead, each column must contain a single, scalar value. In addition, the table must not contain repeating columns that represent a set of values.
- A table in first normal form often has repeating values in its rows. This can be resolved by applying the second normal form.

Figure 9-10 How to apply the first normal form

How to apply the second normal form

Figure 9-11 shows how to apply the second normal form. To be in second normal form, every column in a table that isn't a key column must be dependent on the entire primary key. This form only applies to tables that have composite primary keys, which is often the case when you start with data that is completely unnormalized. The table at the top of this figure, for example, shows the invoice data in first normal form after key columns have been added. In this case, the primary key consists of the invoice_id and invoice_sequence columns. The invoice_sequence column is needed to uniquely identify each line item for an invoice.

Now, consider the three non-key columns shown in this table. Of these three, only one, item_description, depends on the entire primary key. The other two, vendor_name and invoice_number, depend only on the invoice_id column. Because of that, these columns should be moved to another table. The result is a data structure like the second one shown in this figure. Here, all of the information related to an invoice is stored in the Invoices table, and all of the information related to an individual line item is stored in the Invoice_Line_Items table.

Notice that the relationship between these tables is based on the invoice_id column. This column is the primary key of the Invoices table, and it's the foreign key in the Invoice_Line_Items table that relates the rows in that table to the rows in the Invoices table. This column is also part of the primary key of the Invoice_Line_Items table.

When you apply second normal form to a data structure, it eliminates some of the redundant row data in the tables. In this figure, for example, you can see that the invoice number and vendor name are now included only once for each invoice. In first normal form, this information was included for each line item.

The invoice data in first normal form with keys added

	INVOICE_ID	VENDOR_NAME	INVOICE_NUMBER	INVOICE_SEQUENCE	ITEM_DESCRIPTION
1	1	Cahners Publishing	112897	1	VB ad
2	2	Cahners Publishing	112897	2	SQL ad
3	3	Cahners Publishing	112897	3	Library directory
4	4	Zylka Design	97/522	1	Catalogs
5	5	Zylka Design	97/522	2	SQL flyer
6	6	Zylka Design	97/533B	1	Card revision

The invoice data in second normal form

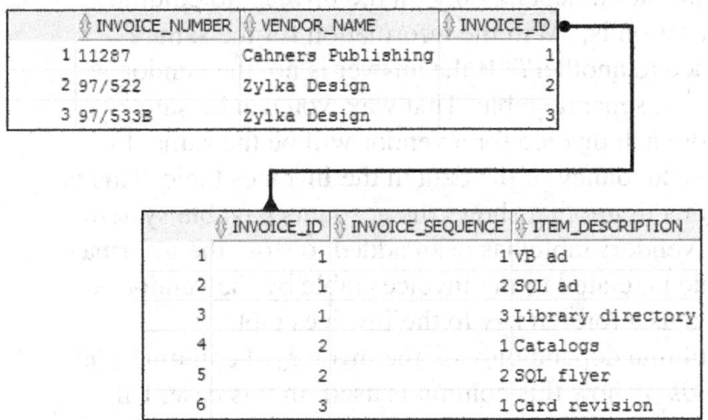

	INVOICE_NUMBER	VENDOR_NAME	INVOICE_ID
1	11287	Cahners Publishing	1
2	97/522	Zylka Design	2
3	97/533B	Zylka Design	3

	INVOICE_ID	INVOICE_SEQUENCE	ITEM_DESCRIPTION
1	1	1	VB ad
2	1	2	SQL ad
3	1	3	Library directory
4	2	1	Catalogs
5	2	2	SQL flyer
6	3	1	Card revision

Description

- For a table to be in second normal form, every non-key column must depend on the entire primary key. If a column doesn't depend on the entire key, it indicates that the table contains information for more than one entity. This is reflected by the table's composite key.
- To apply second normal form, you move columns that don't depend on the entire primary key to another table and then establish a relationship between the two tables.
- Second normal form helps remove redundant row data, which can save storage space, make maintenance easier, and reduce the chance of storing inconsistent data.

Figure 9-11 How to apply the second normal form

How to apply the third normal form

To apply the third normal form, you make sure that every non-key column depends *only* on the primary key. Figure 9-12 illustrates how you can apply this form to the data structure for the accounts payable system. At the top of this figure, you can see all of the columns in the Invoices and Invoice_Line_Items tables in second normal form. Then, you can see a list of questions that you might ask about some of the columns in these tables when you apply third normal form.

First, does the vendor information depend only on the invoice_id column? Another way to phrase this question is, "Will the information for the same vendor change from one invoice to another?" If the answer is no, the vendor information should be stored in a separate table. That way, you can be sure that the vendor information for each invoice for a vendor will be the same. In addition, you will reduce the redundancy of the data in the Invoices table. This is illustrated by the diagram in this figure that shows the accounts payable system in third normal form. Here, a Vendors table has been added to store the information for each vendor. This table is related to the Invoices table by the vendor_id column, which has been added as a foreign key to the Invoices table.

Second, does the terms column depend only on the invoice_id column? The answer to that question depends on how this column is used. In this case, I'll assume that this column is used not only to specify the terms for each invoice, but also to specify the default terms for a vendor. Because of that, the terms information could be stored in both the Vendors and the Invoices tables. To avoid redundancy, however, the information related to different terms can be stored in a separate table, as illustrated by the Terms table in this figure. As you can see, the primary key of this table is an identity column named terms_id. Then, a foreign key column named default_terms_id has been added to the Vendors table, and a foreign key column named terms_id has been added to the Invoices table.

Third, does the account_no column depend only on the invoice_id column? Again, that depends on how this column is used. In this case, it's used to specify the general ledger account number for each line item, so it depends on the invoice_id and the invoice_sequence columns. In other words, this column should be stored in the Invoice_Line_Items table. In addition, each vendor has a default account number, which should be stored in the Vendors table. Because of that, another table named General_Ledger_Accounts has been added to store the account numbers and account descriptions. Then, foreign key columns have been added to the Vendors and Invoice_Line_Items tables to relate them to this table.

Fourth, can the invoice_due_date column in the Invoices table and the line_item_amount column in the Invoice_Line_Items table be derived from other data in the database? If so, they depend on the columns that contain that data rather than on the primary key columns. In this case, the value of the line_item_amount column can always be calculated from the item_quantity and item_unit_price columns. Because of that, this column could be omitted. Alternatively, you could omit the item_quantity and item_unit_price columns

The accounts payable system in second normal form

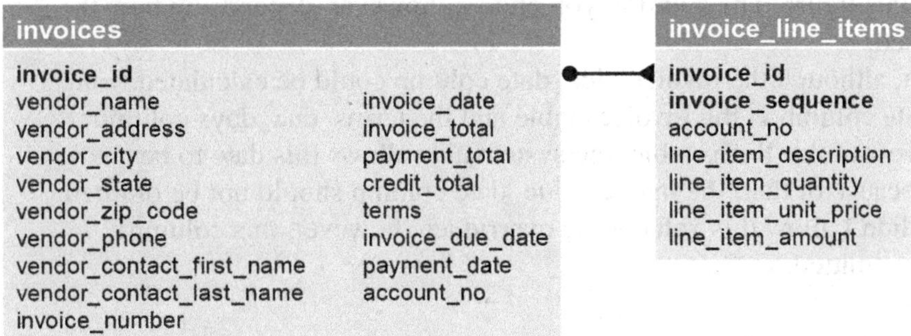

Questions about the structure

1. Does the vendor information (vendor_name, vendor_address, etc.) depend only on the invoice_id column?
2. Does the terms column depend only on the invoice_id column?
3. Does the account_no column depend only on the invoice_id column?
4. Can the invoice_due_date and line_item_amount columns be derived from other data?

The accounts payable system in third normal form

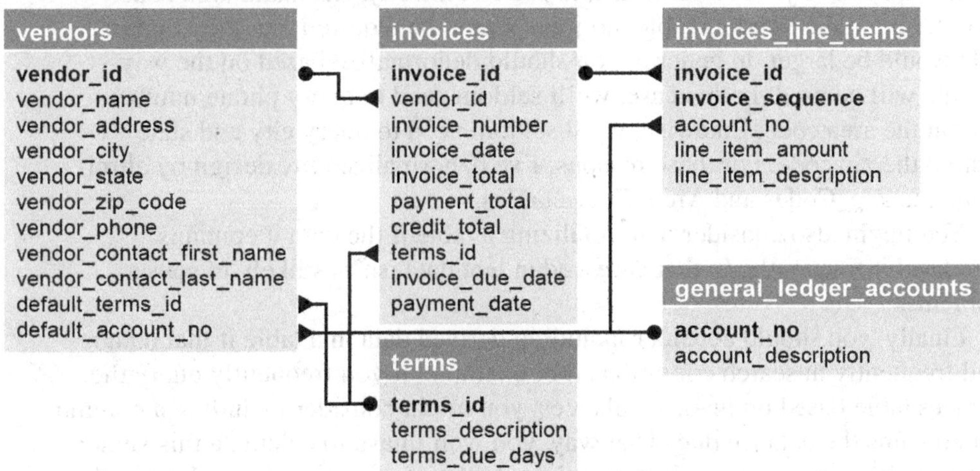

Description

- For a table to be in third normal form, every non-key column must depend *only* on the primary key.
- If a column doesn't depend only on the primary key, it implies that the column is assigned to the wrong table or that it can be computed from other columns in the table. A column that can be computed from other columns contains *derived data*.

Figure 9-12 How to apply the third normal form

and keep just the line_item_amount column. That's what I did in the data structure shown in this figure. The solution you choose, however, depends on how the data will be used.

In contrast, although the invoice_due_date column could be calculated from the invoice_date column in the Invoices table and the terms_due_days column in the related row of the Terms table, the system also allows this date to be overridden. Because of that, the invoice_due_date column should not be omitted. If the system didn't allow this value to be overridden, however, this column could be safely omitted.

When and how to denormalize a data structure

Denormalization is the deliberate deviation from the normal forms. Most denormalization occurs beyond the third normal form. In contrast, the first three normal forms are almost universally applied.

To illustrate when and how to denormalize a data structure, figure 9-13 presents the design of the accounts payable system in fifth normal form. Here, the vendor zip codes are stored in a separate table that contains the city and state for each zip code. In addition, the area codes are stored in a separate table. Because of that, a query that retrieves vendor addresses and phone numbers would require two joins. In contrast, if you left the city, state, and area code information in the Vendors table, no joins would be required, but the Vendors table would be larger. In general, you should denormalize based on the way the data will be used. In this case, we'll seldom need to query phone numbers without the area code. Likewise, we'll seldom need to query city and state without the zip code. For these reasons, I've denormalized my design by eliminating the Zip_Codes and Area_Codes tables.

You might also consider denormalizing a table if the data it contains is updated infrequently. In that case, redundant data isn't as likely to cause problems.

Finally, you should consider including derived data in a table if that data is used frequently in search conditions. For example, if you frequently query the Invoices table based on invoice balances, you might consider including a column that contains the balance due. That way, you won't have to calculate this value each time it's queried. Keep in mind, though, that if you store derived data, it's possible for it to deviate from the derived value. For this reason, you may need to protect the derived column so it can't be updated directly. Alternatively, you could update the table periodically to reset the value of the derived column.

Because normalization eliminates the possibility of data redundancy errors and optimizes the use of storage, you should carefully consider when and how to denormalize a data structure. In general, you should denormalize only when the increased efficiency outweighs the potential for redundancy errors and storage problems. Of course, your decision to denormalize should also be based on your knowledge of the real-world environment in which the system will be used. If you've carefully analyzed the real-world environment as outlined in this chapter, you'll have a good basis for making that decision.

The accounts payable system in fifth normal form

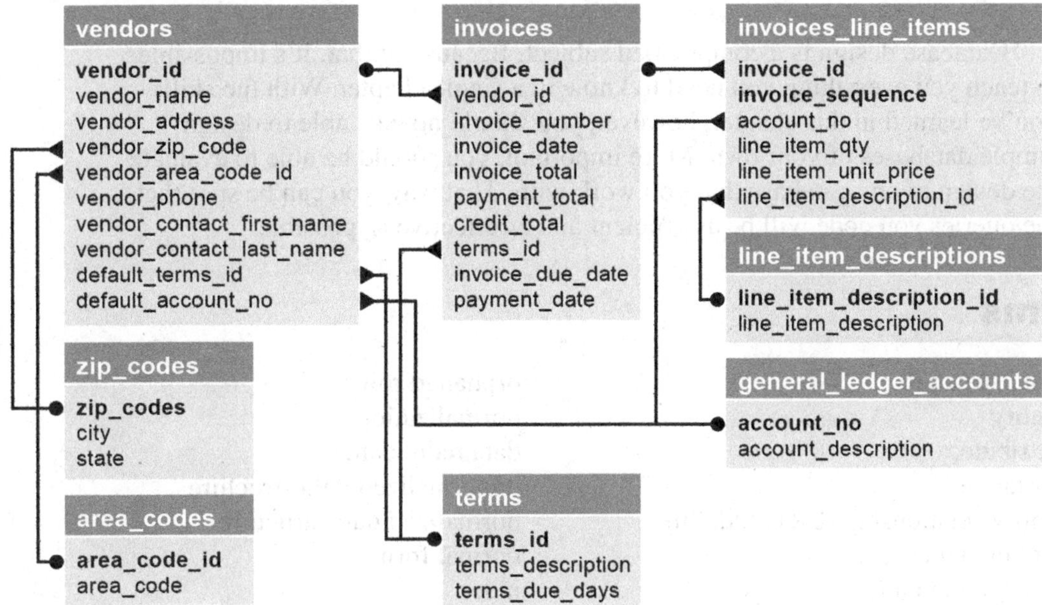

When to denormalize

- When a column from a joined table is used repeatedly in search criteria, you should consider moving that column to the primary key table if it will eliminate the need for a join.
- If a table is updated infrequently, you should consider denormalizing it to improve efficiency. Because the data remains relatively constant, you don't have to worry about data redundancy errors once the initial data is entered and verified.
- Include columns with derived values when those values are used frequently in search conditions. If you do that, you need to be sure that the column value is always synchronized with the values of the columns it's derived from.

Description

- Data structures that are normalized to the fourth normal form and beyond typically require more joins than tables normalized to the third normal form and can therefore be less efficient.
- SQL statements that work with tables that are normalized to the fourth normal form and beyond are typically more difficult to code and debug.
- Most designers *denormalize* data structures to some extent, usually to the third normal form.
- *Denormalization* can result in larger tables, redundant data, and reduced performance.
- Only denormalize when necessary. It is better to adhere to the normal forms unless it is clear that performance will be improved by denormalizing.

Figure 9-13 When and how to denormalize a data structure

Perspective

Database design is a complicated subject. Because of that, it's impossible to teach you everything you need to know in a single chapter. With the skills you've learned in this chapter, however, you should now be able to design simple databases of your own. More important, you should be able to evaluate the design of any database that you work with. That way, you can be sure that the queries you code will be as efficient and as effective as possible.

Terms

data structure
entity
attribute
instance
entity-relationship (ER) modeling
linking table
connecting table
associate table
referential integrity
declarative referential integrity (DRI)
foreign key constraint
orphaned row
normalization
data redundancy
unnormalized data structure
normalized data structure
normal forms
index
composite index
derived data
denormalization

Exercises

Design a database for tracking memberships

1. Draw a database diagram for a database that tracks the memberships for an association and for the groups within the association. Assume that each member can belong to any number of groups and that each group can have any number of members.
2. Modify your design for exercise 2 to keep track of the role served by each individual in each group. Assume that each individual can only serve one role within a group and that each group has its own set of roles that members can fulfill.

10

How to create a database

Now that you've learned how to design a database, you're ready to learn how to implement your design by creating a database and its tables, indexes, and sequences. To do that, you use the set of SQL statements that are known as the data definition language (DDL).

This chapter begins by teaching you how to code DDL statements yourself because it's common to store the DDL statements that create a database in a script. Then, this chapter finishes by showing how to use SQL Developer to work with the tables, indexes, and sequences of a database.

How to work with container databases........................294
An introduction to CDBs and PDBs........................294
How to create and drop a pluggable database........................296
How to work with tables........................298
How to create a table........................298
How to code a primary key constraint........................300
How to code a foreign key constraint........................302
How to code a check constraint........................304
How to alter the columns of a table........................306
How to alter the constraints of a table........................308
How to rename, truncate, and drop a table........................310
How to work with indexes........................312
How to create an index........................312
How to drop an index........................312
How to work with sequences........................314
How to create a sequence........................314
How to use a sequence........................314
How to alter a sequence........................316
How to drop a sequence........................316
How to automatically generate ID values........................318
A script that's used to create a schema........................320
An introduction to scripts........................320
How the DDL statements work........................320
How to use SQL Developer........................326
How to work with tables, indexes, and sequences........................326
How to display an EER diagram for a table........................328
Perspective........................330

How to work with container databases

Before you can begin creating the tables of a database, you need to create the database itself. Then, if multiple databases are running on the server, you need to connect to the correct database before you begin working with it. If you later decide that you no longer need a database, you can drop it, which causes the database and all of its tables and data to be deleted.

An introduction to CDBs and PDBs

With Oracle 12c and later, all databases are part of a multi-tenant container architecture. Within that architecture, databases may be either container databases or pluggable databases.

A *container database (CDB)* contains smaller databases as well as objects and users available, or common, to all databases within the CDB. *Common users*, such as the system and sys users, can access and work with any of the databases and other objects within the container.

Within every CDB is a single CDB$ROOT container. You can think of it as the "root" of the database. This root container can contain one or more *pluggable databases* (*PDBs*). These databases are "pluggable" because they can be easily moved to, or plugged into, another container whenever necessary.

Each PDB has its own *local users*, tables, and other objects. Local users can be administrators for the PDB or end users, but they do not have access to other databases in the CDB.

In addition, each CDB includes a database named PDB$SEED. This database is used as a template, or "seed", for creating new PDBs. It is not used for storing data.

PDBs make it easier to administer a *multi-tenant database*, which is a database that has more than one organization or group using it. This means that multiple organizations can use the same CDB, but each organization has its own PDB. This allows each organization to remain isolated from each other, saves computing resources, and makes it easier to move a PDB from one CDB to another if that becomes necessary.

Each PDB can contain one or more schemas, and each *schema* provides a way to group and organize the tables of the PDB. For instance, this book has been using schemas named AP, OM, and EX to organize its tables in a single PDB.

A container database (CDB) with multiple pluggable databases (PDBs)

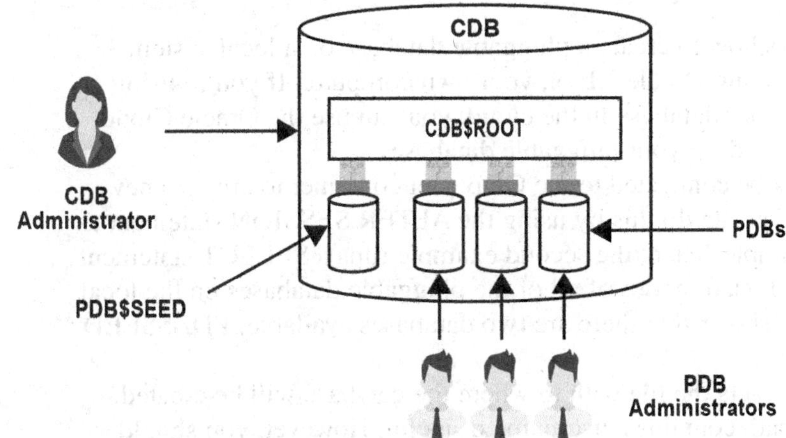

A database with multiple schemas

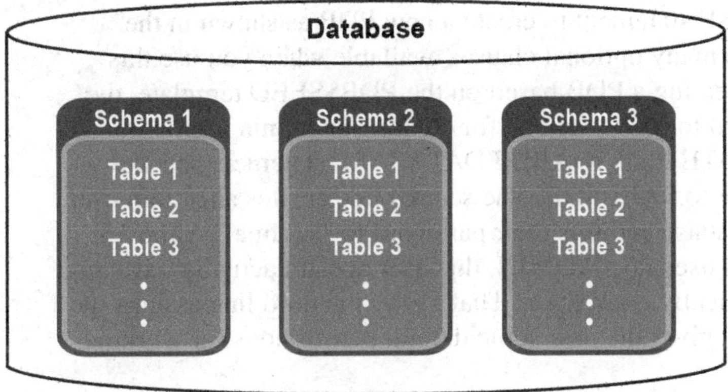

Description

- A *container database* (*CDB*) contains databases as well as objects and users available, or common, to all databases within the CDB.
- A CDB has a single container called CDB$ROOT which can contain *pluggable databases* (*PDBs*). A PDB is a collection of schemas, tables, users, and other objects which can be easily moved, or plugged into, different containers.
- *Common users* are users that have access to, or are common to, the entire container or database. The system and sys users are both common users.
- Each PDB has its own *local users* and tables. Local users can include administrators and regular users for the PDB.
- PDBs make it easier to administer a *multi-tenant database*, which is a database that has more than one organization using it.
- A *schema* provides a way to group and organize the tables for a database. Each PDB can contain one or more schemas.

Figure 10-1 An introduction to CDBs and PDBs

How to create and drop a pluggable database

Figure 10-2 shows how to create a pluggable database on a local system, such when you are running Oracle XE on your own computer. If you're using Oracle Cloud to host your database in the cloud, you can use the Oracle Cloud web pages to create and drop your pluggable databases.

To start, you must be connected to the CDB root container to create a new pluggable database. You can do this by using the ALTER SESSION statement as shown in the first example. Then, the second example runs a SELECT statement to view the names and open modes of all of the pluggable databases on the local system. The result set shows that there are two databases available: PDB$SEED and XEPDB1.

The third example sets the file path to where the database will be created. In this example, the path contains a username of joelm. However, you should change this username and any other parts of this path to match the path to the Oracle database files on your system.

Once you've set the path for the database, you can use the CREATE PLUGGABLE DATABASE statement to create a new PDB as shown in the fourth example. There are many optional clauses available when you use this statement, but if you're creating a PDB based on the PDB$SEED template, the only clause you're required to use is the one for creating an admin user.

The first line in CREATE PLUGGABLE DATABASE statement sets the name for the new database to SALESDB. The second line creates an admin user named salesdb for the database and sets their password to "sesame". Although the clause that creates this user says ADMIN, this user doesn't actually have any privileges until you specifically grant them. That's why the third line assigns the DBA role to the user. This gives the user some default permissions for working with the database.

The fourth line sets the default *tablespace* to the Users tablespace. Oracle uses tablespaces to organize, store, and manage the data files for a database. Typically, you want to store your files in the Users tablespace, which is a tablespace automatically created when you install Oracle Database.

When you create a new PDB, it isn't open initially, so you need to open it as shown in the fifth example before you can work with it. Then, the sixth example shows how to set the container to the new PDB. That's because Oracle runs SQL statements against the current session's container.

Once you've set the container to a PDB, you can grant privileges to a user of the PDB shown in the seventh example. In this example, the GRANT statement grants all privileges to the user named salesdb. As a result, the salesdb user can create tables, indexes, and sequences as described later in this chapter.

The last two statements show how to close and drop a PDB. Before you do that, you need to again set the container to the root container as shown in the first example. Since closing the PDB closes it for all users, you should only do this if you plan to unplug or drop the database.

If you want to drop a PDB, you can use the code in the last example to do that. This deletes the database and all of its data files, so make sure you really want to delete this database before running this statement.

Set the container to the root container

```
-- run using a connection with admin privileges, like Sys
ALTER SESSION SET CONTAINER = CDB$ROOT;
```

View the open mode for all PDBs

```
SELECT name, open_mode FROM v$pdbs;
```

	NAME	OPEN_MODE
1	PDB$SEED	READ ONLY
2	XEPDB1	READ WRITE

Set the location for the database files

```
ALTER SYSTEM SET DB_CREATE_FILE_DEST =
  '/app/joelm/product/21c/oradata/XE';
```

Create a PDB from the PDB$SEED

```
CREATE PLUGGABLE DATABASE salesdb
  ADMIN USER salesdb IDENTIFIED BY sesame
  ROLES=(DBA)
  DEFAULT TABLESPACE users;
```

Open the PDB

```
ALTER PLUGGABLE DATABASE salesdb OPEN;
```

Set the container to the PDB

```
ALTER SESSION SET CONTAINER = salesdb;
```

Grant privileges to the PDB admin user

```
GRANT ALL PRIVILEGES TO salesdb;
```

Close a PDB

```
ALTER PLUGGABLE DATABASE salesdb CLOSE;
```

Drop a PDB

```
DROP PLUGGABLE DATABASE salesdb INCLUDING DATAFILES;
```

Description

- To create a new PDB, the current container must be the CDB$ROOT container and you must have administrative privileges for the CDB.
- The PDB$SEED database is a template, or seed, for creating other PDBs.
- When you create a new PDB using the PDB$SEED, you must create an admin user. However, this user has no privileges until they are specifically granted.
- XEPDB1 is a blank PDB included with Oracle Database XE. If you created the schemas as described in the appendixes, it contains the AP, OM, and EX schemas.
- Oracle uses *tablespaces* to organize, store, and manage the data files for a database

Figure 10-2 How to create and drop a pluggable database

How to work with tables

This section shows how to code the DDL statements that work with the tables of a pluggable database. Because the syntax for these statements is complex, this chapter doesn't present complete syntax diagrams for these statements. Instead, the diagrams present only the commonly used clauses. If you're interested in the complete syntax of any statement, though, you can refer to the Oracle documentation.

As you read this chapter, use the EX connection to run the examples unless otherwise specified. This will allow you to experiment with the code without affecting the tables in the AP schema that are used for other chapters.

How to create a table

Figure 10-3 presents the simplified syntax for the CREATE TABLE statement. In its simplest form, this statement consists of the name of the new table followed by the names and data types of the columns it will contain. This is illustrated by the first example in this figure, which creates a table named Vendors with two columns. The first column, vendor_id, has the NUMBER data type and the second column, vendor_name, has the VARCHAR2(50) data type.

However, in most cases, you'll want to code one or more *column constraints* or *attributes*. Two common constraints and an attribute are listed in the table in this figure.

To indicate that a column doesn't accept null values, you can code the NOT NULL constraint. If you omit this attribute, the column will allow null values.

To indicate that each row in a column must contain a unique value, you can code the UNIQUE constraint. However, a unique column can still contain multiple null values. That's why it's common to use the NOT NULL and UNIQUE attributes together to define a column where each value in the column must be unique. This is illustrated in the second example.

To specify a default value for a column, you can use the DEFAULT attribute. This value is used if another value isn't specified when a row is added to the database. The default value you specify must correspond to the data type for the column. For example, the default value for the invoice_date column in the third example is set to the value that's returned by the SYSDATE function. Similarly, the default value for the payment_total column is set to a literal value of zero.

By default, the CREATE TABLE statement creates a new table in the schema you are currently connected to. If that's not what you want, you can qualify the table name with the desired schema name as shown in the third example.

Finally, if you are using 23ai or later, you can optionally code an IF NOT EXISTS clause before the name of the table as shown here:

```
CREATE TABLE IF NOT EXISTS table_name
...
```

This clause causes the statement to only create the new table if a table with that name does not already exist.

The simplified syntax of the CREATE TABLE statement

```
CREATE TABLE [schema_name.]table_name
(
  column_name_1 data_type [column_attributes]
  [, column_name_2 data_type [column_attributes]]...
  [, table_level_constraints]
)
```

Common column constraints and attributes

Attribute	Description
`NOT NULL`	The column doesn't accept null values.
`UNIQUE`	Each value stored in the column must be unique from every other value or else null.
`DEFAULT default_value`	Specifies a default value for the column.

Create a table with two columns

```
CREATE TABLE vendors
(
  vendor_id      NUMBER,
  vendor_name    VARCHAR2(50)
)
```

Define the columns with constraints

```
CREATE TABLE vendors
(
  vendor_id      NUMBER          NOT NULL    UNIQUE,
  vendor_name    VARCHAR2(50)    NOT NULL    UNIQUE
)
```

Create a table within a specific schema and with default values

```
CREATE TABLE ex.invoices
(
  invoice_id       NUMBER          NOT NULL    UNIQUE,
  vendor_id        NUMBER          NOT NULL,
  invoice_number   VARCHAR2(50)    NOT NULL,
  invoice_date     DATE                        DEFAULT SYSDATE,
  invoice_total    NUMBER(9,2)     NOT NULL,
  payment_total    NUMBER(9,2)                 DEFAULT 0
)
```

Description

- The CREATE TABLE statement creates a table based on the column definitions, constraints, and attributes you specify.
- A *column constraint* restricts the kind of values that can be entered for that column when adding a new row.
- A *column attribute* defines additional information about a column.
- To test the code in this figure and in the figures that follow, you can use the EX schema.

Figure 10-3 How to create a table

How to code a primary key constraint

To declare that a column is the primary key for a table, you can code a *primary key constraint* as shown in figure 10-4. When you identify a column as the primary key, the column automatically gains the NOT NULL and UNIQUE constraints. So, you don't need to specify them. In addition, an index is automatically created for the column.

The easiest way to define a primary key is to code the PRIMARY KEY keywords after the data type for the column as shown in the first example. However, you can also define a primary key as a named constraint. When you code a named constraint, you code the CONSTRAINT keyword, the name of the constraint, and then the constraint itself.

To illustrate, the second example uses the CONSTRAINT keyword to provide names for all three constraints that were defined in the first example. The first constraint provides a name of vendors_pk for the primary key of the Vendors table. The second constraint names the NOT NULL constraint vendor_name_nn, and the third constraint names the UNIQUE constraint vendor_name_uq.

Note how the names provided here begin with the corresponding column name and end with a two-letter suffix to identify the type of constraint. This is a common way to keep track of constraints and helps keep your code organized.

Because the constraints in the first and second examples are coded after columns, they are known as *column-level constraints* or *inline constraints*. However, you can also code a *table-level constraint*, or *out-of-line constraint*. A table-level constraint is coded after all the columns have been defined and names any constrained columns in parentheses. It can apply to a single column or to the table as a whole. Like a column-level constraint, a table-level constraint can optionally be named. To illustrate, the third example creates the vendors_pk and vendor_name_uq constraints at the table level.

Some constraints work the same regardless of whether they are coded at the column level or the table level. As a result, where you decide to code these constraints is largely a matter of personal preference. However, a table-level constraint does provide one capability that isn't available from column-level constraints: it can refer to multiple columns in the table. As a result, if you need to refer to multiple columns, you must use a table-level constraint. For example, to create a primary key based on two or more columns, you must code the PRIMARY KEY constraint at the table level as shown in the fourth example.

Finally, note that when you code a constraint at the table level, you must code a comma at the end of the preceding column definition. If you don't, you will get an error when you try to run the statement.

The syntax of a column-level primary key constraint

```
[CONSTRAINT constraint_name] PRIMARY KEY
```

The syntax of a table-level primary key constraint

```
[CONSTRAINT constraint_name]
PRIMARY KEY (column_name_1 [, column_name_2]...)
```

Column-level constraints

```
CREATE TABLE vendors
(
  vendor_id      NUMBER          PRIMARY KEY,
  vendor_name    VARCHAR2(50)    NOT NULL    UNIQUE
)
```

Named column-level constraints

```
CREATE TABLE vendors
(
  vendor_id      NUMBER          CONSTRAINT vendors_pk PRIMARY KEY,
  vendor_name    VARCHAR2(50)    CONSTRAINT vendor_name_nn NOT NULL
                                 CONSTRAINT vendor_name_uq UNIQUE
)
```

Named table-level constraints

```
CREATE TABLE vendors
(
  vendor_id      NUMBER,
  vendor_name    VARCHAR2(50)    NOT NULL,
  CONSTRAINT vendors_pk PRIMARY KEY (vendor_id),
  CONSTRAINT vendor_name_uq UNIQUE (vendor_name)
)
```

Two-column primary key constraint

```
CREATE TABLE invoice_line_items
(
  invoice_id              NUMBER         NOT NULL,
  invoice_sequence        NUMBER         NOT NULL,
  line_item_description   VARCHAR2(100)  NOT NULL,
  CONSTRAINT line_items_pk PRIMARY KEY (invoice_id, invoice_sequence)
)
```

Description

- A *column-level constraint*, or *inline constraint*, is coded as part of the definition of the column it constrains.
- A *table-level constraint*, or *out-of-line constraint*, is coded after the column definitions. It names the columns it constrains in parentheses.
- A *primary key constraint* indicates that a column is the primary key for the table.
- Only one primary key constraint can be coded per table, and a column designated as a primary key automatically gains the NOT NULL and UNIQUE constraints.

Figure 10-4 How to code a primary key constraint

How to code a foreign key constraint

Figure 10-5 shows how to code a *foreign key constraint*, which is also known as a *referential integrity constraint*. This type of constraint is used to define the relationships between tables and, as you might guess, enforce referential integrity.

To create a foreign key constraint at the column level, you code the REFERENCES keyword followed by the name of the related table and then the name of the related column in parentheses. In this figure, for instance, the first example creates a table with a vendor_id column that includes a REFERENCES clause that identifies the vendor_id column in the Vendors table as the related column.

The second example shows how to code the same primary and foreign key constraints shown in the first example at the table level. When you do this, you code the FOREIGN KEY keywords and the column name in parentheses, as you would for a primary key constraint, and then the references clause.

The third example in this figure shows what happens when you try to insert a row into the Invoices table with a vendor_id value that isn't matched by the vendor_id column in the Vendors table. Because of the foreign key constraint, the system enforces referential integrity by refusing to do the operation. It also displays an error message that indicates the constraint that was violated.

Similarly, if you try to delete a row from the Vendors table that has related rows in the Invoices table, the delete operation will fail and the system will display an error message. Since this prevents rows in the Invoices table from being orphaned, this is usually what you want.

In some cases, though, you may want to automatically delete any related rows in the Invoices table when the related row in the Vendors table is deleted. To do that, you can code the ON DELETE clause after the foreign key constraint as illustrated by the fourth example.

Here, the ON DELETE clause is coded with the CASCADE option. This means that when you delete a row from the primary key table, the delete operation is *cascaded* to the related rows in the foreign key table, causing those rows to be deleted as well. For example, if you deleted a row from the Vendors table, all related rows in the Invoices table would also be deleted. Because a cascading delete operation makes it easy to delete data that you didn't intend to delete, you should use this clause with caution.

The other option for the ON DELETE clause is SET NULL. Then, when you delete a row from the primary key table, the corresponding values in the foreign key column of the foreign key table are set to null. Since this creates rows in the foreign key table that aren't related to the primary key table, you'll rarely want to use this option.

The syntax of a column-level foreign key constraint

```
[CONSTRAINT constraint_name]
  REFERENCES table_name (column_name)
  [ON DELETE {CASCADE|SET NULL}]
```

The syntax of a table-level foreign key constraint

```
[CONSTRAINT constraint_name]
  FOREIGN KEY (column_name_1 [, column_name_2]...)
  REFERENCES table_name (column_name_1 [, column_name_2]...)
  [ON DELETE {CASCADE|SET NULL}]
```

A column-level foreign key constraint

```
CREATE TABLE invoices
(
  invoice_id       NUMBER          PRIMARY KEY,
  vendor_id        NUMBER          REFERENCES vendors (vendor_id),
  invoice_number   VARCHAR2(50)    NOT NULL    UNIQUE
)
```

A table-level foreign key constraint

```
CREATE TABLE invoices
(
  invoice_id       NUMBER          NOT NULL,
  vendor_id        NUMBER          NOT NULL,
  invoice_number   VARCHAR2(50)    NOT NULL    UNIQUE,
  CONSTRAINT invoices_pk PRIMARY KEY (invoice_id),
  CONSTRAINT invoices_fk_vendors
    FOREIGN KEY (vendor_id) REFERENCES vendors (vendor_id)
)
```

An INSERT statement that fails because a related row doesn't exist

```
INSERT INTO invoices
VALUES (1, 1, '1')
```

The response from the system

```
SQL Error: ORA-02291: integrity constraint (EX.INVOICES_FK_VENDORS) violated
- parent key not found
*Cause:     A foreign key value has no matching primary key value.
*Action:    Delete the foreign key or add a matching primary key.
```

A constraint that uses the ON DELETE clause

```
CONSTRAINT invoices_fk_vendors
  FOREIGN KEY (vendor_id) REFERENCES vendors (vendor_id)
  ON DELETE CASCADE
```

Description

- A *foreign key constraint* requires values in one table to match values in another table. This defines the relationship between two tables and enforces referential integrity.
- To define a foreign key that consists of two or more columns, you must define the constraint at the table level.

Figure 10-5 How to code a foreign key constraint

How to code a check constraint

Figure 10-6 shows how to code a *check constraint*. A check constraint allows you to code a Boolean expression that checks a value before it is stored in a column and returns an error if the value does not satisfy the expression.

The first example in this figure uses a column-level check constraint to limit the values in the invoice_total column to numbers greater than or equal to zero. After the NOT NULL constraint on the column, the CHECK constraint defines a Boolean condition within parentheses. A similar CHECK constraint is coded for the payment_total column after its DEFAULT constraint.

The second example shows how to code the same constraints at the table level. Because any single constraint violation will prevent a row from being inserted, you can usually combine multiple column-level check constraints into one table-level constraint. This is done by using the AND keyword to combine the Boolean expressions into a single Boolean expression.

In general, you should use check constraints to restrict the values in a column whenever feasible to keep your data clean and useful. In some situations, however, check constraints can be too restrictive. As an example, consider a telephone number that's constrained to the "(000) 000-0000" format used for US phone numbers. The problem with this constraint is that it doesn't let you store phone numbers with extensions (although you could store extensions in a separate column) or phone numbers with an international format.

For this reason, check constraints aren't used by all database designers. That way, the database can store values with formats that weren't predicted when the database was designed. However, this flexibility comes at the cost of allowing some invalid data. For some systems, this tradeoff is acceptable.

The syntax of a check constraint

```
[CONSTRAINT constraint_name] CHECK (condition)
```

Two column-level check constraints

```
CREATE TABLE invoices
(
  invoice_id        NUMBER          PRIMARY KEY,
  invoice_total     NUMBER(9,2)     NOT NULL      CHECK (invoice_total >= 0),
  payment_total     NUMBER(9,2)     DEFAULT 0     CHECK (payment_total >= 0)
)
```

The same check constraints coded at the table level

```
CREATE TABLE invoices
(
  invoice_id        NUMBER          PRIMARY KEY,
  invoice_total     NUMBER(9,2)     NOT NULL,
  payment_total     NUMBER(9,2)     DEFAULT 0,
  CONSTRAINT invoices_ck CHECK (invoice_total >= 0 AND payment_total >= 0)
)
```

An INSERT statement that fails due to the check constraint

```
INSERT INTO invoices
VALUES (1, 99.99, -10)
```

The response from the system

```
SQL Error: ORA-02290: check constraint (EX.INVOICES_CK) violated
02290. 00000 -  "check constraint (%s.%s) violated"
*Cause:    The values being inserted do not satisfy the named check
*Action:   do not insert values that violate the constraint.
```

Description

- *Check constraints* limit the values that can be stored in the columns of a table.
- The condition you specify for a check constraint is evaluated as a Boolean expression. If the expression is true, the insert or update operation proceeds. Otherwise, it fails.
- A check constraint coded at the column level can refer only to that column. A check constraint coded at the table level can refer to any column in the table, and may refer to more than one column.

Figure 10-6 How to code a check constraint

How to alter the columns of a table

After you create a table, you may need to change the columns of that table. For example, you may need to add, modify, or drop a column. To do that, you can use the ALTER TABLE statement shown in figure 10-7. When you use the ALTER TABLE statement, you code the table name followed by any number of ADD, DROP, or MODIFY statements.

The first example in this figure shows how to add a new column to an existing table. As you can see, you code the column definition the same way as when you create a new table. To start, you specify the column name. Then, you code the data type and its attributes, including any constraints.

The second example shows how to drop an existing column. Note that Oracle will prevent you from dropping some columns. For example, you can't drop a column if it's the primary key column.

The third example shows how to modify the length of the data type for an existing column. In this case, a column that was defined as VARCHAR2(50) is changed to VARCHAR2(100). Since the new data type is bigger than the old data type, you can be sure that all existing data will still fit.

The fourth example shows how to change the data type to a different data type. In this case, a column that was defined as VARCHAR2(100) is changed to CHAR(100). Since these data types both store the same type of characters, you can again be sure no data will be lost.

The fifth example shows how to change the default value for a column. In this case, a default value of "New Vendor" is assigned to the vendor_name column.

In the first five statements, Oracle can alter the table without losing any data. As a result, these statements execute successfully and alter the table. However, if the change will result in a loss of data, Oracle prevents the change. For example, the sixth statement attempts to change the length of the vendor_name column to a length that's too small for existing data that's stored in this column. As a result, Oracle doesn't modify the column and the system instead returns an error message like the one shown in this figure.

With Oracle 23ai or later, the ALTER TABLE statement has an optional IF EXIST clause that you can use to check if a table with that name exists before altering the database. You use it as shown here:

```
ALTER TABLE IF EXISTS table_name
...
```

The simplified syntax for modifying the columns of a table

```
ALTER TABLE [schema_name.]table_name
{  ADD          column_name data_type [column_attributes]
 | DROP COLUMN column_name |
 | MODIFY       column_name data_type [column_attributes]}...
```

Add a new column

```
ALTER TABLE vendors
ADD last_transaction_date DATE;
```

Drop a column

```
ALTER TABLE vendors
DROP COLUMN last_transaction_date;
```

Change the length of a column

```
-- vendor_name is defined with VARCHAR2(50)
ALTER TABLE vendors
MODIFY vendor_name VARCHAR2(100);
```

Change the data type of a column

```
ALTER TABLE vendors
MODIFY vendor_name CHAR(100);
```

Change the default value of a column

```
ALTER TABLE vendors
MODIFY vendor_name DEFAULT 'New Vendor';
```

A statement that fails because it would cause data to be lost

```
ALTER TABLE vendors
MODIFY vendor_name VARCHAR2(10);
```

The response from the system

```
SQL Error: ORA-01441: cannot decrease column length because some value is
too big
```

Description

- You can use the ALTER TABLE statement to modify the columns of an existing table.
- Oracle won't allow you to change a column if that change would cause data to be lost.

Warning

- You should never alter a table or other database object in a production database without first consulting the DBA.

Figure 10-7 How to alter the columns of a table

How to alter the constraints of a table

After you create a table, you may need to change the constraints. For example, you may need to add or drop a constraint. Or, you may want to disable or enable a constraint. To do that, you can use the ALTER TABLE statement as shown in figure 10-8.

The first example shows how to add a check constraint to a table. However, if the invoice_total column contained existing values that didn't satisfy the check constraint, Oracle wouldn't alter the table and would instead return an error.

The second example shows how to drop the check constraint that was added in the first example. To drop a constraint, you have to know its name. Since the first example supplied a name for the constraint, it's easy to drop. If you don't name your constraints, though, Oracle will automatically generate a name for the constraint, such as SYS_C0012769. In that case, before you can drop a constraint, you'll need to look up its name as shown later in this chapter.

The third example shows how to use the DISABLE keyword to add a new check constraint that's *disabled*. A disabled constraint can still be modified, but does not constrain any data. Since the check constraint in this example is disabled when it's added, Oracle won't check to make sure existing values satisfy this constraint. As a result, the constraint will be created even if there are existing values in the column that don't satisfy the constraint.

The fourth example shows how to use the ENABLE and NOVALIDATE keywords to *enable* a constraint for new values only. When you use NOVALIDATE, Oracle doesn't validate whether existing data meets the newly enabled constraint. In this case, the code enables the check constraint that was added in the third example without validating existing values against the constraint. Of course, if you do want to validate values when you enable the constraint, you can omit the NOVALIDATE keyword.

The fifth example shows how to use the DISABLE keyword to disable an existing constraint. In some cases, this is useful if you need to temporarily disable a constraint while you update the data in the column. Then, you can re-enable the constraint when you're done.

The last three examples show how to add other types of constraints. The sixth example shows how to add a foreign key constraint, the seventh example shows how to add a unique constraint, and the eighth example shows how to add a not null constraint. Note that the not null constraint uses the syntax for modifying a column rather than adding the constraint at the table level.

The simplified syntax for modifying the constraints of a table

```
ALTER TABLE table_name
{ ADD        CONSTRAINT   constraint_name constraint_definition [DISABLE]
| DROP       CONSTRAINT   constraint_name |
| ENABLE   [NOVALIDATE] constraint_name |
| DISABLE                 constraint_name }...
```

Add a new check constraint

```
ALTER TABLE invoices
ADD CONSTRAINT invoice_total_ck CHECK (invoice_total >= 0);
```

Drop a check constraint

```
ALTER TABLE invoices
DROP CONSTRAINT invoice_total_ck;
```

Add a disabled constraint

```
ALTER TABLE invoices
ADD CONSTRAINT invoice_total_ck CHECK (invoice_total >= 1) DISABLE;
```

Enable a constraint for new values only

```
ALTER TABLE invoices
ENABLE NOVALIDATE CONSTRAINT invoice_total_ck;
```

Disable a constraint

```
ALTER TABLE invoices
DISABLE CONSTRAINT invoice_total_ck;
```

Add a foreign key constraint

```
ALTER TABLE invoices
ADD CONSTRAINT invoices_fk_vendors
FOREIGN KEY (vendor_id) REFERENCES vendors (vendor_id);
```

Add a unique constraint

```
ALTER TABLE vendors
ADD CONSTRAINT vendor_name_uq UNIQUE (vendor_name);
```

Add a not null constraint by modifying a column

```
ALTER TABLE vendors
MODIFY vendor_name CONSTRAINT vendor_name_nn NOT NULL;
```

Description

- You use the ALTER TABLE statement to add or drop the constraints of an existing table.
- To drop a constraint, you must know its name.
- By default, Oracle verifies that existing data satisfies a new constraint. To prevent this, you can add a *disabled* constraint, then *enable* it later without validating existing data.

Figure 10-8 How to alter the constraints of a table

How to rename, truncate, and drop a table

Figure 10-9 shows how to use the RENAME, TRUNCATE TABLE, and DROP TABLE statements. When you use these statements, use them cautiously, especially when you're working on a production database.

To start, you can use the RENAME statement to rename an existing table. This is useful if you want to change the name of a table without modifying its column definitions or the data that's stored in the table. In this figure, for instance, the first example changes the name of the Vendors table to Vendor. However, if you rename a table, you should probably update the names of any constraints that use the name of the table.

You can use the TRUNCATE TABLE statement to delete all of the data from a table without deleting the column definitions for the table. In this figure, the second example deletes all rows from the newly renamed Vendor table.

Finally, you can use the DROP TABLE statement to delete all of the data from a table and also delete the definition of the table, including the constraints for the table. The third and fourth examples both drop the Vendor table. However, the fourth example explicitly specifies that it is dropping the Vendor table that's stored in the EX schema.

When you execute a DROP TABLE statement, Oracle first checks to see if other tables depend on the table you're trying to delete. If one does, Oracle won't allow the deletion. For example, you couldn't delete the Vendor table if a foreign key constraint in the Invoices table referred to the Vendor table. In that case, you would need to drop the Invoices table before you could drop the Vendor table.

If you're using Oracle 23ai or later, the DROP TABLE statement provides an IF EXISTS clause that you can use to check if a table exists before dropping it like this:

```
DROP TABLE IF EXISTS table_name
...
```

This is useful if you want to suppresses the error message that's displayed when you attempt to drop a table that doesn't exist.

Rename a table

```
RENAME vendors TO vendor
```

Delete all data from a table

```
TRUNCATE TABLE vendor
```

Delete a table

```
DROP TABLE vendor
```

Specify the schema for the table to be deleted

```
DROP TABLE ex.vendor
```

Description

- You can use the RENAME statement to change the name of an existing table.
- You can use the TRUNCATE TABLE statement to delete all data from a table without deleting the definition for the table.
- You can use the DROP TABLE statement to delete a table.
- You can't truncate or drop a table if a foreign key constraint in another table refers to that table.
- When you drop a table, all of its data, constraints, indexes, and triggers are deleted.

Warnings

- You can't roll back a TRUNCATE TABLE statement.
- You should not use these statements on a production database without first consulting the DBA.

Figure 10-9 How to rename, truncate, and drop a table

How to work with indexes

An *index* speeds up joins and searches by providing a way for a database management system to find a row without having to search through all the rows first. By default, Oracle creates indexes for the primary keys and unique constraints of a table, which is usually what you want. In addition, you may want to create indexes for foreign keys and other columns that are used frequently in search conditions or joins. However, you'll want to avoid creating indexes on columns that are updated frequently since this slows down insert, update, and delete operations.

How to create an index

Figure 10-10 presents the simplified syntax of the CREATE INDEX statement. To create an index, you use the ON clause to specify the name the table followed by the columns or expression that the index will be based on.

For each column or expression, you can specify the ASC or DESC keyword to indicate whether you want the index sorted in ascending or descending sequence. If you don't specify a sort order, ASC is the default. In addition, you can use the UNIQUE keyword to specify that an index can contain only unique values. This is effectively the same as coding a UNIQUE constraint for a column.

In these examples, the names follow a standard naming convention that's used by many developers. The index name specifies the name of the table, followed by the name of the column or columns, followed by a suffix of IX. This naming convention makes it easy to see which columns of which tables have been indexed.

The fifth and sixth examples show how to create a *function-based index*, which is an index that's based on an expression. This expression may include one or more functions or operators. When you create an index like this, it's usually for use with search conditions in a WHERE clause.

The syntax and examples for creating indexes omit many optional clauses that you can use for tuning indexes for better performance. This tuning is often done by DBAs working with large databases, and usually isn't necessary for small databases. However, you should be aware that there are many ways to optimize indexes that depend on the specific needs of your application.

How to drop an index

The seventh example shows how to use the DROP INDEX statement to delete an index. You may want to drop an index if you suspect that it isn't speeding up your joins and searches or that it's slowing down your insert, update, or delete operations.

If you're using Oracle 23ai or later, both the CREATE INDEX and DROP INDEX statements allow you to check if an index exists before executing the statement. As a result, you can use the IF NOT EXISTS and IF EXISTS clauses as you would if you were creating or dropping a table.

The simplified syntax of the CREATE INDEX statement

```
CREATE [UNIQUE] INDEX [schema_name.]index_name
  ON table_name (column_name_1 [ASC|DESC] [, column_name_2 [ASC|DESC]]...)
```

Create an index based on a single column

```
CREATE INDEX invoices_vendor_id_ix ON invoices (vendor_id);
```

Create a composite index based on two columns

```
CREATE INDEX invoices_vendor_id_invoice_number_ix
  ON invoices (vendor_id, invoice_number);
```

Create a unique index

```
CREATE UNIQUE INDEX vendors_vendor_phone_ix
  ON vendors (vendor_phone);
```

Create an index that's sorted in descending order

```
CREATE INDEX invoices_invoice_total_ix
  ON invoices (invoice_total DESC);
```

Create a function-based index

```
CREATE INDEX vendors_vendor_name_upper_ix
  ON vendors (UPPER(vendor_name));
```

Another way to create a function-based index

```
CREATE INDEX invoices_balance_due_ix
  ON invoices (invoice_total - payment_total - credit_total DESC);
```

Drop an index

```
DROP INDEX vendors_vendor_state_ix
```

Description

- An *index* allows Oracle to directly or near-directly find the row for a specific value instead of having to search for it.
- Oracle automatically creates an index for primary keys and for columns with the UNIQUE constraint.
- If a column is most frequently accessed in descending order, you should create the index for that column in descending order to speed up row retrieval.
- When creating a *composite index* based on more than one column, the order matters. List the most-frequently accessed columns first.
- A *function-based index* is based on an expression that includes functions or operators.
- Because indexes take up storage space and Oracle must update a table's indexes every time that table is modified, you shouldn't create more indexes than you need.
- There are many ways to optimize indexes. The larger your tables are, the more important having optimized indexes is. Consult the Oracle documentation for more information.

Figure 10-10 How to work with indexes

How to work with sequences

In Oracle, a *sequence* automatically generates a sequence of integer values. Typically, a sequence is used to generate values for the primary key of a table.

How to create a sequence

Figure 10-11 shows how to create a sequence. Most of the time, you can create a sequence by simply coding the CREATE SEQUENCE statement followed by the name of the sequence. In the first example, for instance, the CREATE SEQUENCE statement creates a sequence named vendor_id_seq that will be used to get a value for the primary key of the Vendors table.

This creates a sequence that starts with 1, is incremented by a value of 1, has no maximum or minimum value, and caches the next 20 generated numbers. In addition, this sequence doesn't guarantee that sequence numbers are generated in order of request. In most cases, these settings are adequate.

If you need to create a sequence that works differently, however, you can use the optional clauses of the CREATE SEQUENCE statement to modify the sequence. Most of these are listed in the simplified syntax at the top of this figure. For instance, you can use the STARTS WITH clause to start the sequence at a specific value. The highest vendor ID in the Vendors table is 123. As a result, it makes sense to use a sequence that starts at 124 to generate any new values for the vendor_id column.

The third example illustrates how the other clauses work. This example generates a sequence that begins with 100, is incremented each time by a value of 10, has a minimum value of 0, has a maximum value of 1,000,000, cycles back to the beginning of the sequence if and when it reaches the end, and guarantees that sequence numbers are generated in order of request.

How to use a sequence

Once you've created a sequence, you typically use it within an INSERT statement as shown in the fourth example. Here, the NEXTVAL pseudocolumn increments the sequence and returns the newly incremented value so it can be inserted into the vendor_id column of the Vendors table. Then, the fifth example shows how to use the CURRVAL pseudocolumn to check the current, or most recently generated, value of the sequence.

When you use these pseudocolumns, be aware that Oracle doesn't initialize the sequence until you call NEXTVAL for the first time. As a result, you can't call CURRVAL until you've called the NEXTVAL pseudocolumn at least once. If you do, Oracle will return an error.

The same sequence can be used in multiple tables and by multiple users. Additionally, it is possible for a sequence to be incremented within a transaction that is later rolled back. Because of that, you may notice that some numbers seem to have been "skipped" by the sequence. If it is important that a sequence always increments exactly as you expect, consider restricting access to the sequence.

The simplified syntax of the CREATE SEQUENCE statement

```
CREATE SEQUENCE [schema_name.]sequence_name
  [START WITH starting_integer]
  [INCREMENT BY increment_integer]
  [{MAXVALUE maximum_integer | NOMAXVALUE}]
  [{MINVALUE minimum_integer | NOMINVALUE}]
  [{CYCLE|NOCYCLE}]
  [{CACHE cache_size|NOCACHE}]
  [{ORDER|NOORDER}]
```

Create a sequence of integers that starts with 1 and is incremented by 1

```
CREATE SEQUENCE vendor_id_seq
```

Specify a starting integer

```
CREATE SEQUENCE vendor_id_seq
  START WITH 124
```

Specify multiple parameters

```
CREATE SEQUENCE test_seq
  START WITH 100 INCREMENT BY 10
  MINVALUE 0 MAXVALUE 1000000
  CYCLE CACHE 10 ORDER;
```

Use the NEXTVAL pseudocolumn to get the next value

```
INSERT INTO vendors
VALUES (vendor_id_seq.NEXTVAL, 'Acme Co.', '123 Main St.', NULL,
  'Fresno', 'CA', '93711', '(800) 221-5528',
  'Wiley' , 'Coyote');
```

Use the CURRVAL pseudocolumn to get the current value

```
SELECT vendor_id_seq.CURRVAL from dual;
```

Description

- By default, the CREATE SEQUENCE statement creates a *sequence* of integers that starts with 1, is incremented 1, has no maximum or minimum value, caches the next 20 generated numbers, and doesn't guarantee that numbers are generated in order of request.
- Sequences are typically used to generate integer values for primary keys.
- The NEXTVAL pseudocolumn increments the sequence and returns the new value.
- The CURRVAL pseudocolumn returns the current value in the sequence.
- If another user is also incrementing the sequence, CURRVAL may return a result you don't expect.

Figure 10-11 How to work with sequences (part 1 of 2)

How to alter a sequence

Once you've created a sequence, you can use the ALTER SEQUENCE statement to alter the attributes of a sequence. This works similarly to the CREATE SEQUENCE statement. However, there are a few notable differences.

When you alter a sequence, you can choose to use the RESTART keyword to restart the sequence. By default, a restarted sequence starts again at 1. However, when used with an ascending sequence that has a minimum value specified, RESTART sets the next value to the minimum value. Similarly, when a descending sequence has a maximum value, RESTART sets the next value to the maximum value. And if RESTART is used with START WITH, the sequence restarts with the specified starting value. This can be useful if you need to "rewind" a sequence that was used in a transaction that was rolled back.

Additionally, Oracle performs some validation when you alter a sequence and prevents some changes. For example, you can't set the minimum and maximum attributes to values that don't make sense. If the current value of the sequence is 99, for instance, you can't set the maximum value to 98.

How to drop a sequence

To drop a sequence, you must use the DROP SEQUENCE statement. In this figure, the last example drops the sequence named test_seq that was created in part 1 of this figure. When you drop a sequence, it doesn't affect any values already generated by that sequence.

With Oracle 23ai or later, both the CREATE SEQUENCE and DROP SEQUENCE statements allow you to check if a sequence exists before executing the statement. As a result, you can use the IF NOT EXISTS and IF EXISTS clauses as you would if you were creating or dropping a table.

The simplified syntax of the ALTER SEQUENCE statement

```
ALTER SEQUENCE sequence_name
  [sequence_attributes]
  [RESTART]
```

Alter a sequence and restart it

```
ALTER SEQUENCE test_seq
  INCREMENT BY 9
  MINVALUE 99 MAXVALUE 999999
  NOCYCLE CACHE 9 NOORDER
  RESTART;
```

Drop a sequence

```
DROP SEQUENCE test_seq;
```

Description

- The RESTART keyword restarts a sequence. By default, the sequence restarts with 1 unless any of START WITH, MINVALUE, or MAXVALUE have been set.
- If MINVALUE has been set and the sequence is ascending, RESTART sets NEXTVAL to the minimum value. For a descending sequence with MAXVALUE, RESTART sets NEXTVAL the maximum value.
- When altering a sequence, you can't specify a new MINVALUE or MAXVALUE that doesn't make sense with the current values of the sequence.
- You can use the DROP SEQUENCE statement to drop a sequence.

Figure 10-11 How to work with sequences (part 2 of 2)

How to automatically generate ID values

Figure 10-12 begins by showing the simplified syntax of the IDENTITY clause, which you can use when you create or alter a column. This clause creates a sequence used exclusively for the column it is defined on. To use it, you code the GENERATED keyword and then optionally either ALWAYS or BY DEFAULT. When you use ALWAYS, the sequence always supplies the value for the column and attempting to insert a literal value for a row will return and error. When you use BY DEFAULT, the sequence is used only if another value isn't supplied. If you don't specify either, ALWAYS is the default.

Then, you code the AS IDENTITY keywords followed optionally by any specifications for the sequence in parentheses. These are largely the same as the options for sequences you saw in figure 10-11 and are coded the same way.

As the name implies, the IDENTITY clause is most commonly used with ID columns and primary key columns. The first three examples in this figure illustrate this.

The first example shows how to use the NEXTVAL pseudocolumn of a sequence as a default column value. If you're working with a column that already has a sequence defined for it, you may want to use this technique.

The second example, on the other hand, uses the IDENTITY clause to create and use a sequence that starts at 1 and is incremented by 1. This sequence will be incremented only when a row is inserted into the Vendors table.

The third example shows how to modify the sequence created by the IDENTITY clause. It uses START WITH to start the sequence at 124. Although you can use any of the clauses for defining a sequence, you typically only need to use the START WITH clause to specify a starting value for the sequence.

When you use the IDENTITY clause, the sequence is dropped if a table is dropped. Conversely, the sequence is recreated if the table is recreated. This resets the values supplied by the sequence, which is usually what you want.

This figure finishes by showing two INSERT statements that use automatically generated ID values. The first INSERT statement uses the DEFAULT keyword to specify the value for the vendor_id column. As a result, the vendor_id column uses the ID value that's automatically generated. The second INSERT statement uses a column list that doesn't include the vendor_id column. As a result, the vendor_id column uses the ID value that's automatically generated.

The simplified syntax of the IDENTITY clause

```
GENERATED [{ALWAYS|BY DEFAULT}]
AS IDENTITY [(identity_options)]
```

Three ways to automatically generate an ID value

Use a sequence as the default value for the column

```
CREATE TABLE vendors
(
  vendor_id     NUMBER          DEFAULT vendor_id_seq.NEXTVAL PRIMARY KEY,
  vendor_name   VARCHAR2(50)    NOT NULL       UNIQUE
)
```

Use a simple IDENTITY clause

```
CREATE TABLE vendors
(
  vendor_id     NUMBER          GENERATED AS IDENTITY PRIMARY KEY,
  vendor_name   VARCHAR2(50)    NOT NULL       UNIQUE
)
```

Use a more complex IDENTITY clause

```
CREATE TABLE vendors
(
  vendor_id     NUMBER          GENERATED BY DEFAULT
                                AS IDENTITY (START WITH 124) PRIMARY KEY,
  vendor_name   VARCHAR2(50)    NOT NULL       UNIQUE
)
```

Two ways to use the generated ID value

Use the DEFAULT keyword

```
INSERT INTO vendors VALUES (DEFAULT, 'default test');
```

Use a column list

```
INSERT INTO vendors (vendor_name) VALUES ('column list test');
```

Description

- You can use the NEXTVAL pseudocolumn of a sequence as the default value for a column.
- You can use the IDENTITY clause to automatically create a sequence and use it for that column. By default, this sequence starts with 1 and is incremented by 1.
- When you set the identity's options, you can specify any of the sequence clauses, such as START WITH, INCREMENT BY, MAXVALUE, and so on.
- When you use the IDENTITY clause, the sequence is dropped if the table is dropped. Conversely, the sequence is recreated if the table is recreated. This resets the values of the sequence.

Figure 10-12 How to automatically generate ID values

A script that's used to create a schema

Figure 10-13 presents the DDL statements that are used to create a schema named AP2 in a pluggable database named murach_db. This AP2 schema is a simplified version of the AP schema that's used throughout this book. In this figure, these statements are coded as part of a script.

This script is designed to be run against Oracle XE running on a local system. If you're using Oracle Cloud to host a database as described in chapter 13, you would use the Oracle Cloud web pages to create the database and the AP2 user. However, the statements for creating the tables and indexes in the AP2 schema would work the same.

An introduction to scripts

As you learned in chapter 2, a script is a file that contains one or more SQL statements. In addition, a script can store PL/SQL code, which can be used to conditionally run SQL statements. Scripts are often used to create the objects for a database as shown in this figure.

When you code a script, you code a semicolon at the end of each SQL statement. In addition, you code a front slash at the end of each block of PL/SQL code that's stored in the script. At any point in the script, you can code a COMMIT statement to commit the changes to the database. However, DDL statements such as CREATE and DROP statements are automatically committed immediately after they're executed. As a result, you don't need to code a COMMIT statement after DDL statements like the ones shown in this figure.

How the DDL statements work

The ALTER SESSION statement that begins this script sets the container for the current session to the CDB$ROOT. This allows you to create the murach_db pluggable database in the CDB$ROOT container.

After the session container is set, a block of PL/SQL code is used to close and drop the murach_db PDB if it exists. This makes it possible to run this script again without manually removing the database first yourself. If the database exists, this code drops it. If not, this code suppresses any error messages that would be displayed when you try to drop an object that doesn't exist.

In the PL/SQL code, the ALTER PLUGGABLE DATABASE and DROP PLUGGABLE DATABASE statements are coded within single quotes after the EXECUTE IMMEDIATE keywords. This is necessary to execute DDL statements from within a block of PL/SQL code.

After the block of PL/SQL, the script sets the DB_CREATE_FILE_DEST variable to the path to the folder that stores the database files. Then, the code uses the CREATE PLUGGABLE DATABASE statement to create the murach_db database. Once the database has been created, the code opens the database so that the script can run other DDL statements against that database.

A script that creates a pluggable database and a schema — Page 1

```
-- Set container to CDB
ALTER SESSION SET CONTAINER = CDB$ROOT;

-- Use an anonymous PL/SQL script to close and drop the PDB
-- and suppress any error messages
BEGIN
  EXECUTE IMMEDIATE 'ALTER PLUGGABLE DATABASE murach_db CLOSE';
  EXECUTE IMMEDIATE 'DROP PLUGGABLE DATABASE murach_db INCLUDING DATAFILES';
EXCEPTION
  WHEN OTHERS THEN
    DBMS_OUTPUT.PUT_LINE('');
END;
/

-- Enable Oracle Managed Files (OMF)
ALTER SYSTEM SET DB_CREATE_FILE_DEST = '/app/joelm/product/21c/oradata/XE';

-- Create the PDB
CREATE PLUGGABLE DATABASE murach_db
  ADMIN USER ap2 IDENTIFIED BY sesame
  ROLES=(DBA)
  DEFAULT TABLESPACE users;

-- Open the PDB
ALTER PLUGGABLE DATABASE murach_db OPEN;

-- Grant permissions to the PDB admin
ALTER SESSION SET CONTAINER = murach_db;
GRANT ALL PRIVILEGES TO ap2;

--  Connect as the admin user for the AP2 schema
CONNECT ap2/sesame@localhost:1521/murach_db;

/****************************************************
* Create the tables for the AP2 schema of the PDB
****************************************************/

CREATE TABLE general_ledger_accounts
(
  account_number          NUMBER          NOT NULL,
  account_description     VARCHAR2(50)    NOT NULL,
  CONSTRAINT gl_accounts_pk
    PRIMARY KEY (account_number),
  CONSTRAINT gl_account_description_uq
    UNIQUE (account_description)
);

CREATE TABLE terms
(
  terms_id                NUMBER          NOT NULL,
  terms_description       VARCHAR2(50)    NOT NULL,
  terms_due_days          NUMBER          NOT NULL,
  CONSTRAINT terms_pk
    PRIMARY KEY (terms_id)
);
```

Figure 10-13 A script that creates a pluggable database and a schema (part 1 of 3)

After opening the database, the code sets the container for the session to the murach_db database. Then, it grants all privileges to the admin user named ap2 that was created earlier in the script.

Next, the script uses a CONNECT statement to connect to the murach_db database as the ap2 user with a password of "sesame". This database is running on a local server on port 1521. This CONNECT statement only works when you run the entire script at once. If you try to run these statements one at a time, SQL Developer resets the connection to the one currently selected for the worksheet immediately after the CONNECT statement.

After connecting as the ap2 user, the script executes the CREATE TABLE statements for the five main tables of the AP2 schema. Since this code connects as the ap2 user, it creates the tables in the AP2 schema. For each CREATE TABLE statement, the code specifies the primary key column (or columns) first.

Although this isn't required, this is a good programming practice. Since the order in which you declare the columns defines the default order for the columns, the code defines these columns in a logical order. That way, when you use a SELECT * statement to retrieve all of the columns, they're returned in this order.

When you create tables, you must create the tables that don't have foreign keys first. That way, the other tables can define foreign keys that refer to them. In this figure, for example, the script creates the General_Ledger_Accounts and Terms tables first since they don't have foreign keys. Then, it creates the Vendors table, which has foreign keys that refer to those tables. And so on.

Conversely, when you drop tables, you must drop the last table that was created first. Then, you can work back to the first table that was created. Otherwise, the foreign keys might not allow you to drop the tables.

For most of the columns in these tables, the script includes a NOT NULL constraint or a DEFAULT attribute. In general, it's a good practice to only allow a column to accept null values when you want to allow for unknown values. If, for example, a vendor doesn't supply an address, that address is unknown. In that case, you can store a null value in the vendor_address1 and vendor_address2 columns.

Another option is to store an empty string for these columns. To do that, you could define the vendor address columns like this:

```
vendor_address1      VARCHAR2(50)      DEFAULT '',
vendor_address2      VARCHAR2(50)      DEFAULT '',
```

In this case, Oracle would store empty strings for these columns unless other values were assigned to them.

In practice, a null value is a more intuitive representation of an unknown value than a default value is. Conversely, it makes sense to use a default value like an empty string to indicate that a value is known but the column is empty. For example, an empty string might indicate that a vendor hasn't provided its street address. Although how you use nulls and empty strings is largely a matter of personal preference, it does of course affect the way you query a table.

A script that creates a pluggable database and a schema **Page 2**

```
CREATE TABLE vendors
(
  vendor_id                     NUMBER          GENERATED BY DEFAULT
                                                AS IDENTITY (START WITH 124),
  vendor_name                   VARCHAR2(50)    NOT NULL,
  vendor_address1               VARCHAR2(50),
  vendor_address2               VARCHAR2(50),
  vendor_city                   VARCHAR2(50)    NOT NULL,
  vendor_state                  CHAR(2)         NOT NULL,
  vendor_zip_code               VARCHAR2(20)    NOT NULL,
  vendor_phone                  VARCHAR2(50),
  vendor_contact_last_name      VARCHAR2(50),
  vendor_contact_first_name     VARCHAR2(50),
  default_terms_id              NUMBER          NOT NULL,
  default_account_number        NUMBER          NOT NULL,
  CONSTRAINT vendors_pk
    PRIMARY KEY (vendor_id),
  CONSTRAINT vendors_vendor_name_uq
    UNIQUE (vendor_name),
  CONSTRAINT vendors_fk_terms
    FOREIGN KEY (default_terms_id)
    REFERENCES terms (terms_id),
  CONSTRAINT vendors_fk_accounts
    FOREIGN KEY (default_account_number)
    REFERENCES general_ledger_accounts (account_number)
);

CREATE TABLE invoices
(
  invoice_id          NUMBER          GENERATED BY DEFAULT
                                      AS IDENTITY (START WITH 115),
  vendor_id           NUMBER          NOT NULL,
  invoice_number      VARCHAR2(50)    NOT NULL,
  invoice_date        DATE            NOT NULL,
  invoice_total       NUMBER(9,2)     NOT NULL,
  payment_total       NUMBER(9,2)                 DEFAULT 0,
  credit_total        NUMBER(9,2)                 DEFAULT 0,
  terms_id            NUMBER          NOT NULL,
  invoice_due_date    DATE            NOT NULL,
  payment_date        DATE,
  CONSTRAINT invoices_pk
    PRIMARY KEY (invoice_id),
  CONSTRAINT invoices_fk_vendors
    FOREIGN KEY (vendor_id)
    REFERENCES vendors (vendor_id),
  CONSTRAINT invoices_fk_terms
    FOREIGN KEY (terms_id)
    REFERENCES terms (terms_id)
);
```

Figure 10-13 A script that creates a pluggable database and a schema (part 2 of 3)

Since Oracle automatically creates an index for each primary key and unique constraint, this script provides names for all primary key and unique constraints. It also follows a strict convention for naming these constraints that begins with the name or abbreviated name of the table.

In addition to the indexes that are created automatically, this script uses CREATE INDEX statements to create six more indexes to improve the performance of the database. Here again, the script follows the naming conventions used for the primary key and unique constraints. As a result, if you view the indexes for the schema as shown later in this chapter, it's easy to determine how the indexes relate to the tables.

The first five of these indexes are based on the foreign keys that each referring table uses to relate to another table. For example, since the vendor_id column in the Invoices table refers to the vendor_id column in the Vendors table, the script creates an index on vendor_id in the Invoices table. Finally, it creates an index for the invoice_date column in the Invoices table because this column is frequently used to search for rows in this table.

After the indexes, this script includes a SET DEFINE statement to disable substitution variable prompting. Substitution variable prompting checks for any variables prefixed with an ampersand (&). If it finds any, the execution of statements is halted while you are prompted to provide a value. Turning it off isn't necessary for the INSERT statements that follow, but it's still a good practice.

After the SET DEFINE statement, the script includes two INSERT statements for the General_Ledger_Accounts table. To keep this script simple, only two INSERT statements are shown. However, it's common to add INSERT statements that add the starting data to all of the tables in the database.

The final two statements commit the data from the INSERT statements and test that the data was inserted successfully. At this point, the General_Ledger_Accounts table is the only table with data, but if you insert data into any of the other tables and commit your changes it will work the same.

A script that creates a pluggable database and a schema **Page 3**

```
CREATE TABLE invoice_line_items
(
  invoice_id              NUMBER          NOT NULL,
  invoice_sequence        NUMBER          NOT NULL,
  account_number          NUMBER          NOT NULL,
  line_item_amt           NUMBER(9,2)     NOT NULL,
  line_item_description   VARCHAR2(100)   NOT NULL,
  CONSTRAINT line_items_pk
    PRIMARY KEY (invoice_id, invoice_sequence),
  CONSTRAINT line_items_fk_invoices
    FOREIGN KEY (invoice_id) REFERENCES invoices (invoice_id),
  CONSTRAINT line_items_fk_acounts
    FOREIGN KEY (account_number)
    REFERENCES general_ledger_accounts (account_number)
);

-- Create the indexes
CREATE INDEX vendors_terms_id_ix
  ON vendors (default_terms_id);
CREATE INDEX vendors_account_number_ix
  ON vendors (default_account_number);

CREATE INDEX invoices_invoice_date_ix
  ON invoices (invoice_date DESC);
CREATE INDEX invoices_vendor_id_ix
  ON invoices (vendor_id);
CREATE INDEX invoices_terms_id_ix
  ON invoices (terms_id);
CREATE INDEX line_items_account_number_ix
  ON invoice_line_items (account_number);

-- Disable substitution variable prompting
-- (This isn't necessary for the next two INSERT statements,
-- but still a good practice.)
SET DEFINE OFF;

-- INSERT INTO general_ledger_accounts
INSERT INTO general_ledger_accounts VALUES (100,'Cash');
INSERT INTO general_ledger_accounts VALUES (110,'Accounts Receivable');

-- Other insert statements could go here

-- Commit INSERT statements
COMMIT;

-- Check table
SELECT * FROM general_ledger_accounts;
```

Figure 10-13 A script that creates a pluggable database and a schema (part 3 of 3)

How to use SQL Developer

When working with a database, you often use a script to create database objects such as tables, indexes, and sequences. As a result, it's important to understand the DDL statements presented in this chapter. Once you understand these statements, it's easy to learn how to use a graphical user interface such as SQL Developer to work with database objects. In particular, it's often useful to view these database objects before writing SELECT, INSERT, UPDATE, or DELETE statements that use them.

How to work with tables, indexes, and sequences

Figure 10-14 shows how to use SQL Developer to work with some of the database objects of a database. For example, you can view the column definitions for a table by expanding the node for the schema, expanding the Tables folder, and clicking on the name of the table. In this figure, the first screen shows the column definitions for the Invoices table in the AP2 schema of the murach_db database. This table and its database and schema were created by the script from the previous figure.

Viewing the column definitions for a table allows you to examine more detailed information about the table. This information includes information about each column such as its name, its data type, whether it can contain null values, and its default value if it has one.

The same skills apply to other database objects such as indexes and sequences. For example, to view all indexes for a schema, you expand the Indexes folder. Then, you can right-click on an index to view more information about it as shown in the second screen in this figure.

If you want to use SQL Developer to modify a database object, you can right-click on it and use the resulting menu to work with it. However, it's more common to modify a database object by running a SQL statement than it is to use SQL Developer. That's because you can record the changes that have been made to a database by saving any SQL statements that modify a database in a script. By contrast, if you use SQL Developer to modify a database, you don't have a record of what changes you made to the database. Generally, SQL Developer is an excellent tool for quickly viewing the objects of a database or for quickly creating temporary tables or other objects that won't need to be recreated later.

The column definitions for the Invoices table

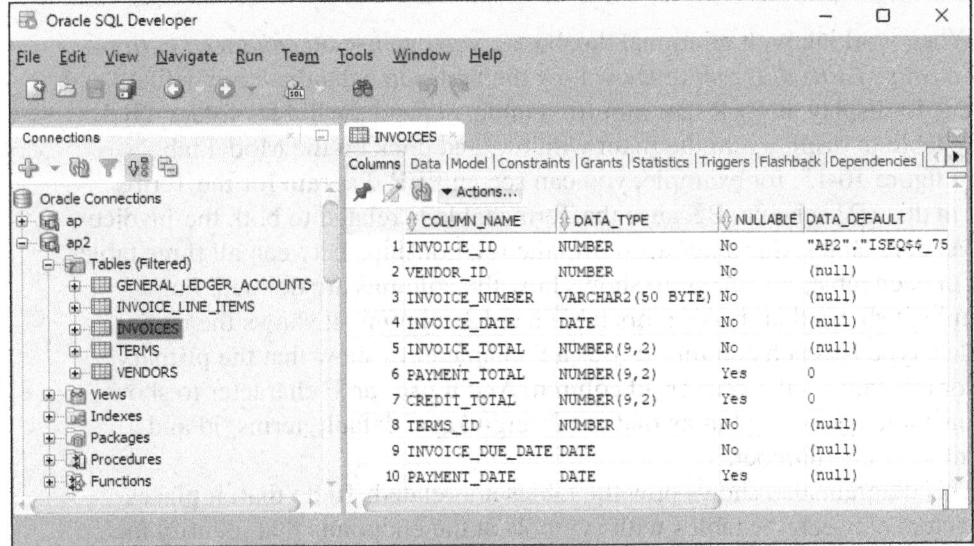

The indexes for a schema

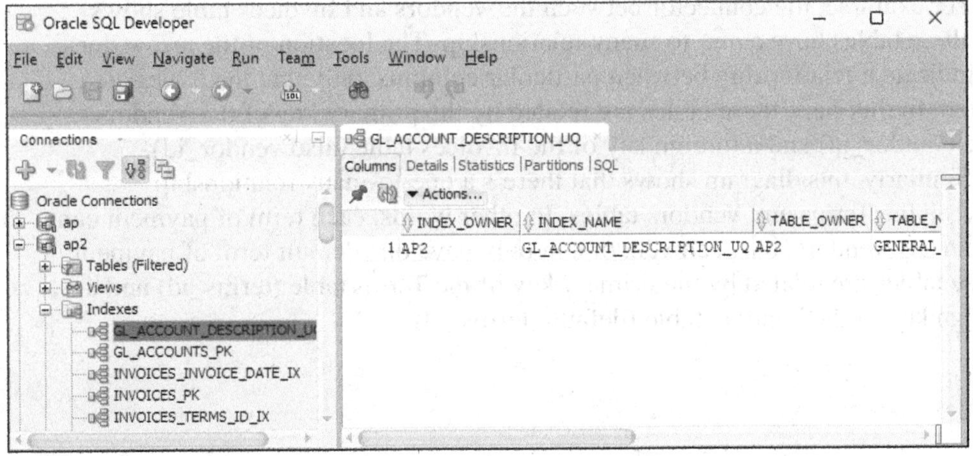

Description

- To display a table in the main window, expand the Tables folder and click on the table. Then, you can click on the tabs above the table to view its column definitions, its data, its constraints, and its indexes.
- To view all indexes for a database, expand the Indexes folder. Then, to get more information about a particular index, click on it to display it in the main window.
- To view the sequences for a database, expand the Sequences folder. Then, to get more information about a sequence, click on its name to display it in the main window.
- You can use SQL Developer to add, edit, and drop database objects. However, you typically use SQL statements to do that.

Figure 10-14 How to work with tables, indexes, and sequences

How to display an EER diagram for a table

When working with relational databases, you can use an *enhanced entity-relationship* (*EER*) *diagram* to show how the tables in a database are defined and related. To display an EER diagram for a table, expand the Tables folder, click on the table to display it in the main window, and click on the Model tab.

In figure 10-15, for example, you can see an EER diagram for the Terms table in the AP2 schema. Because the Terms table is related to both the Invoices and Vendors tables, this diagram shows the relationships between all three tables.

For each table, this diagram shows how the columns are defined. For example, it shows that the Vendors table has 12 columns. It shows the name and data type for each column. It uses a P character to show that the primary key for this table is the vendor_id column. And it uses an F character to show that the table has two columns that are foreign keys: default_terms_id and default_account_number.

This diagram also shows how the tables are related. To do that, it places connectors between the tables with symbols at the endpoints that identify the type of relationship. The two most common symbols are shown in the table in this figure. These symbols identify the "one" and "many" sides of a relationship.

For example, the connector between the Vendors and Invoices table shows that these tables have a one-to-many relationship. The location of the arrow does not indicate a relationship between particular columns, only that the tables are related. In this case, these tables are related by the primary key of the Vendors table (vendor_id) and a foreign key of the Invoices table (also vendor_id).

Similarly, this diagram shows that there's a one-to-many relationship between the Terms and Vendors tables. In other words, each term of payment can have many vendors, but each vendor can only have one default term of payment. These tables are related by the primary key of the Terms table (terms_id) and a foreign key of the Vendors table (default_terms_id).

A database diagram for the Terms table

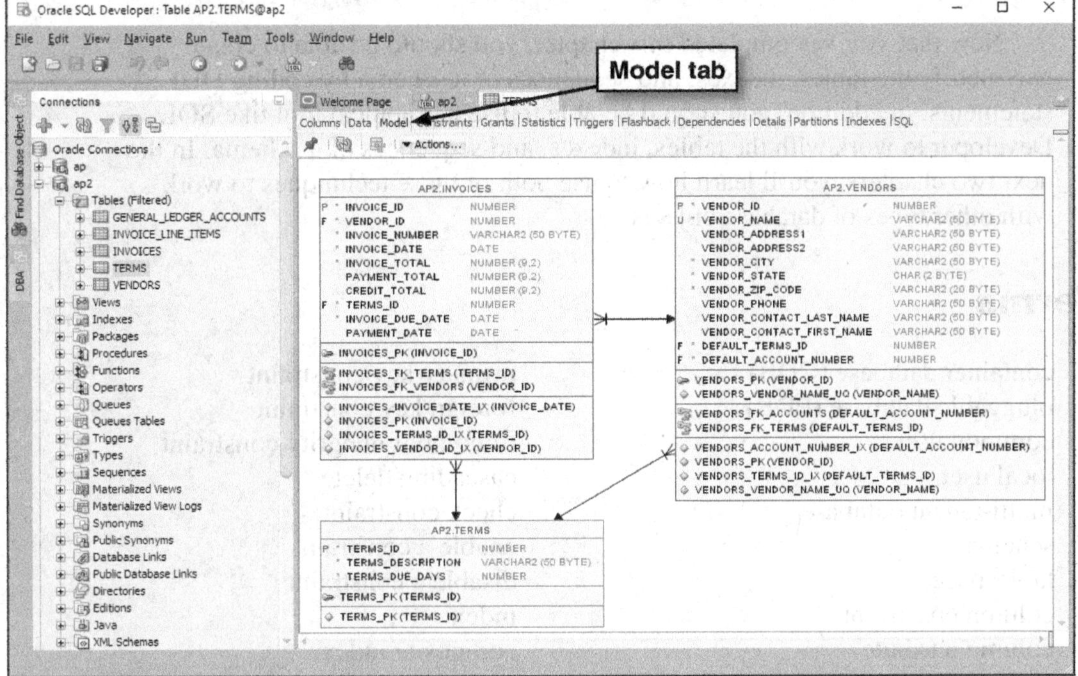

Common relationship symbols

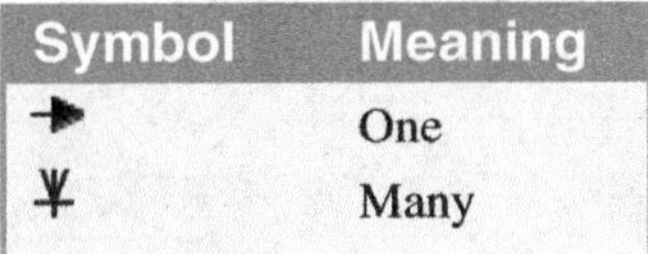

Symbol	Meaning
	One
	Many

Description

- SQL Developer can automatically generate an *enhanced entity-relationship (EER) diagram* that shows how the tables in a database are defined and related.
- To display an EER diagram for a table, expand the Tables folder, click on the table to display it in the main window, and click on the Model tab.

Figure 10-15 How to display an EER diagram for a table

Perspective

Now that you've completed this chapter, you should be able to create and modify the tables, indexes, and sequences of a schema by coding DDL statements. In addition, you should be able to use a graphical tool like SQL Developer to work with the tables, indexes, and sequences of a schema. In the next two chapters, you'll learn how to use both of these techniques to work with other types of database objects.

Terms

container database (CDB)
pluggable database (PDB)
common user
local user
multi-tenant database
schema
tablespace
column constraint
column attribute
column-level constraint
inline constraint
table-level constraint
out-of-line constraint
primary key constraint
foreign key constraint
referential integrity constraint
cascading delete
check constraint
enable a constraint
disable a constraint
index
composite index
function-based index
sequence
enhanced entity-relationship (EER) diagram

Exercises

Add constraints and an index to the AP schema

1. Write an ALTER TABLE statement that adds two new check constraints to the Invoices table of the AP schema. The first should allow (1) payment_date to be null only if payment_total is zero and (2) payment_date to be not null only if payment_total is greater than zero. The second constraint should prevent the sum of payment_total and credit_total from being greater than invoice_total.
2. Add an index to the AP schema for the zip code field in the Vendors table.

Implement a schema design

3. Write the CREATE TABLE statements needed to implement the following design in the EX schema:

These tables provide for members of an association, and each member can be registered in one or more groups within the association. There should be one row for each member in the Members table and one row for each group in the Groups table. The member ID and group ID columns are the primary keys for the Members and Groups tables. And the Members_Groups table relates each member to one or more groups.

When you create the tables, be sure to include the key constraints. Also, include any null or default constraints that you think are appropriate.

4. Write INSERT statements that add two rows to the Members table for member IDs 1 and 2, two rows to the Groups table for group IDs 1 and 2, and three rows to the Members_Groups table: one row for member 1 and group 2; one for member 2 and group 1; and one for member 2 and group 2. Use any values you like for the other columns.

   Then, write a SELECT statement that joins the three tables and retrieves the group name, member last name, and member first name columns.

5. Create two sequences that can be used to number the member ID and group ID values starting with 3 (since you already added members and groups for IDs 1 and 2).

6. Write an INSERT statement that adds another row to the Groups table. This statement should use the NEXTVAL pseudocolumn to get the value for the next group ID from the sequence that you created in exercise 5.

   Then, write a SELECT statement that gets all of the data for all of the rows in the Groups table to make sure your sequence worked correctly.

7. Write an ALTER TABLE statement that adds two new columns to the Members table: one column for annual dues that provides for three digits to the left of the decimal point and two to the right; and one column for the payment date. The annual dues column should have a default value of 52.50.

8. Write an ALTER TABLE statement that modifies the Groups table so the group name in each row has to be unique. Then, re-run the INSERT statement that you used in exercise 6 to make sure this works.

11

How to create views

As you've seen throughout this book, SELECT queries can be complicated, particularly if they use multiple joins, subqueries, or complex functions. Because of that, you may want to save the queries you use regularly so you don't have to keep re-creating them. One way to do that is to store the statement in a script file. Another way is to create a view.

Unlike scripts, which are stored in files, views are stored in a schema. As a result, they can be used by SQL programmers and by custom applications that have access to the schema. This provides some advantages over using tables directly.

An introduction to views....................................334
How views work....................................334
Benefits of using views....................................336
How to work with views....................................338
How to create a view....................................338
How to create an updatable view....................................342
How to create a read-only view....................................342
How to use WITH CHECK OPTION....................................344
How to insert or delete rows through a view....................................346
How to alter or drop a view....................................348
How to use SQL Developer with views....................................350
Perspective....................................352

An introduction to views

Before you learn the details for working with a view, it's helpful to get a general idea of how a view works. In addition, it's helpful to consider some of the benefits of views so you can determine whether you want to use them.

How views work

A *view* is a SELECT statement that's stored in the database as an object. To create a view, you use a CREATE VIEW statement like the one shown in figure 11-1. This statement creates a view named vendors_min that retrieves the vendor_name, vendor_state, and vendor_phone columns from the Vendors table.

You can think of a view as a virtual table that consists only of the rows and columns specified in its CREATE VIEW statement. The table or tables that are listed in the FROM clause are called the *base tables* for the view. Since the view refers back to the base tables, it doesn't store any data itself, and it always reflects the most current data in the base tables.

To use a view, you refer to it from another SQL statement. In the second example in this figure, for instance, the SELECT statement uses the vendors_min view in the FROM clause instead of a table. As a result, this SELECT statement extracts its result set from the virtual table that the view represents. In this case, all the rows for vendors in California are retrieved from the view.

When you create a view like the one in this figure, the view is updateable. As a result, it's possible to use the view in an INSERT, UPDATE, or DELETE statement. In this figure, for example, the UPDATE statement uses the vendors_min view to update the vendor_phone column in the Vendors table for the specified vendor.

To drop a view, you can use the DROP VIEW statement as shown in the last example in this figure. This works similarly to the DROP statements for tables, indexes, and sequences that you learned about in the previous chapter.

Because a view is stored as an object in a database, it can be used by anyone who has access to the database. That includes users who have access to the database through applications that provide for ad hoc queries and report generation. In addition, that includes custom applications that are written specifically to work with the data in the database. In fact, views are often designed to be used with these types of applications.

Create a view named Vendors_Min

```
CREATE VIEW vendors_min AS
  SELECT vendor_name, vendor_state, vendor_phone
  FROM vendors;
```

The virtual table for the view

	VENDOR_NAME	VENDOR_STATE	VENDOR_PHONE
1	US Postal Service	WI	(800) 555-1205
2	National Information Data Ctr	DC	(301) 555-8950
3	Register of Copyrights	DC	NULL
4	Jobtrak	CA	(800) 555-8725
5	Newbrige Book Clubs	NJ	(800) 555-9980

(122 rows)

Use the Vendors_Min view

```
SELECT * FROM vendors_min
WHERE vendor_state = 'CA'
ORDER BY vendor_name
```

The result set

	VENDOR_NAME	VENDOR_STATE	VENDOR_PHONE
1	ASC Signs	CA	NULL
2	Abbey Office Furnishings	CA	(559) 555-8300
3	American Express	CA	(800) 555-3344
4	Aztek Label	CA	(714) 555-9000
5	BFI Industries	CA	(559) 555-1551

(75 rows)

Use a view to update the base table

```
UPDATE vendors_min
SET vendor_phone = '(800) 555-3941'
WHERE vendor_name = 'Register of Copyrights'
```

The response from the system

```
1 row updated.
```

Drop a view

```
DROP VIEW vendors_min
```

Description

- A *view* consists of a SELECT statement that's stored as an object in the database. The tables referenced in the SELECT statement are called the *base tables* for the view.
- When you create a view, you can refer to the view anywhere you would normally use a table in any of the DML statements: SELECT, INSERT, UPDATE, and DELETE.
- Although a view behaves like a virtual table, it doesn't store any data. Instead, a view always refers back to its base tables.
- A view can also be referred to as a *viewed table* because it provides a view to the underlying base tables.

Figure 11-1 How views work

Benefits of using views

Figure 11-2 describes some of the advantages of using views. To start, you can use views to limit the exposure of the tables in your database to external users and applications. To illustrate, suppose a view refers to a table that you've decided to divide into two tables. To accommodate this change, you simply modify the view. In other words, you don't have to modify any statements that refer to the view. That means that users who query the database using the view don't have to be aware of the change in the database structure, and applications that use the view don't have to be modified.

You can also use views to restrict access to a database. To do that, you include just the columns and rows you want a user or an application to have access to in the view. Then, you let the user or application access the data only through the views. For example, let's assume you have an Employees table that has a salary column that contains information about each employee's salary. In this case, you can create a view that doesn't include the salary column for the users who need to view and maintain this table, but who should not be able to view salaries. Then, you can create another view that includes the salary column for the users who need to view and maintain salary information.

In addition, you can use views to hide the complexity of a SELECT statement. For example, if you have a long and unwieldy SELECT statement that joins multiple tables, you can create a view for that statement. This makes it easier for you and other database users to work with this data.

Finally, when you create a view, you can allow data in the base table to be updated through the view. To do that, you use INSERT, UPDATE, or DELETE statements to work with the view.

Some of the benefits provided by views

Benefit	Description
Design independence	Views can limit the exposure of tables to external users and applications. As a result, if the design of the tables changes, you can modify the view as necessary so the users and applications that use the view don't need to be modified.
Data security	Views can restrict access to the data in a table by not including all columns or rows of a table.
Simplified queries	Views can be used to hide the complexity of retrieval operations. Data can be retrieved using simple SELECT statements that specify a view in the FROM clause.
Updatability	With certain restrictions, views can be used to update, insert, and delete data from a base table.

Description

- You can create a view based on almost any SELECT statement. That means that you can code views that join tables, summarize data, and use subqueries and functions.

Figure 11-2 Benefits of using views

How to work with views

Now that you have a general understanding of how views work and of the benefits that they provide, you're ready to learn the details for working with them.

How to create a view

Figure 11-3 presents the CREATE VIEW statement that you use to create a view. In its simplest form, you code the CREATE VIEW keywords, followed by the name of the view, followed by the AS keyword and a SELECT statement that defines the view. In this figure, for instance, the first statement creates a view named vendors_phone_list. This view includes four columns from the Vendors table for all vendors with invoices.

When you code a CREATE VIEW statement, you can specify that you want to automatically drop an existing view that has the same name if one exists and then replace it. To do that, you specify the OR REPLACE keywords after the CREATE keyword as shown in all of the examples in this figure except for the first.

Conversely, like CREATE TABLE, with Oracle 23ai or later you can use the IF NOT EXISTS clause to only create the view if a view with that name doesn't already exist. However, you can't use IF NOT EXISTS if you include the OR REPLACE keywords.

The SELECT statement for a view can use most of the features of a normal SELECT statement. In this figure, for instance, the second example creates a view that joins data from two tables. Similarly, the third statement creates a view that uses a subquery and row limiting clause.

The simplified syntax of the CREATE VIEW statement

```
CREATE [OR REPLACE] [{FORCE|NOFORCE}] VIEW [schema.]view_name
  [(column_alias_1[, column_alias_2]...)]
AS
  select_statement
  [WITH {READ ONLY|CHECK OPTION} [CONSTRAINT constraint_name]]
```

Create a view of vendors that have invoices

```
CREATE VIEW vendors_phone_list AS
  SELECT vendor_name, vendor_contact_last_name,
         vendor_contact_first_name, vendor_phone
  FROM vendors
  WHERE vendor_id IN (SELECT vendor_id FROM invoices)
```

Create a view that uses a join

```
CREATE OR REPLACE VIEW vendor_invoices AS
  SELECT vendor_name, invoice_number, invoice_date, invoice_total
  FROM vendors v JOIN invoices i
    ON v.vendor_id = i.vendor_id
```

Create a view that uses a subquery

```
CREATE OR REPLACE VIEW top5_invoice_totals AS
  SELECT vendor_id, invoice_total
  FROM (SELECT vendor_id, invoice_total FROM invoices
        ORDER BY invoice_total DESC)
  FETCH FIRST 5 ROWS ONLY
```

Description

- If you include the OR REPLACE keyword, the CREATE VIEW statement will replace any existing view that has the same name.

Figure 11-3 How to create a view (part 1 of 2)

By default, the columns in a view are given the same names as the columns in the base tables. If a view contains a calculated column, however, you'll want to name that column just as you do in other SELECT statements. In addition, you'll need to rename columns from different tables that have the same name. To do that, you can code the column names in the CREATE VIEW clause as shown in the first example in part 2 of this figure. Or, you can use the AS clause as shown in the second example.

Note that you have to name all of the columns in the first example. In contrast, in the second example, you only have to name the columns you need to rename. As a result, you'll typically want to use the technique presented in the fifth example.

The third example creates a view that summarizes the rows in the Invoices table by vendor. This shows that a view can use aggregate functions and the GROUP BY clause to summarize data. In this case, the rows are grouped by vendor name, and a count of the invoices and the invoice total are calculated for each vendor.

If you attempt to create a view with a base table that doesn't exist, Oracle will display an error message and not create the view. Since this prevents you from creating views for tables that don't exist, this is usually what you want. However, if you want to create a view first and the table later, you can use the FORCE option to create a view as shown in the fourth example. Of course, this view won't display any data until a table named Products is created and the product_description and product_price columns are populated with some data.

When you code views, the SELECT statement you code within the definition of a view can refer to another view. In other words, views can be nested. In theory, *nested views* can make it easier to present data to your users. In practice, using nested views can make the dependencies between tables and views confusing, which can make your code difficult to maintain. As a result, if you use nested views, you should use them carefully.

Finally, note that if the SELECT statement for a view uses the * operator to select all of the columns in a table and then the definition of that table later changes to include more columns, those columns won't appear in the view. To make them appear, you'll have to either recreate the view or alter it.

Name all the view columns in the CREATE VIEW clause

```
CREATE OR REPLACE VIEW invoices_outstanding
  (invoice_number, invoice_date, invoice_total, balance_due)
AS
  SELECT invoice_number, invoice_date, invoice_total,
         invoice_total - payment_total - credit_total
  FROM invoices
  WHERE invoice_total - payment_total - credit_total > 0
```

Name just the calculated column in the SELECT clause

```
CREATE OR REPLACE VIEW invoices_outstanding AS
  SELECT invoice_number, invoice_date, invoice_total,
         invoice_total - payment_total - credit_total AS balance_due
  FROM invoices
  WHERE invoice_total - payment_total - credit_total > 0
```

Create a view that summarizes invoices by vendor

```
CREATE OR REPLACE VIEW invoice_summary AS
  SELECT vendor_name,
    COUNT(*) AS invoice_count,
    SUM(invoice_total) AS invoice_total_sum
  FROM vendors v JOIN invoices i
    ON v.vendor_id = i.vendor_id
  GROUP BY vendor_name
```

Create a view even if the underlying tables don't exist

```
CREATE FORCE VIEW products_list AS
  SELECT product_description, product_price
  FROM products
```

Description

- If you name the columns of a view in the CREATE VIEW clause, you have to name all of the columns. In contrast, if you name the columns in the SELECT clause, you can name just the columns you need to rename.
- If you include the FORCE keyword, the CREATE VIEW statement will create the view even if the underlying tables don't exist.
- You can create a view that's based on another view. This is known as a *nested view*.
- If the SELECT statement of a view gets all columns of a table using the * operator and the table later gets a new column, that column will not be in the view unless you alter or re-create the view.

Figure 11-3 How to create a view (part 2 of 2)

How to create an updatable view

Once you create a view, you can refer to it in a SELECT statement. And if it's an *updatable view*, you can refer to it in INSERT, UPDATE, and DELETE statements to modify the data that's stored in an underlying table. Figure 11-4 lists the primary requirements for creating updatable views.

The first requirement is the most important. To be updatable, Oracle must be able to unambiguously determine which base table and columns are affected by the view.

Then, the next two requirements have to do with what you can code in the select list of the SELECT statement that defines the view. In particular, the select list can't include the DISTINCT clause, and it can't include aggregate functions. In addition, the SELECT statement can't include a GROUP BY or HAVING clause, and two SELECT statements can't be joined by a union operation.

The first CREATE VIEW statement in this figure creates a view that's updatable. As a result, you can refer to it in an INSERT, UPDATE, or DELETE statement. For example, you can use the first UPDATE statement shown in this figure to update the credit_total column in the Invoices base table.

However, you can't update any calculated columns that are used by the view. For example, you can't use the second UPDATE statement shown in this figure to update the balance_due column.

When you update data through a view, you can only update the data in a single base table, even if the view refers to two or more tables. In this figure, for instance, the view includes data from two base tables, the Vendors and Invoices tables. However, since the first UPDATE statement only refers to columns in the Invoices table, it is able to update data in that table.

How to create a read-only view

If you don't follow the requirements for creating an updateable view, you won't be able to use a view to update any columns in the base table. If you attempt to do this, you'll get an error message.

If you want to make sure that a view is a read-only view, you can add a subquery restriction clause with the read-only constraint to the CREATE VIEW statement. To do that, you code the keywords WITH READ ONLY after the SELECT statement. In this figure, for instance, the last example creates a read-only view. As a result, your users won't be able to use this view to update data in the Invoices table.

Primary requirements for creating updatable views

- Each column in the view must correspond to a column of a single table
- The select list can't include a DISTINCT clause.
- The select list can't include an aggregate function.
- The SELECT statement can't include a GROUP BY or HAVING clause.
- The view can't include the UNION operator.

Create an updatable view

```
CREATE OR REPLACE VIEW balance_due_view AS
  SELECT vendor_name, invoice_number,
         invoice_total, payment_total, credit_total,
         invoice_total - payment_total - credit_total AS balance_due
  FROM vendors v JOIN invoices i ON v.vendor_id = i.vendor_id
  WHERE invoice_total - payment_total - credit_total > 0
```

Use the view to update data

```
UPDATE balance_due_view
SET credit_total = 300
WHERE invoice_number = '989319-497'
```

The response from the system

```
1 row updated.
```

A failed attempt to use the view to update a calculated column

```
UPDATE balance_due_view
SET balance_due = 0
WHERE invoice_number = '989319-497'
```

The response from the system

```
SQL Error: ORA-01733: virtual column not allowed here
```

Create a read-only view

```
CREATE OR REPLACE VIEW balance_due_view AS
  SELECT vendor_name, invoice_number,
         invoice_total, payment_total, credit_total,
         invoice_total - payment_total - credit_total AS balance_due
  FROM vendors v JOIN invoices i ON v.vendor_id = i.vendor_id
  WHERE invoice_total - payment_total - credit_total > 0
WITH READ ONLY
```

Description

- An *updatable view* is a view that can be used in an INSERT, UPDATE, or DELETE statement to update the data in a base table.
- A *read-only view* is a view that cannot be used to update the data in the base table. To create a read-only view, code WITH READ ONLY in the subquery restriction clause.
- The requirements for coding updatable views are more restrictive than for coding read-only views. Oracle must be able to unambiguously determine which base tables and columns are affected.
- You can't use WITH READ ONLY if your view contains an ORDER BY clause.

Figure 11-4 How to create updatable and read-only views

How to use WITH CHECK OPTION

When you create an updatable view, it is possible to use that view to modify rows such that they are no longer selected by the view. To avoid this, you can use the subquery restriction clause with the check-option constraint. To do that, you code the keywords WITH CHECK OPTION after the SELECT statement. Figure 11-5 shows an example of an updateable view that uses this constraint to prevent an update if it causes the row to be excluded from the view.

To illustrate how the check-option constraint works, the CREATE VIEW statement creates an updatable view named vendor_payment that joins data from the Vendors and Invoices tables and displays all invoices that have a balance due that's greater than zero. Then, the first UPDATE statement uses this view to modify the payment_date and payment_total columns for a specific invoice. This works because this UPDATE statement doesn't exclude the row from the view.

However, the second UPDATE statement would cause the balance due to become less than zero. That would violate the check-option constraint, so Oracle prevents this statement from running and returns an error instead. This can be a helpful way to provide an extra layer of data validation.

An updatable view that uses WITH CHECK OPTION

```
CREATE OR REPLACE VIEW vendor_payment AS
  SELECT vendor_name, invoice_number, invoice_date, payment_date,
         invoice_total, credit_total, payment_total
  FROM vendors v JOIN invoices i
    ON v.vendor_id = i.vendor_id
  WHERE invoice_total - payment_total - credit_total >= 0
WITH CHECK OPTION
```

Display a row from the view

```
SELECT * FROM vendor_payment
WHERE invoice_number = 'P-0608'
```

The result set

	VENDOR_NAME	INVOICE_NUMBER	INVOICE_DATE	PAYMENT_DATE	INVOICE_TOTAL	CREDIT_TOTAL	PAYMENT_TOTAL
1	Malloy Lithographing Inc	P-0608	11-APR-24	(null)	20551.18	1200	0

Use the view to update a row

```
UPDATE vendor_payment
SET payment_total = 400.00,
    payment_date = '01-AUG-24'
WHERE invoice_number = 'P-0608'
```

The response from the system

```
1 row updated.
```

The same row data after the update

	VENDOR_NAME	INVOICE_NUMBER	INVOICE_DATE	PAYMENT_DATE	INVOICE_TOTAL	CREDIT_TOTAL	PAYMENT_TOTAL
1	Malloy Lithographing Inc	P-0608	11-APR-24	01-AUG-24	20551.18	1200	400

A failed attempt to use the view to update a row

```
UPDATE vendor_payment
SET payment_total = 30000.00,
    payment_date = '01-AUG-24'
WHERE invoice_number = 'P-0608'
```

The response from the system

```
SQL Error: ORA-01402: view WITH CHECK OPTION where-clause violation
```

Description

- A change you make through a view can cause the modified rows to no longer be included in the view.
- If you specify WITH CHECK OPTION in the subquery restriction clause when you create a view, an error will occur if you try to modify a row in such a way that it would no longer be included in the view.
- You can't use WITH CHECK OPTION if your view contains an ORDER BY clause.

Figure 11-5 How to use WITH CHECK OPTION

How to insert or delete rows through a view

In the previous figures, you learned how to use a view to update data in the underlying tables. Now, figure 11-6 shows how to use a view to insert or delete data in an underlying view. In general, this works the same as it does for a table. However, due to constraints, using a view to insert or delete rows often results in errors like the ones shown in this figure. As a result, it's generally more common to work directly with base tables when inserting or deleting rows.

At the top of this figure, you can see a CREATE VIEW statement for a view named ibm_invoices. This view retrieves columns and rows from the Invoices table for the vendor with a vendor_id of 34. Then, the INSERT statement in the second example attempts to insert a row into the Invoices table through this view.

This insert operation fails, though, because the view and the INSERT statement don't include all of the required columns for the Invoices table. This illustrates that to be able to use a view to insert rows, you must design a view that includes all required columns for the underlying table.

In addition, an INSERT statement that uses a view can insert rows into only one table. That's true even if the view is based on two or more tables and all of the required columns for those tables are included in the view. In that case, you could use separate INSERT statements to insert rows into each table in the view.

The third example shows that foreign key constraints work the same way with views as they do with tables. The statement attempts to delete an invoice from the Invoices table through the ibm_invoices view. However, it fails because the invoice contains line items. This causes an error message like the one in this figure to be displayed. Instead, to get this DELETE statement to work you must first delete the related line items for the specified invoice before you use the ibm_invoices view to delete the invoice. This is shown by the last two DELETE statements in this figure.

Create an updatable view

```
CREATE OR REPLACE VIEW ibm_invoices AS
  SELECT invoice_number, invoice_date, invoice_total
  FROM invoices
  WHERE vendor_id = 34;
```

The virtual table for the view

	INVOICE_NUMBER	INVOICE_DATE	INVOICE_TOTAL
1	QP58872	25-FEB-24	116.54
2	Q545443	14-MAR-24	1083.58

An INSERT statement that fails due to columns with null values

```
INSERT INTO ibm_invoices
  (invoice_number, invoice_date, invoice_total)
VALUES
  ('RA23988', '31-JUL-24', 417.34)
```

The response from the system

```
SQL Error: ORA-01400: cannot insert NULL into ("AP"."INVOICES"."INVOICE_ID")
```

A DELETE statement that fails due to a foreign key constraint

```
DELETE FROM ibm_invoices
WHERE invoice_number = 'Q545443'
```

The response from the system

```
SQL Error: ORA-02292: integrity constraint (AP.LINE_ITEMS_FK_INVOICES)
violated - child record found
```

Two DELETE statements that succeed

```
DELETE FROM invoice_line_items
WHERE invoice_id = (SELECT invoice_id FROM invoices
                    WHERE invoice_number = 'Q545443');

DELETE FROM ibm_invoices
WHERE invoice_number = 'Q545443';
```

The response from the system

```
1 row deleted.
1 row deleted.
```

Description

- You can use the INSERT statement to insert rows into a base table through a view. Both the view and the INSERT statement must include all of the columns from the base table that require a value.
- If the view names more than one base table, an INSERT statement can insert data into only one of those tables.
- You can use the DELETE statement to delete rows from a base table through a view. For this to work, the view must be based on a single table.

Figure 11-6 How to insert or delete rows through a view

How to alter or drop a view

Although Oracle has an ALTER VIEW statement, it's not used to alter the definition of an existing view. Instead, it's used to recompile the view or to modify constraints. That's why if you want to change a view, you'll use the CREATE VIEW statement with the OR REPLACE keywords to replace the existing view with a new one.

In figure 11-7, for instance, the first example uses a CREATE VIEW statement to create a view named vendors_sw that retrieves rows from the Vendors table for vendors located in four states. Then, the second example uses OR REPLACE to modify this view so it includes vendors in two additional states and is read-only.

Alternately, you can first drop the view and then create a new one with the same name. To drop a view, you use the DROP VIEW statement to name the view you want to drop. In this figure, for instance, the third example drops the view named vendors_sw. Like the other statements for dropping database objects, this statement permanently deletes the view. As a result, you should be careful when you use it. And like the other drop statements, you can optionally code an IF EXISTS clause to tell Oracle to only execute the statement if a view with that name exists.

Create a view

```
CREATE VIEW vendors_sw AS
SELECT *
FROM vendors
WHERE vendor_state IN ('CA','AZ','NV','NM')
```

Replace the view with a read-only view

```
CREATE OR REPLACE VIEW vendors_sw AS
SELECT *
FROM vendors
WHERE vendor_state IN ('CA','AZ','NV','NM','UT','CO')
WITH READ ONLY
```

Drop the view

```
DROP VIEW vendors_sw
```

Description

- To change the definition of a view, use the CREATE VIEW statement with the OR REPLACE keywords to replace the existing view with a new one.
- To delete a view from the database, use the DROP VIEW statement.

Figure 11-7 How to alter or drop a view

How to use SQL Developer with views

Once you understand how to write SQL code that creates and drops views, it's easy to learn how to use SQL Developer to work with views. Figure 11-8 shows how.

To get information about an existing view, you can expand the Views folder in the Connections window to see a list of all views that are stored in a schema. In this figure, for example, the Views folder shows all of the views in the AP schema that were created by the SQL statements in this chapter. Here, the balance_due_view is selected and displayed in the main window, and the Columns tab presents information about the columns of this view.

If you want to view the data that's retrieved by a view, you can click on the Data tab. Then, if the view is updateable, you can use the buttons at the top of this tab to insert, update, and delete rows.

If you want to get other information about a view, you can click on the appropriate tab. In particular, you can click on the Grants tab to see the users that have been granted privileges for working with the view. You can click on the Dependencies tab to view the tables that the view depends on. You can view the Details tab to get miscellaneous information about the view. And you can click on the SQL tab to see the SQL code that was used to create the view.

To alter the design of an existing view, you can right-click on the view and select Edit. Or, to create a new view, you can right-click on the Views folder and select New View. The resulting dialog box allows you to directly modify the SQL code or to generate SQL code by selecting tables and columns from the graphical user interface.

Using SQL Developer to drop a view works the same as using SQL Developer to drop any other type of database object. Right-click the view and select Drop. Then, use the resulting dialog box to confirm the drop.

SQL Developer with the views for the AP schema displayed

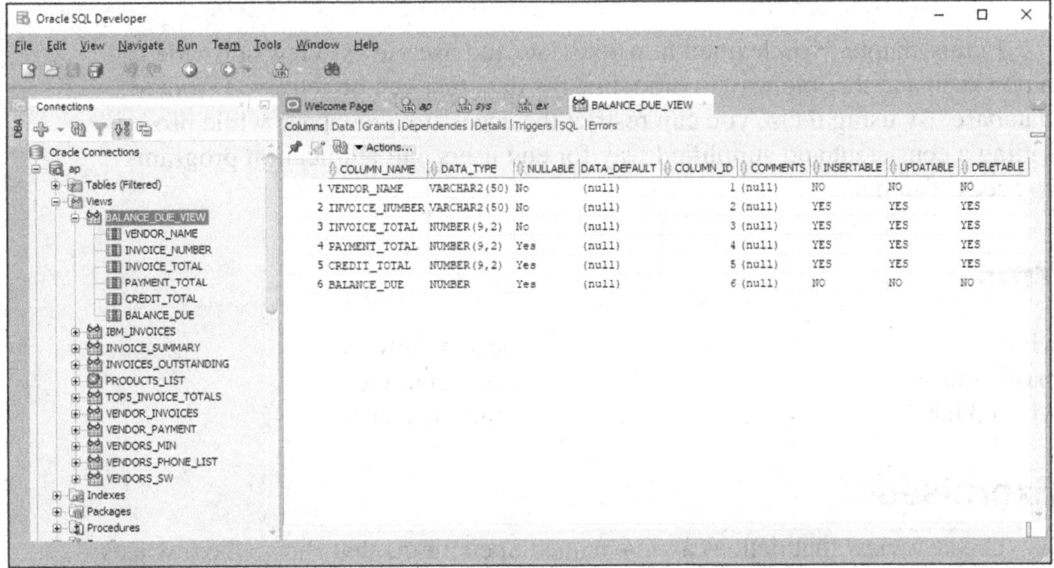

Description

- To examine the columns of a view, expand the Views folder and click on the view to display it in the main window. Then, if necessary, click on the Columns tab.
- To work with the data for a view, click on the view to display it in the main window, and click on the Data tab. Then, you can view the data for the table, and you can insert, update, and delete rows if the view is updateable.
- To get other information about a view, click on the view to display it in the main window. Then, click on the Grants, Dependencies, Details, or SQL tabs.
- To drop a view, right-click on the view and select Drop.
- To modify the design of an existing view, right-click on the view and select Edit.
- To create a new view, right-click on the Views folder and select New View.

Figure 11-8 How to use SQL Developer to work with views

Perspective

In this chapter, you learned how to create and use views. Views provide a powerful and flexible way to predefine the data that can be retrieved from a database. By using them, you can restrict the access to a database while providing a consistent and simplified way for end users and application programs to access that data.

Terms

view
base table
viewed table
nested view
updatable view
read-only view

Exercises

1. Create a view that defines a view named open_items that shows the invoices that haven't been paid. This view should return four columns from the Vendors and Invoices tables: vendor_name, invoice_number, invoice_total, and balance_due (invoice_total - payment_total - credit_total). However, a row should only be returned when the balance due is greater than zero, and the rows should be in sequence by vendor_name. Then, run the script to create the view, and use SQL Developer to review the data that it returns. (You may have to click on the Refresh button in the Connections window after you click on the Views node to show the view you just created.)
2. Write a SELECT statement that returns all of the columns in the open_items view that you created in exercise 1, with one row for each invoice that has a balance due of $1000 or more.
3. Create a view named open_items_summary that returns one summary row for each vendor that contains invoices with unpaid balances due. Each row should include vendor_name, open_item_count (the number of invoices with a balance due), and open_item_total (the total of the balance due amounts), and the rows should be sorted by the open item totals in descending sequence. Then, run the script to create the view, and use SQL Developer to review the data that it returns.
4. Write a SELECT statement that returns just the first 5 rows in the open_items_summary view that you created in exercise 3.
5. Create an updatable view named vendor_address that returns the vendor_id, both address columns, and the city, state, and zip code columns for each vendor. Then, use SQL Developer to review the data in this view.
6. Write an UPDATE statement that changes the address for the row with vendor ID 4 so the suite number (Ste 260) is stored in vendor_address2 instead of vendor_address1. Then, use SQL Developer to verify the change (you may need to click the Refresh button at the top of the Data tab to see the change). If this works correctly, go back to the tab for the UPDATE statement and click the Commit button to commit the change.

12

How to manage database security

If you are only working with sample databases like the ones in this book, security isn't a big concern. However, when you use Oracle in a production environment, it's imperative that you configure security to prevent misuse of your data. In this chapter, you'll learn how to do that by using SQL statements to create users and roles that have restricted access to your database. In addition, you'll learn how to use SQL Developer to perform many of these same security-related tasks.

How to work with users and roles 354
How to create a user.................................354
How to create an admin user.................................356
How to alter and drop a user.................................358
How to create and drop a role.................................360
How to work with privileges and synonyms.................................362
System privileges and object privileges.................................362
How to grant privileges.................................364
How to revoke privileges.................................366
How to work with synonyms.................................368
A script that creates roles and users.................................370
How to view the privileges for users and roles.................................374
How to use SQL Developer.................................376
How to create an admin connection for a PDB.................................376
How to work with users.................................378
How to grant and revoke roles.................................380
How to grant and revoke system privileges.................................382
Perspective.................................384

How to work with users and roles

The heart of database security is making sure authorized users can access the data they need to while preventing unauthorized users from seeing any data at all. To achieve that, you can create users and roles with specific levels of access to the database.

How to create a user

Figure 12-1 begins with the simplified syntax for creating a new database user. You begin by coding the CREATE USER keywords followed by the name of the new user. Then, you can code the IDENTIFIED clause.

The IDENTIFIED clause can specify authentication one of three ways. First, it can use a password. Second, it can specify authentication externally via an SSL certificate or Kerberos principal. Third, it can specify authentication globally through an enterprise directory service, Microsoft Azure login, or Amazon IAM identity. This chapter shows how to use password authentication since it doesn't require any additional services, but you should be aware that other options are available.

After the IDENTIFIED clause, you can code several optional clauses to specify attributes for the new user. These clauses allow you to force password expiration, lock or unlock the user's account, and set default and temporary tablespaces. In addition, they allow you to specify a profile and a disk storage quota.

Creating a user can be simple, as shown by the first example in this figure. This statement creates a user named John with a password of 'sesame'. Because no other clauses are included, the defaults are applied. Unfortunately, that's often not what you want. For example, the default tablespace on many systems is the System tablespace, which should be reserved for Oracle's internal use only. Furthermore, the user has not been allocated any disk storage, so they can't create any objects like tables.

The second statement in this figure also shows how to create a user named Joel with a password of 'sesame'. However, this password will expire the first time he uses it, forcing him to change it. His default tablespace is Users and his temporary tablespace is Temp. Within the Users tablespace, his *quota* specifies that he can use a maximum of 100 megabytes of storage.

If you are connected to the PDB when you run a CREATE USER statement, Oracle creates a *local user*, which means that the user is stored in the PDB and can only access objects in that PDB, not objects in other PDBs or the CDB. However, if you are connected to the CDB root when you run the CREATE USER statement, Oracle creates a *common user*, which means that the user is stored in the CDB and can access its objects including objects in its PDBs.

With Oracle 23ai or later, you can code an IF NOT EXISTS clause to only create the new user if a user with that name doesn't already exist like this:

```
CREATE USER IF NOT EXISTS username IDENTIFIED BY password
```

The simplified syntax of the CREATE USER statement

```
CREATE USER username
{IDENTIFIED {BY password|EXTERNALLY|GLOBALLY}}
[PASSWORD EXPIRE]
[ACCOUNT {LOCK|UNLOCK}]
[DEFAULT  TABLESPACE tablespace_name]
[TEMPORARY TABLESPACE tablespace_name]
[PROFILE profile_name]
[QUOTA {quota_size|UNLIMITED} ON tablespace_name]
```

Three tablespaces created when Oracle is installed

Tablespace	Stores
SYSTEM	Permanent objects that are used internally by Oracle
USERS	Users and other permanent objects such as tables and views
TEMP	Temporary objects

Create a user

```
CREATE USER john IDENTIFIED BY sesame
```

Create a user with additional specifications

```
CREATE USER joel IDENTIFIED BY sesame
PASSWORD EXPIRE
DEFAULT TABLESPACE users
TEMPORARY TABLESPACE temp
QUOTA 100M ON users
```

Description

- If you're connected to the CDB root, users you create are *common* by default, which means they are stored in the container database and can access its objects, including its PDBs and their objects.
- If you're connected to a PDB, users you create are *local* by default, which means they are stored in that PDB and can only access its objects, not objects from other PDBs or the container database.
- To force a user to change their password the first time they log in, use PASSWORD EXPIRE.
- If you don't specify a default tablespace or a temporary tablespace for a user, the database default tablespaces are used.
- A *quota* limits the amount of disk storage the user can use in the tablespace. If you don't want to place a limit on the user's disk space, you can specify UNLIMITED.

Figure 12-1 How to create a user

How to create an admin user

Figure 12-2 begins by presenting a script that creates an administrative (or admin) user who has all privileges for working with the objects of a pluggable database. This script is designed to work with Oracle XE running on a local system, but the concepts are similar for a database running in the cloud. To run this script, you can use the sys connection or another connection with the sysdba role.

In this script, the first statement sets the container for the current session to the PDB named XEPDB1. As a result, the CREATE USER statement creates the user as a local user in this PDB. Then, the script grants all privileges to this user. As a result, this user will be able to execute DDL statements to create tables, views, and other database objects. In addition, this user will be able to execute DML statements such as the SELECT, INSERT, UPDATE, and DELETE statements.

After this script grants all privileges to the AR user, it connects as the new user to the same PDB using a CONNECT statement. However, there are two caveats to using CONNECT this way. The first is that if you're using SQL Developer, the worksheet's connection will revert to the selected connection after the script finishes executing. So, if you run the statements in a script one by one, SQL Developer uses the worksheet's connection, not the connection specified by the CONNECT statement.

The second caveat to using CONNECT is that anyone who has access to the script can see the name and password of the user. That's usually something you want to avoid for obvious security reasons. However, for learning purposes, CONNECT is a convenient way to switch between users in a single script.

After connecting as the AR user, the script uses the CREATE TABLE statement to create a table named Customers. Since this statement doesn't specify a schema for the table, the table is stored in a schema with the same name as the user. As a result, the Customers table is stored in the AR schema, which is stored in the Users tablespace in the PDB named XEPDB1.

After creating the table, this script uses three INSERT statements to insert three rows into the table, followed by a SELECT statement to select all data from the Customers table. Since the script is still connected as the AR user, there's no need to qualify the table name with the schema name in this statement.

A script that creates an admin user and a table

```
-- use the sys connection to run this script

-- set the container for the current session to the XEPDB1 PDB
ALTER SESSION SET CONTAINER = xepdb1;

-- Create the AR user and grant it all privileges
CREATE USER ar IDENTIFIED BY sesame DEFAULT TABLESPACE users;
GRANT ALL PRIVILEGES TO ar;

-- Connect as the AR user
CONNECT ar/sesame@localhost:1521/xepdb1;

-- Create the Customers table in the AR schema
CREATE TABLE customers
(
  customer_id            NUMBER          NOT NULL,
  customer_first_name    VARCHAR2(50)    NOT NULL,
  customer_last_name     VARCHAR2(50)    NOT NULL,
  CONSTRAINT customers_pk
    PRIMARY KEY (customer_id)
);

INSERT INTO customers VALUES (1, 'Karen', 'Beecher');
INSERT INTO customers VALUES (2, 'Stephanie', 'Brown');
INSERT INTO customers VALUES (3, 'Jessica', 'Cruz');

SELECT * FROM customers;      -- select from AR schema
```

Description

- In a script, you can use the ALTER SESSION command to set the container for the session to the PDB that you want to work with.
- In a script, you can use the CONNECT statement to connect to a database. However, the connection will revert to the worksheet's selected connection after the script finishes.
- You can use the GRANT ALL PRIVILEGES statement to grant all privileges to an administrative, or admin, user. You should not do this for regular users.
- If a user adds a table or other database object without specifying a schema, the database object is stored in a schema with the same name as the user.

Figure 12-2 How to create an admin user

How to alter and drop a user

You may want to change a user account after you create it. To do so, you use the ALTER USER statement. As the examples in figure 12-3 show, this statement works much like the CREATE USER statement.

The first and second statements in this figure show how to change the amount of disk space that's allocated to a user. Here, the first statement changes the amount of disk space for Joel from 100 megabytes to 10. Then, the second statement changes the amount of disk space for Joel to 0 bytes. As a result, Joel won't be able to create any more database objects. However, any existing objects that were created by Joel will remain in the database.

The third and fourth statements show how to work with a user's password. The third statement changes the password for Joel to 'secret'. Then, the fourth statement causes the password for Joel to expire. As a result, the next time Joel logs in, he'll have to enter the old password and then immediately create a new password.

The fifth and sixth statements show how to lock or unlock a user's account. If, for example, you want to temporarily disable Joel's access and lock him out of the database, you can use the fifth statement. Then, to re-enable his access, you can use the sixth.

The seventh statement shows how to change the profile assigned to a user. A *profile* is a set of resource limits that specify things like how long a user may be connected to a database before being timed out, the maximum number of failed login attempts allowed before automatically locking an account, the maximum number of simultaneous sessions allowed per user, and so on.

This figure summarizes three of Oracle's built-in profiles that you can assign to users. The DEFAULT profile contains the system defaults, the ORA_CIS_PROFILE complies with the Center for Internet Security's (CIS) guidelines, and the ORA_STIG_PROFILE complies with the Security Technical Implementation Guide (STIG). If you don't specify a profile when you create a user, Oracle assigns the DEFAULT profile to the user.

Finally, this figure shows how to drop a user. If the user hasn't created any tables or other database objects, you can use the DROP USER statement without the CASCADE keyword as shown by the eighth statement. However, if the user has created any tables or other database objects, Oracle would display an error message and not drop the user. Since this prevents you from accidentally dropping database objects, this is usually what you want. But, if you're sure you want to drop the user and all database objects stored in the user's schema, you can specify the CASCADE keyword as shown in the last statement.

With Oracle 23ai or later, you can code an IF EXISTS clause to only alter the user if a user with that name exists like this:

```
ALTER USER IF EXISTS username
[new_specifications]
```

Built-in profiles

Profile	Description
DEFAULT	Use the system profile defaults.
ORA_CIS_PROFILE	Complies with the Center for Internet Security's (CIS) requirements.
ORA_STIG_PROFILE	Complies with the Security Technical Implementation Guide's (STIG) requirements.

The simplified syntax of the ALTER USER statement

```
ALTER USER username
[new_specifications]
```

Change a user's tablespace allocation quota

```
ALTER USER joel QUOTA 10M ON users
```

Don't allow the user to use more space in the tablespace

```
ALTER USER joel QUOTA 0 ON users
```

Change a user's password

```
ALTER USER joel IDENTIFIED BY secret
```

Force the user to change their password

```
ALTER USER joel PASSWORD EXPIRE
```

Lock a user's account to disable access

```
ALTER USER joel ACCOUNT LOCK
```

Unlock a user's account to enable access

```
ALTER USER joel ACCOUNT UNLOCK
```

Change a user's profile

```
ALTER USER joel PROFILE ORA_CIS_PROFILE
```

Drop a user

```
DROP USER joel
```

Drop a user and all of their created objects

```
DROP USER ar CASCADE
```

Description

- Most clauses of the ALTER USER statement are identical to the clauses of the CREATE USER statement.
- A *profile* specifies resource limits for a user. A user may only be assigned one profile.
- You use the DROP USER statement to drop a user. If you use the optional CASCADE keyword, it first deletes any objects created by the user.

Figure 12-3 How to alter and drop a user

How to create and drop a role

In addition to individual users, you can create roles in Oracle Database. A *role* is a set of privileges you can grant to users. If, for example, you're working with a database that contains multiple users, you can use roles to grant similar privileges to similar users. You can create roles for developers, administrators, managers, application users, and so on. A user can have more than one role, and a role can even be assigned to another role.

To create a role, you use the CREATE ROLE statement that's shown in figure 12-4. After you create a role, you can grant privileges to it, and you can grant the role to a user as shown in the next figure. Later, if you want to drop a role, you can use the DROP ROLE statement shown in this figure.

The syntax of the CREATE ROLE statement

```
CREATE ROLE role_name
[CONTAINER = {CURRENT|ALL}]
```

Create a role

```
CREATE ROLE ap_user
```

Drop a role

```
DROP ROLE ap_user
```

Description

- You can use the CREATE ROLE statement to create a role. A *role* is a set of privileges you can grant to users.
- A user can have multiple roles.
- A role can be assigned to another role.
- A newly created role does not have any privileges.
- Like users, a role may be common (stored in the container database) or local (stored in a specific PDB).
- You can use the DROP ROLE statement to drop roles.

Figure 12-4 How to create and drop a role

How to work with privileges and synonyms

Now that you know how to create users and roles, you're ready to learn the details for coding SQL statements that assign privileges to users and roles. In addition, you're ready to learn how to use synonyms to make it easier to access tables that are stored in another user's schema.

System privileges and object privileges

Figure 12-5 lists some of the privileges that you can grant to or revoke from users or roles. To start, *system privileges* allow a user to connect to a database and create, alter, or drop database objects such as tables, sequences, views, and stored procedures.

Although this figure doesn't show them all, Oracle provides system privileges for most types of DDL statements. For example, Oracle provides a CREATE INDEX privilege, a CREATE TRIGGER privilege, and so on. To see a complete list of privileges, you can look up the GRANT statement in the Oracle documentation.

For most privileges, you can use the ANY keyword to expand the privilege to any schema in the database except SYS and AUDSYS. For example, if you grant the CREATE TABLE privilege to a user while connected to the AR schema, the user can only create tables in the AR schema. However, if you grant the CREATE ANY TABLE privilege, the user can create a table in any schema on the database except SYS and AUDSYS.

For most CREATE privileges, there are corresponding DROP and ALTER privileges. For example, Oracle provides a DROP TABLE privilege and an ALTER TABLE privilege. Like the CREATE TABLE privilege, you can use the ANY keyword to apply these privileges to any schema in the database except SYS and AUDSYS. Otherwise, they will only apply to the schema of the user granting the privilege.

While system privileges allow a user to create schema objects, *object privileges* allow a user to use database objects such as tables, views, sequences, and stored procedures after they have been created. In other words, object privileges allow a user to execute DML statements such as the SELECT, UPDATE, INSERT, and DELETE statements. In addition, the EXECUTE privilege lets the user run a stored procedure or a function.

When you work with object privileges, you'll find that some privileges can be granted only for certain types of objects. For example, you can grant a SELECT privilege only to an object from which you can select data, such as a table or view. Likewise, you can grant an EXECUTE privilege only to an object that you can execute, such as a stored procedure or function.

Some of the system privileges

Privilege	Lets the user...
`CREATE SESSION`	Connect to the database.
`CREATE TABLE`	Create a table in the current schema.
`CREATE ANY TABLE`	Create a table in any schema.
`DROP TABLE`	Drop a table in the current schema.
`DROP ANY TABLE`	Drop a table in any schema.
`UNLIMITED TABLESPACE`	Create objects such as tables without regard to tablespace quotas.
`CREATE SEQUENCE`	Create a sequence in the current schema.
`CREATE VIEW`	Create a view in the current schema.
`CREATE PROCEDURE`	Create a stored procedure, function, or package in the current schema.
`CREATE PUBLIC SYNONYM`	Create a synonym in the current schema that's available to all users.

Some of the object privileges

Privilege	Lets the user...
`SELECT`	Select data from the specified object. This applies to tables, sequences, and views.
`UPDATE`	Update data. This applies to tables and views.
`INSERT`	Insert data. This applies to tables and views.
`DELETE`	Delete data. This applies to tables and views.
`EXECUTE`	Execute a stored procedure or a function.

Description

- *System privileges* allow the user to connect to a database and to create, alter, or drop database objects such as tables, views, sequences, and stored procedures.
- *Object privileges* allow the user to work with database objects such as tables, views, sequences, and stored procedures.
- The privileges available for an object depend on the type of object.
- For a complete list of privileges, look up the GRANT statement in the Oracle documentation.

Figure 12-5 System privileges and object privileges

How to grant privileges

Figure 12-6 shows how to use the GRANT statement to grant system and object privileges to a role or a user. To start, the first two statements show how to grant system privileges. Here, the first statement grants the CREATE SESSION privilege to the role named ap_user. Then, the second statement grants the DROP ANY VIEW privilege to the role named ap_developer. In addition, since this statement includes the WITH ADMIN OPTION clause, any user with the ap_developer role can grant the specified privilege to other users.

The third and fourth statements show how to grant object privileges. Here, the third statement grants the SELECT privilege on the Vendors table to the ap_user role, and the fourth statement grants the same privilege to the user named John. However, since this fourth statement includes the WITH GRANT OPTION clause, John can grant the specified privilege to other users.

The fifth statement shows that you can grant multiple privileges by separating the privileges with commas. This statement grants the SELECT, INSERT, UPDATE, and DELETE privileges on the Invoices table to the ap_user role. Although it isn't shown, you can use a similar technique for granting multiple system privileges. You can also grant privileges to multiple roles and users by separating them with commas.

The sixth statement shows how to use the ALL keyword to grant all related privileges for the specified database object. This statement grants all privileges on the Invoices table to the ap_user role, including all of the privileges that are granted by the fifth example: SELECT, INSERT, UPDATE, and DELETE. In addition, this statement grants several other privileges such as FLASHBACK and DEBUG.

The advantage of using the ALL keyword is that you write less code to grant all privileges on an object. The disadvantage is that you may accidentally grant a user more privileges than the user needs, which could open a hole in the security of your database.

Granting ALL may cause unintended side effects. For example, if you grant ALL permissions for a user on a table and then revoke SELECT permissions, the user may still be able to select data from the table. This is due to the permissions granted by ALL superseding the revoked SELECT permission.

Before you can grant privileges, you must connect as an appropriate user. In this figure, for instance, the examples assume that you are connected as the AP user for two reasons. First, this user has the necessary privileges to grant all system and object privileges to other roles and users. Second, the database objects shown in this figure are stored in the schema for this user. As a result, you don't need to qualify the names of these objects with the schema name. In this figure, that means that you don't need to qualify the Vendors and Invoices tables with the AP schema. However, if you weren't connected as the AP user, you would need to qualify these object names with the schema name.

Finally, note that you can grant privileges to specific columns for an object that has columns, such as a table. However, it's usually easier to create a view that provides access to the columns that you want to grant to the user. Then, you can grant privileges on the view but not on the table.

The simplified syntax of the GRANT statement for system privileges

```
GRANT system_privilege[, ...]
TO user_or_role [, ...]
[WITH ADMIN OPTION]
```

The simplified syntax of the GRANT statement for object privileges

```
GRANT object_privilege[, ...]
ON [schema_name.]object_name [(column [, ...])]
TO user_or_role [, ...]
[WITH GRANT OPTION]
```

Grant a system privilege to a role

```
GRANT CREATE SESSION TO ap_user
```

Grant a system privilege with the admin option

```
GRANT DROP ANY VIEW TO ap_developer WITH ADMIN OPTION
```

Grant an object privilege to a role

```
GRANT SELECT ON vendors TO ap_user
```

Grant an object privilege to a user with the grant option

```
GRANT SELECT ON vendors TO john WITH GRANT OPTION
```

Grant multiple object privileges for a table to a role

```
GRANT SELECT, INSERT, UPDATE, DELETE ON invoices TO ap_user
```

Grant all object privileges for a table to a role

```
GRANT ALL ON invoices TO ap_user
```

Grant a role to multiple users

```
GRANT ap_user TO john, juanita
```

Grant a role to another role

```
GRANT ap_user TO ap_developer
```

Description

- You can use the GRANT statement to grant system or object privileges to users or roles.
- The WITH ADMIN OPTION clause allows the user or role to grant the specified system privileges to other users or roles.
- The WITH GRANT OPTION clause allows the user or role to grant the specified object privileges to other users or roles.

Figure 12-6 How to grant privileges

How to revoke privileges

Figure 12-7 shows how to use the REVOKE statement to revoke system or object privileges. As you can see, this statement is similar to the GRANT statement, but it reverses the action of a GRANT statement.

To start, the first REVOKE statement in this figure shows how to revoke a system privilege. This statement revokes the DROP ANY VIEW privilege from the role named ap_developer.

In contrast, the second statement shows how to revoke an object privilege. This statement revokes the SELECT privilege on the Invoices table from the ap_user role.

The third statement shows that you can revoke multiple object privileges by separating each privilege with a comma. This statement revokes the INSERT, UPDATE, and DELETE privileges on the Invoices table from the ap_user role.

The fourth statement shows how to use the ALL keyword to revoke all privileges on a database object. This works similarly to the previous two statements. However, it also revokes any other privileges that may have been granted on the Invoices table from the ap_user role. If you need to make sure that you've revoked all privileges on an object, this is the way to do it.

The fifth statement shows how to revoke a role from multiple users. In particular, it revokes the ap_users role from the users named John and Juanita.

The sixth statement shows how to revoke one role from another role. Specifically, it revokes the ap_users role from the ap_developer role. As a result, the ap_developer role will no longer have the privileges that have been granted to the ap_user role.

The simplified syntax of the REVOKE statement for system privileges

```
REVOKE system_privilege [, ...]
FROM user_or_role [, ...]
```

The simplified syntax of the REVOKE statement for object privileges

```
REVOKE object_privilege
ON [schema_name.]object_name [(column [, ...])]
FROM user_or_role [, ...]
```

Revoke a system privilege from a role

```
REVOKE DROP ANY VIEW FROM ap_developer
```

Revoke an object privilege from a role

```
REVOKE SELECT ON invoices FROM ap_user
```

Revoke multiple object privileges from a role

```
REVOKE INSERT, UPDATE, DELETE ON invoices FROM ap_user
```

Revoke all object privileges for a table from a role

```
REVOKE ALL ON invoices FROM ap_user
```

Revoke a role from multiple users

```
REVOKE ap_user FROM john, juanita
```

Revoke a role from another role

```
REVOKE ap_user FROM ap_developer
```

Description

- You can use the REVOKE statement to revoke privileges from a user or role.

Figure 12-7 How to revoke privileges

How to work with synonyms

Figure 12-8 shows how to work with *synonyms*. A synonym is an alias for a table, view, or other database object. A synonyms may be *private*, which means it's only available to the user who creates it, or *public*, which means that it's available to all users of the database.

The primary benefit of using a synonym is that it allows applications to work without modification regardless of which schema a table or view is stored in. To illustrate, the first example in this figure shows what happens when a user named John tries to use the Vendors table in the AP schema without qualifying the table name. In this case, an error message is displayed and the statement fails.

To solve this problem, John creates a private synonym as shown by the second example. This example assumes John already has the CREATE SYNONYM privilege and the SELECT privilege on the Vendors table. After creating the synonym, the statement in the third example succeeds because "vendors" is now a synonym for "ap.vendors". However, if another user attempted to use the synonym, the statement would still fail.

The fourth example shows how to drop a private synonym. This works much like dropping any other database object.

Alternately, instead of a private synonym, you can create a public synonym by using the PUBLIC keyword as shown in the fifth example. This example assumes you are using the AP connection and thus have all object privileges for the schema. Then, in the fifth example, the EX user is also able to access the Vendors table through the synonym. Note that this works because the EX user already has the SELECT privilege for the Vendors table.

The last statement in this figure shows how to drop a public synonym. This is the same as dropping a private synonym except you include the PUBLIC keyword.

The simplified syntax of the CREATE SYNONYM statement

```
CREATE [PUBLIC] SYNONYM synonym_name
FOR [schema.]object_name
```

A statement that fails because it doesn't specify the schema

```
-- connect as John
SELECT vendor_name FROM vendors WHERE vendor_id = 1
```

The error message that's displayed

```
SQL Error: ORA-00942: table or view does not exist
```

Create a private synonym

```
-- connect as John
CREATE SYNONYM vendors FOR ap.vendors
```

A statement that succeeds due to the private synonym

```
-- connect as John
SELECT vendor_name FROM vendors WHERE vendor_id = 1
```

The result set

	VENDOR_NAME
1	US Postal Service

Drop a private synonym

```
DROP SYNONYM vendors
```

Create a public synonym

```
-- connect as AP
CREATE PUBLIC SYNONYM vendors FOR ap.vendors
```

A statement that succeeds due to the public synonym

```
-- connect as EX
SELECT vendor_name FROM vendors WHERE vendor_id = 1
```

Drop a public synonym

```
-- connect as AP
DROP PUBLIC SYNONYM vendors
```

Description

- A *synonym* is an alias for a table, view, or other schema object.
- Synonyms allow applications to work without modification regardless of which user owns the table or view.
- A *private synonym* is a synonym that's only available to the user who creates it.
- *A public synonym* is a synonym that's available to all users.
- Before a user can use a synonym, the user must have appropriate privileges on the underlying object.

Figure 12-8 How to work with synonyms

A script that creates roles and users

Figure 12-9 shows a script that creates roles and users that can work with the tables and other database objects in the AP schema. To start, this script connects as the AP user. Then, it uses an anonymous PL/SQL block to drop any existing users, roles, or synonyms that are created by the script. This PL/SQL drops the users, roles, and synonyms if they exist and also suppresses any error messages that would be displayed if they don't exist.

With Oracle 23ai or later, you don't need to use the PL/SQL block to suppress the error messages. Instead, you can code regular DROP statements with the IF EXISTS clause like this:

```
DROP USER IF EXISTS john;
DROP USER IF EXISTS juanita;
...
```

After the script drops any objects that will be created by the script, it creates three roles: ap_user, ap_manager, and ap_developer. Then, the script grants privileges to these three roles.

The ap_user role has the fewest privileges. It can connect to the database. It can select and modify data in the Vendors, Invoices, and Invoice_Line_Items tables. It can select data from the Terms and General_Ledger_Accounts tables. And it can use the sequences for the Vendors and Invoices tables. In short, this role has all of the privileges needed for a user to work with an accounts payable application.

The ap_manager role has all of the privileges of the ap_user role and a few more. It can grant the ap_user role to other users, and it can modify data in the Terms and General_Ledger_Accounts tables. In short, this role has all of the privileges needed for a manager to perform managerial functions with an accounts payable application.

A script that creates roles and users

Page 1

```
-- connect as AP user

-- use an anonymous PL/SQL block to
-- drop all end users, roles, and synonyms in the current database
-- and suppress any error messages that may be displayed
-- if these objects don't exist
BEGIN
  EXECUTE IMMEDIATE 'DROP USER john';
  EXECUTE IMMEDIATE 'DROP USER juanita';
  EXECUTE IMMEDIATE 'DROP USER jiro';
  EXECUTE IMMEDIATE 'DROP USER joel CASCADE';

  EXECUTE IMMEDIATE 'DROP ROLE ap_user';
  EXECUTE IMMEDIATE 'DROP ROLE ap_manager';
  EXECUTE IMMEDIATE 'DROP ROLE ap_developer';

  EXECUTE IMMEDIATE 'DROP PUBLIC SYNONYM vendors';
  EXECUTE IMMEDIATE 'DROP PUBLIC SYNONYM invoices';
  EXECUTE IMMEDIATE 'DROP PUBLIC SYNONYM invoice_line_items';
  EXECUTE IMMEDIATE 'DROP PUBLIC SYNONYM general_ledger_accounts';
  EXECUTE IMMEDIATE 'DROP PUBLIC SYNONYM terms';
EXCEPTION
  WHEN OTHERS THEN
    DBMS_OUTPUT.PUT_LINE('');
END;
/

-- create the roles
CREATE ROLE ap_user;
CREATE ROLE ap_manager;
CREATE ROLE ap_developer;

-- grant privileges to the ap_user role
GRANT CREATE SESSION TO ap_user;
GRANT ALL ON vendors TO ap_user;
GRANT SELECT, INSERT, UPDATE, DELETE ON invoices TO ap_user;
GRANT SELECT, INSERT, UPDATE, DELETE ON invoice_line_items TO ap_user;
GRANT SELECT ON general_ledger_accounts TO ap_user;
GRANT SELECT ON terms TO ap_user;
GRANT SELECT ON invoice_id_seq TO ap_user;
GRANT SELECT ON vendor_id_seq TO ap_user;

-- grant privileges to the ap_manager role
GRANT ap_user TO ap_manager WITH ADMIN OPTION;
GRANT ALL ON general_ledger_accounts TO ap_manager;
GRANT ALL ON terms TO ap_manager;
```

Figure 12-9 A script that creates roles and users (part 1 of 2)

The ap_developer role has all of the privileges of the ap_manager role and a few more. It can create and drop tables, views, and sequences from any schema in the PDB. Although this role doesn't have all the privileges that are available to the AP user, it has enough privileges for a developer to work with the tables, views, and sequences of the Accounts Payable application.

After this script grants privileges to the roles, it creates some users named John, Juanita, Jiro, and Joel. It assigns a password of sesame to all of them, and it sets the Users tablespace as the default tablespace. As a result, these users and any objects that they create will be stored in the Users tablespace.

After this script creates the users, it assigns roles to each. Here, John and Juanita are assigned the ap_user role, Jiro is assigned the ap_manager role, and Joel is assigned the ap_developer role.

After this script assigns roles for the users, it alters the user named Joel so he can allocate up to 10MB of disk space in the Users tablespace. Without this allocation, Joel wouldn't be able to create tables or views or other objects that allocate disk space even though the ap_developer role has the CREATE ANY TABLE and CREATE ANY VIEW privileges.

After this script sets the quota for Joel, it creates public synonyms for the tables in the AP schema. Since these synonyms are public, they are available to Joel, John, Juanita, Jiro, and any other users of the PDB that are created later.

Finally, this script alters the password for each user so it is expired. This forces each user to change their password when they connect to the database for the first time.

Generally speaking, you don't need to use ALTER USER statements in the same script that creates the users you want to alter. This is because you can include options like QUOTA and PASSWORD EXPIRE in the CREATE USER statement. However, using ALTER USER can make the statements in a script easier to follow, and it can make it easier to comment out, or skip, options for testing purposes.

For example, in this script, you may want to comment out the ALTER USER statements that cause the password to expire. Then, you can use the passwords specified by the CREATE USER statement to test the new user accounts. When you code your own scripts, you can use whichever method makes the most sense for your purposes.

A script that creates roles and users **Page 2**

```
-- grant privileges to the ap_developer role
GRANT
  ap_manager,
  CREATE ANY TABLE,
  DROP ANY TABLE,
  CREATE ANY VIEW,
  DROP ANY VIEW,
  CREATE ANY SEQUENCE,
  DROP ANY SEQUENCE
TO ap_developer;

-- create the users
CREATE USER john IDENTIFIED BY sesame DEFAULT TABLESPACE users;
CREATE USER juanita IDENTIFIED BY sesame DEFAULT TABLESPACE users;
CREATE USER jiro IDENTIFIED BY sesame DEFAULT TABLESPACE users;
CREATE USER joel IDENTIFIED BY sesame DEFAULT TABLESPACE users;

-- assign roles to the users
GRANT ap_user TO john, juanita;
GRANT ap_manager TO jiro;
GRANT ap_developer TO joel;

-- allow joel to create tables
ALTER USER joel QUOTA 10M ON users;

-- create synonyms for all users
CREATE PUBLIC SYNONYM vendors FOR ap.vendors;
CREATE PUBLIC SYNONYM invoices FOR ap.invoices;
CREATE PUBLIC SYNONYM invoice_line_items FOR ap.invoice_line_items;
CREATE PUBLIC SYNONYM general_ledger_accounts FOR
                      ap.general_ledger_accounts;
CREATE PUBLIC SYNONYM terms FOR ap.terms;

-- require the users to change their passwords when they log in
ALTER USER john PASSWORD EXPIRE;
ALTER USER juanita PASSWORD EXPIRE;
ALTER USER jiro PASSWORD EXPIRE;
ALTER USER joel PASSWORD EXPIRE;
```

Description

- When you create a database, you can use a script to create the users and roles and assign privileges to them.
- A script like this provides a record of the privileges initially assigned to each user and role, and can be modified or reused as convenient.

Figure 12-9 A script that creates roles and users (part 2 of 2)

How to view the privileges for users and roles

Oracle provides a number of views that you can use to view information about users, roles, and privileges. Some of these views are summarized at the top of figure 12-10.

In the first example, the user connects as the AP user and grants one system privilege and one object privilege directly to the user named John. That way, the second and third examples return rows for these privileges when you view the privileges for John.

The second example shows how to view system privileges that have been granted directly to the current user, but not any privileges granted via roles. Assuming you are connected to the database as John, this statement returns only the CREATE PROCEDURE privilege that was granted by the first example.

The third example shows how to view object privileges that have been granted to the current user, again excluding privileges granted through roles. When connected as John, this statement returns the SELECT privilege on the Vendors table that was granted to John by the first example.

The fourth example shows how to view the roles assigned to the current user. In this case, it shows that John has the ap_user role. Then, the fifth example shows how to view the system privileges granted by a specific role by using the ROLE_SYS_PRIVS view with a WHERE clause. Note that for the WHERE clause to work, you must use all caps when coding the name of the role. Otherwise, Oracle won't return any data.

Finally, the sixth example displays all object privileges John has been granted via roles. Here, the result set shows the first five rows of the object privileges that were granted to the ap_user role by the script shown in the previous figure.

Some of Oracle's default views for user information

View	Shows data for...
DBA_USERS	All database users
USER_USERS	The current user only
USER_RESOURCE_LIMITS	Resource limits for the current user
USER_SYS_PRIVS	System privileges for the current user (not including those from roles)
USER_TAB_PRIVS	Object privileges for the current user (not including those from roles)
USER_ROLE_PRIVS	Roles for the current user
ROLE_SYS_PRIVS	System privileges for roles (limited to roles the user has access to)
ROLE_TAB_PRIVS	Object privileges for roles (limited to roles the user has access to)

Grant system and object privileges to a user

```
-- connect as AP user
GRANT CREATE PROCEDURE TO john;      -- system privilege
GRANT SELECT ON vendors TO john;     -- object privilege
```

View system privileges granted directly to the current user

```
-- connect as John
SELECT * FROM user_sys_privs
```

	USERNAME	PRIVILEGE	ADMIN_OPTION	COMMON	INHERITED
1	JOHN	CREATE PROCEDURE	NO	NO	NO

View object privileges granted directly to the current user

```
SELECT * FROM user_tab_privs
```

	GRANTEE	OWNER	TABLE_NAME	GRANTOR	PRIVILEGE	GRANTABLE	HIERARCHY	COMMON	TYPE	INHERITED
1	JOHN	AP	VENDORS	AP	SELECT	NO	NO	NO	TABLE	NO
2	PUBLIC	SYS	JOHN	JOHN	INHERIT PRIVILEGES	NO	NO	NO	USER	NO

View roles for the current user

```
SELECT * FROM user_role_privs
```

	USERNAME	GRANTED_ROLE	ADMIN_OPTION	DELEGATE_OPTION	DEFAULT_ROLE	OS_GRANTED	COMMON	INHERITED
1	JOHN	AP_USER	NO	NO	YES	NO	NO	NO

View all system privileges for the ap_user role

```
SELECT * FROM role_sys_privs WHERE role = 'AP_USER'
```

	ROLE	PRIVILEGE	ADMIN_OPTION	COMMON	INHERITED
1	AP_USER	CREATE SESSION	NO	NO	NO

View all object privileges granted through roles to the current user

```
SELECT * FROM role_tab_privs
```

	ROLE	OWNER	TABLE_NAME	COLUMN_NAME	PRIVILEGE	GRANTABLE	COMMON	INHERITED
1	AP_USER	AP	VENDORS	(null)	ALTER	NO	NO	NO
2	AP_USER	AP	VENDORS	(null)	DELETE	NO	NO	NO
3	AP_USER	AP	INVOICES	(null)	DELETE	NO	NO	NO
4	AP_USER	AP	INVOICE_LINE_ITEMS	(null)	DELETE	NO	NO	NO
5	AP_USER	AP	VENDORS	(null)	INSERT	NO	NO	NO

(22 rows)

Figure 12-10 How to view the privileges for users and roles

How to use SQL Developer

Since you often use a script to set up the security for a database or to view the privileges that have been granted to a user, it's important to understand the SQL statements presented in this chapter. However, once the security for a database has been set up, you may want to use SQL Developer to work with security. For example, you can use SQL Developer to drop or alter an existing user or to grant or revoke the privileges for a role.

How to create an admin connection for a PDB

Before you can use SQL Developer to work with users, you need to create a connection to a PDB for a user that has sysdba privileges. To do this, you can create a new connection using the sys login as shown in figure 12-11. Here, the connection is named "sys - pdb1" to distinguish it from the sys connection for the CDB root. Note that this creates a connection for the user named sys to the pluggable database named XEPDB1, not to the container database named XE. In addition, it uses the Role dropdown to specify the sysdba role.

If you connect to the PDB as a user that doesn't have appropriate privileges, you won't be able to use SQL Developer to work with the users and roles for the database. For example, the AP, EX, and OM users that you have been using throughout this book don't have appropriate privileges to work with other users because they haven't been granted the sysdba role.

If you're using a database hosted in Oracle Cloud, you already have an admin login for your pluggable database. As a result, you just need to create a connection for the admin user as shown in chapter 13.

A sysdba connection to a PDB named XEPDB1

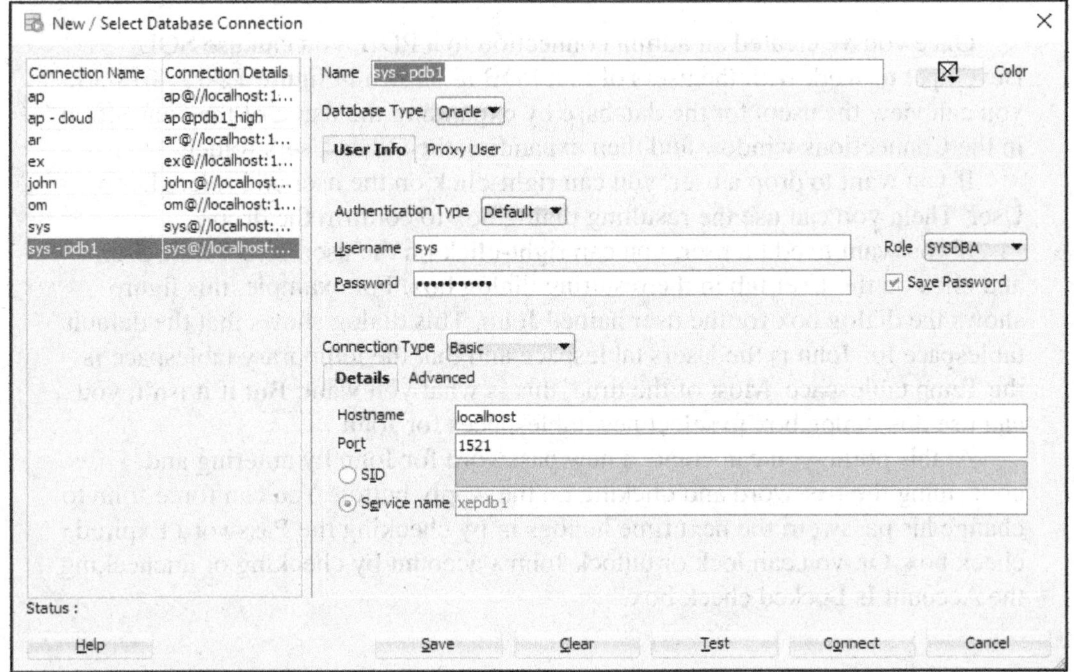

Description

- To use SQL Developer to work with security for a PDB, you can create a connection to the PDB using the sys user's credentials.

Figure 12-11 How to create an admin connection for a PDB

How to work with users

Once you've created an admin connection to a PDB, you can use SQL Developer to work with the users of that PDB as shown in figure 12-12. To start, you can view the users for the database by expanding the name of the connection in the Connections window and then expanding the Other Users folder.

If you want to drop a user, you can right-click on the user and select Drop User. Then, you can use the resulting dialog box to confirm the drop.

If you want to edit a user, you can right-click on the user, select Edit User, and click on the User tab in the resulting dialog box. For example, this figure shows the dialog box for the user named John. This dialog shows that the default tablespace for John is the Users tablespace and that the temporary tablespace is the Temp tablespace. Most of the time, this is what you want. But if it isn't, you can use this dialog box to select new tablespaces for John.

At this point, you can create a new password for John by entering and confirming the password and clicking on the Apply button. You can force John to change his password the next time he logs in by checking the Password Expired check box. Or, you can lock or unlock John's account by checking or unchecking the Account Is Locked check box.

The dialog box for working with a user

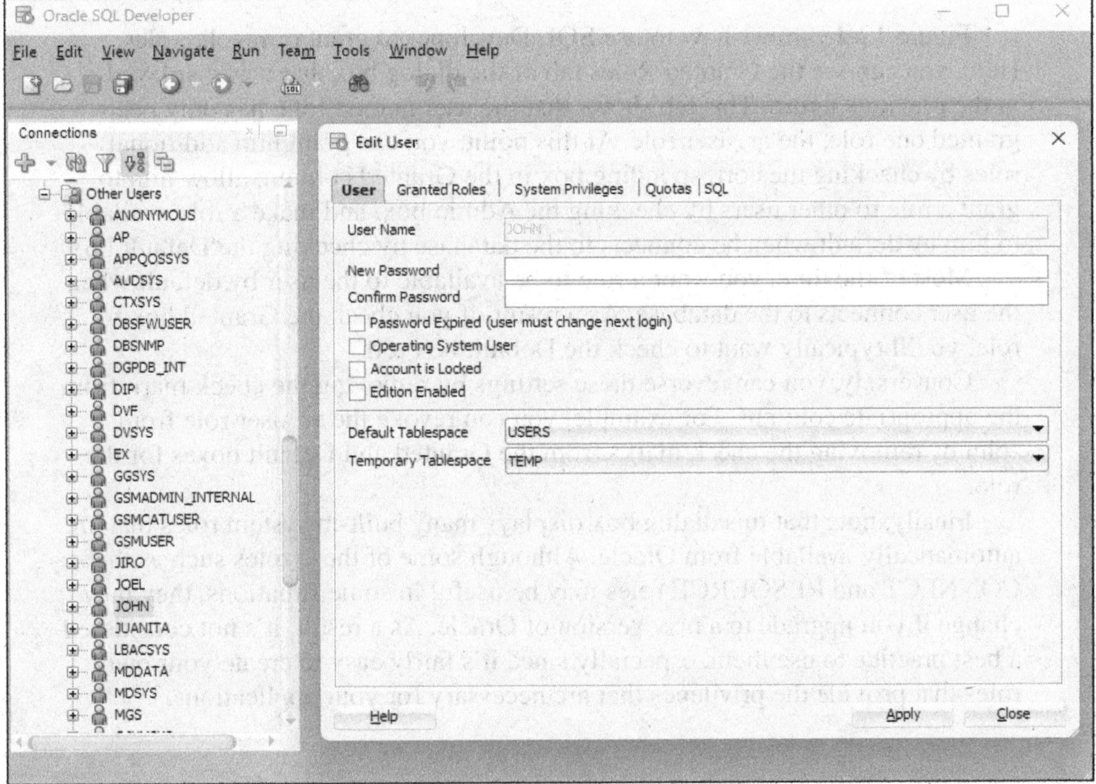

Description

- To view all users for a database or schema, you can expand a connection and then the Other Users folder in the Connections window.
- To drop a user, right-click on the user and select Drop User.
- To edit a user, right-click on the user and select Edit User. Then, you can work with the user's password, granted roles, system privileges, and so on.

Figure 12-12 How to work with users

How to grant and revoke roles

Figure 12-13 shows how to use SQL Developer to grant or revoke roles. Here, you can see the Granted Roles tab of the dialog box that was displayed in the previous figure. This tab shows that the user named John has only been granted one role, the ar_user role. At this point, you can grant him additional roles by checking the corresponding box in the Granted column, allow him to grant a role to other users by checking the Admin box, and make a role available to him by default when he connects to the database by checking the Default box.

Most of the time, you want a role to be available to the user by default when the user connects to the database. As a result, if you check the Granted box for a role, you'll typically want to check the Default box too.

Conversely, you can reverse these settings by removing the check mark from the appropriate columns. For example, you can revoke the ar_user role from John by removing the check marks from the Granted and Default boxes for the role.

Finally, note that this dialog box displays many built-in system roles that are automatically available from Oracle. Although some of these roles such as the CONNECT and RESOURCE roles may be useful in some situations, they may change if you upgrade to a new version of Oracle. As a result, it's not considered a best practice to use them, especially since it's fairly easy to create your own roles that provide the privileges that are necessary for your applications.

The dialog box for working with roles

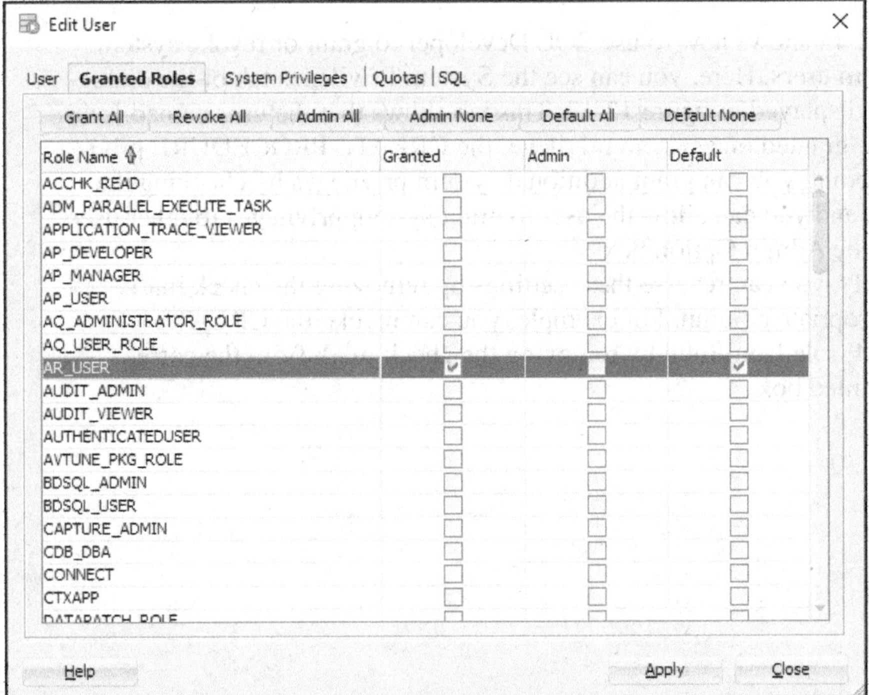

Description

- To view, grant, or revoke roles, right-click on a user in the Other Users folder, select Edit User, and click on the Granted Roles tab in the resulting dialog box.
- You can also grant or revoke the admin option for a role, and control whether a user has access to the role by default.
- If you don't see any roles listed in the tab, make sure that you have selected the user from the Other Users folder of an admin connection.

Figure 12-13 How to grant and revoke roles

How to grant and revoke system privileges

Figure 12-14 shows how to use SQL Developer to grant or revoke system privileges from users. Here, you can see the System Privileges tab of the dialog box that was displayed in figure 12-12. This tab shows that the user named John has only been granted one system privilege, the CREATE PROCEDURE privilege. At this point, you can grant additional system privileges by checking the Granted box, and you can allow the user to grant system privileges to other users by checking the Admin Option box.

Conversely, you can reverse these settings by removing the check mark from the appropriate column. For example, you can revoke the CREATE PROCEDURE role from John by removing the check mark from the corresponding Granted box.

The dialog box for working with system privileges

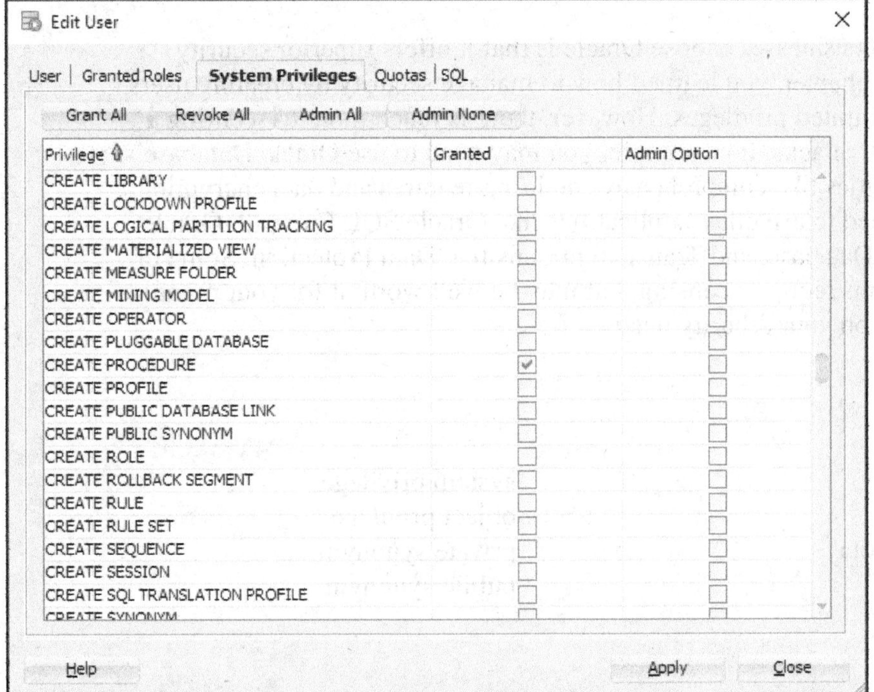

Description

- To view, grant, or revoke the system privileges assigned to a user, right-click on a user in the Other Users folder, select Edit User, and click on the System Privileges tab in the resulting dialog box.
- You can also grant or revoke the admin option for a privilege.

Figure 12-14 How to grant and revoke system privileges

Perspective

One reason businesses choose Oracle is that it offers superior security options. In this chapter, you learned how to manage security by creating users and roles with limited privileges. However, there is much more to securing a database against attacks. For example, you may need to use Oracle Database's built-in capabilities like comprehensive auditing features and data encryption. Or, you may need to use other applications like Oracle SQL Firewall, Oracle Virtual Private Database, and Transparent Sensitive Data Protection. Some of these applications require licensing, but may be well worth it for your organization depending on your security needs.

Terms

common user
local user
disk storage quota
profile
role
system privilege
object privilege
private synonym
public synonym

Exercises

1. Write a script that creates a database role named payment_entry in the AP schema and run it using the AP connection. This new role should have the following privileges:

   The privilege to create a session
   SELECT and UPDATE privileges for the Vendors table
   SELECT and UPDATE privileges for the Invoices table
   SELECT, UPDATE, and INSERT privileges for the Invoice_Line_Items table

2. Write a script that creates a user named Tom with a password of "temp" and grants the new payment_entry role to him. Run the script using thc AP connection.

3. Write SELECT statements that view the user and role privileges that are available to Tom. Create a connection for Tom and use it to run these statements.

4. Write a script that creates a public synonym for each of the tables used by the payment_entry role. Then, use the AP connection to run this script.

5. Write SELECT statements that test each of the public synonyms created by the previous step. Then, use the connection for Tom to run these statements. They should succeed.

   After the SELECT statements, write a DELETE statement that attempts to delete the first row in the Vendors table. Then, use the connection for Tom to run this statement. It should fail due to insufficient privileges.

13

How to host a database in the cloud

For years now, the tech industry has been moving away from using local servers to host databases and websites, preferring instead to use servers in the cloud. The reasons for this trend include increased affordability, flexibility, and scalability. In this chapter, you'll learn how to use Oracle's cloud computing platform, Oracle Cloud, to host a database in the cloud. In particular, you'll learn how to use Oracle's Autonomous Database service to host Oracle databases.

How to get started with Oracle Cloud 386

The Oracle Cloud portal 386

How to create a database in the cloud 388

How to create a user for a schema 390

How to create the tables for a schema 392

How to view the database objects for a schema 394

How to use SQL Developer with a cloud database 396

How to connect to a cloud database 396

How to run SQL against a cloud database 398

More skills for working with a cloud database 400

How to restore and delete a cloud database 400

How to restart a cloud database 402

Perspective 404

How to get started with Oracle Cloud

In this section, you'll learn how to use Oracle Cloud to create and configure an Oracle Autonomous Database instance for hosting a relational database in the cloud. The Autonomous Database service is presented because it can be used for free with some resource restrictions. This means that you can learn and experiment with Oracle Cloud without committing to any paid services.

The Oracle Cloud portal

Before you can use any of Oracle's cloud services, you must first create an Oracle Cloud account. To do that, you can go to the URL for the Oracle Cloud portal shown at the top of figure 13-1.

When you create a new account, you will need to download Oracle's Mobile Authenticator app or otherwise enable two-factor authentication (2FA). In addition, you will need to enter payment information. However, you can still use Oracle Cloud in a limited capacity for free. Specifically, many introductory services are available for free or as free trials, including Autonomous Database.

Once you've created your account and logged in, you'll see the Oracle Cloud home page shown in the first screen of this figure. From this page, you can use the search bar or the Navigation Menu button to navigate to any of the services that Oracle Cloud provides. For example, you can click the Navigation Menu button, the Databases link, and the Autonomous Database link to get to the Autonomous Database page shown in the second screen.

Oracle Autonomous Database is a *managed database service*. This means that Oracle handles most aspects of server and database maintenance for you, including creating backups.

A managed database service is usually a good choice for beginners because it allows you to focus on learning to use SQL and relational databases. Then, when you have a good grasp on the basics, you can look into Oracle's other cloud services, or you can set up a local server and database.

The Oracle Cloud portal

cloud.oracle.com

The portal after logging in

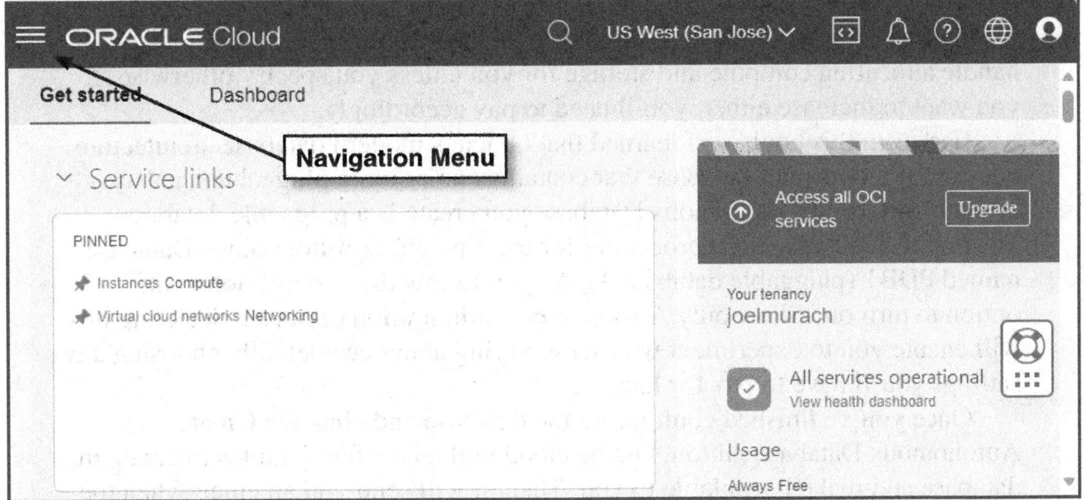

The Databases page

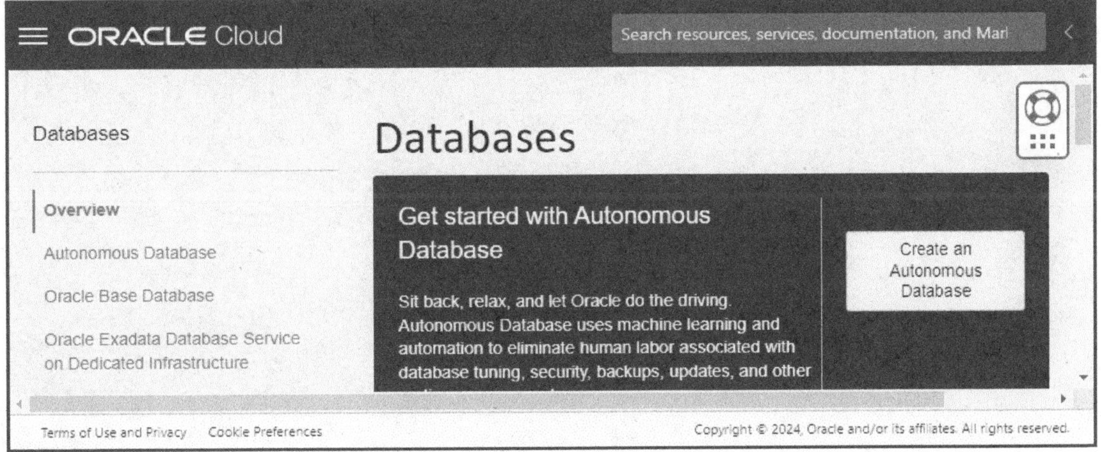

Description

- The Oracle Cloud portal allows you to manage Oracle web services, including databases.
- Oracle Autonomous Database is a *managed database service*. Oracle handles most of the server and database maintenance for you.

Figure 13-1 The Oracle Cloud portal

How to create a database in the cloud

Oracle's Autonomous Database service provides the compute and storage that are used to physically host your database. *Compute and storage* refers to the physical aspects of a server: memory, processing power, and disk storage.

Because Autonomous Database is a managed database service, Oracle will handle allocating compute and storage for you unless you specify otherwise. If you want to increase either, you'll need to pay accordingly.

Earlier in this book, you learned that Oracle's modern database architecture consists of a container database that contains one or more pluggable databases. In the cloud, each Autonomous Database you create is a pluggable database.

Figure 13-2 presents a procedure for creating a free Autonomous Database named PDB1 (pluggable database 1). As you follow these steps, notice the option to turn on "Show only Always Free configuration options". Selecting this will enable you to experiment without worrying about accidentally choosing any settings you'll have to pay for later.

Once you've finished configuring the database and click the Create Autonomous Database button, Oracle cloud will take a few minutes to create the database and make it available to you. Then, it will send you an email when the database is ready.

The Create Autonomous Database page

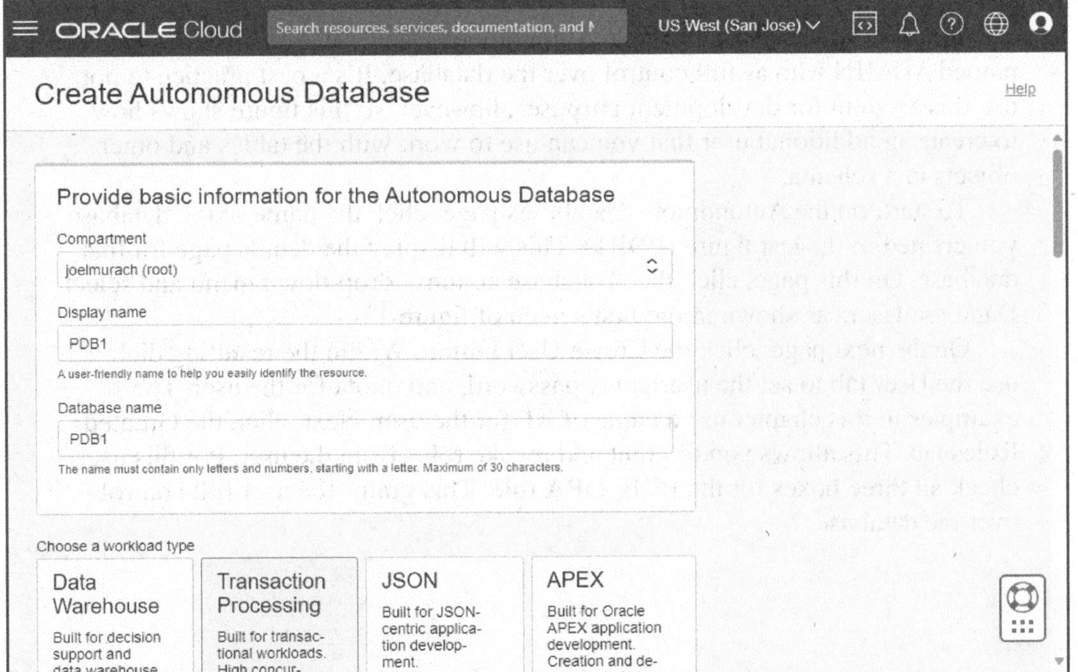

How to create a database in the cloud

1. Navigate to the Autonomous Database page and click Create Autonomous Database.
2. Enter a name and display name for the database. For this chapter, you can enter "PDB1" for both.
3. Select the workload type. For this chapter, you can select the Transaction Processing workload.
4. Select the deployment type. This chapter uses Serverless.
5. In the "Configure the database" section, turn on "Show only Always Free configuration options".
6. Create a password for the user named ADMIN. Make sure to remember this password!
7. Select an access type. This chapter uses "Secure access from IPs and VCNs only".
8. Click the "Add my IP address" button. (You can add more IP addresses later.)
9. Click the Create Autonomous Database button.

Description

- This procedure creates a pluggable database in the Oracle Cloud.
- When in doubt, you can accept the default options.

Figure 13-2 How to create a database in the cloud

How to create a user for a schema

By default, when you create an Autonomous Database, you also create a user named ADMIN who as full control over the database. It's a best practice to not use this account for development purposes, however, so this figure shows how to create an additional user that you can use to work with the tables and other objects in a schema.

To start, on the Autonomous Databases page, click the name of the database you created in the last figure (PDB1). This will display the details page for that database. On this page, click the "Database actions" drop-down menu and select Database Users as shown in the first screen of figure 13-3.

On the next page, click the Create User button. Within the resulting dialog, use the User tab to set the username, password, and quota for the user. The examples in this chapter use a name of AP for the user. Next, click the Granted Roles tab. This allows you to grant and revoke roles from the user. For this user, check all three boxes for the PDB_DBA role. This grants the user full control over the database.

The details page for an Autonomous Database

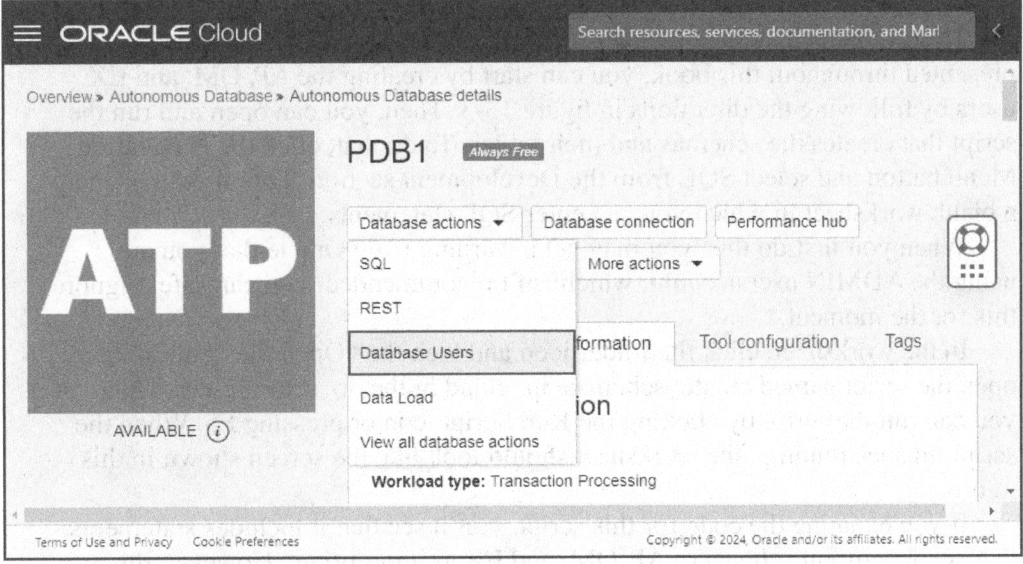

How to create a user for a schema

1. On the details page for a database, click the "Database actions" drop-down list and select Database Users.
2. On the Database Users page, click the Create User button.
3. Enter a username and password for the user, such as AP.
4. Select UNLIMITED from the Quota drop-down.
5. Click the Granted Roles tab and check all three boxes for the PDB_DBA role.

The ADMIN and AP users

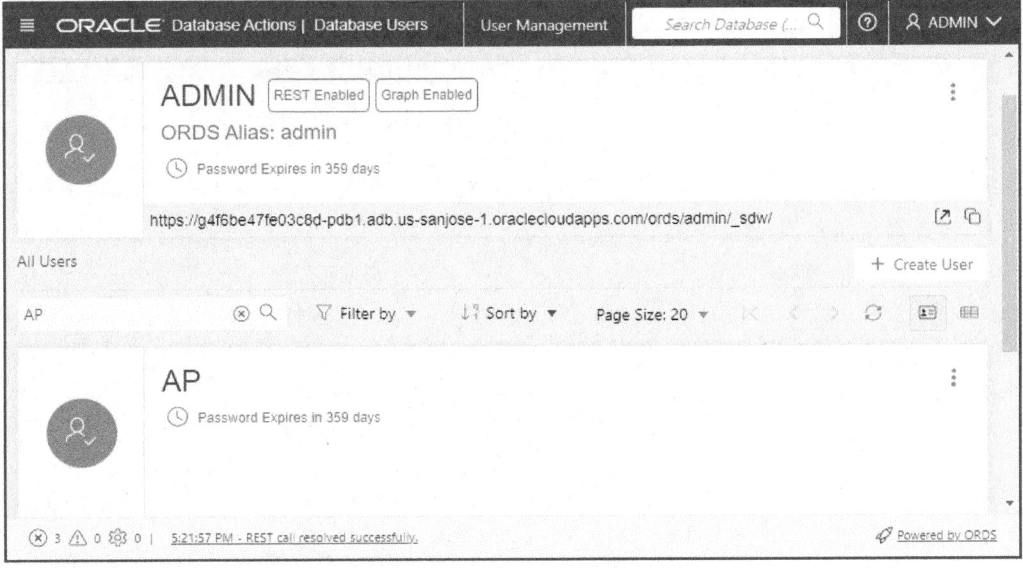

Figure 13-3 How to create a user for a schema

How to create the tables for a schema

To automatically create the tables for the AP, OM, and EX schemas presented throughout this book, you can start by creating the AP, OM, and EX users by following the directions in figure 13-3. Then, you can open and run the script that creates the schemas and their tables. To do that, click the Navigation Menu button and select SQL from the Development section. This should open up a blank worksheet in which you can enter SQL statements.

When you first do this, you may get a warning from Oracle that you are using the ADMIN user account, which isn't recommended. You can safely ignore this for the moment.

In the worksheet, click the folder icon and then the "Open file" button to open the script named create_schemas_in_cloud in the db_setup folder. Then, you can run the script by clicking the Run Script icon or pressing F5. When the script finishes running, the worksheet should look like the screen shown in this figure.

If you examine the code for this script, you'll see that it includes statements that set the current schema to AP, OM, and EX as appropriate. However, this only works if you already created the AP, OM, and EX users.

An Oracle Cloud worksheet after running a script

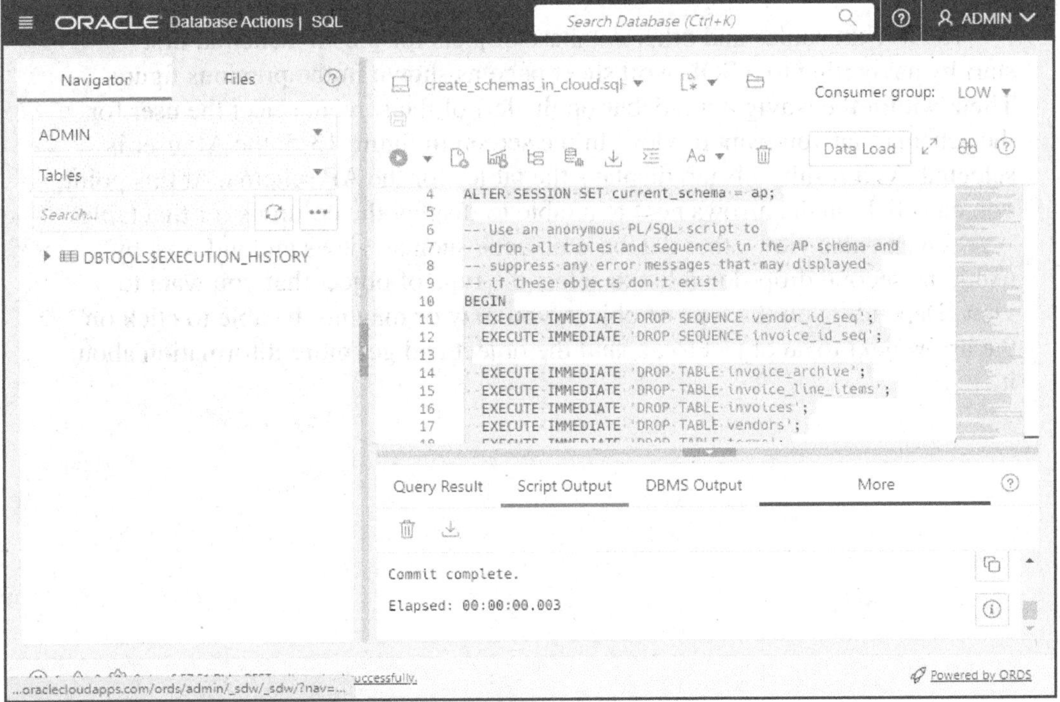

How to run a script that creates the tables for three schemas

1. Create users named AP, OM, and EX as shown in the previous figure.
2. Click the Navigation Menu and select SQL from the Development section. This should open a blank worksheet.
3. Click the folder icon or press Ctrl+O and use the resulting dialog to select the create_schemas_in_cloud.sql file that's in the db_setup folder.
4. Click the Run Script button or press F5 to run all of the statements in the script. This should create the three schemas and their tables.

Figure 13-4 How to create the tables for a schema

How to view the database objects for a schema

To view the tables and other database objects for the AP schema, first start by navigating to a SQL worksheet page as shown in the previous figure. Then, within the Navigator sidebar on the left of the screen, select the user for the schema that you want to view. In the screen in figure 13-5, the AP user is selected. As a result, this tab displays the tables for the AP schema. At this point, you can click on the arrows next to a table to display the columns for that table.

You can also view other database objects, such as views and indexes, by using the second drop-down list to select the type of object that you want to view. Depending on the type of object, you may or may not be able to click on the arrow next to an object to expand the object and get more information about it.

The tables for the AP schema

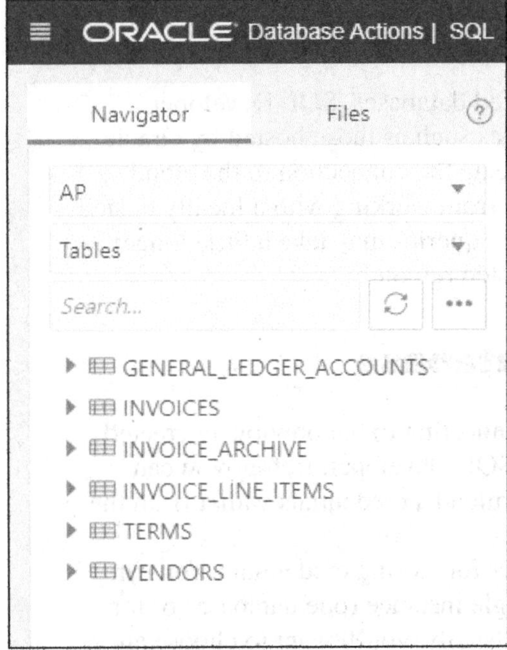

How to view the tables for a schema

1. Open a blank worksheet as described in the previous figure.
2. In the Navigator tab, select the user for the schema from the first drop-down list. This should display the tables at the bottom of the tab.

Description

- To view database objects other than tables, select an option other than Tables from the second drop-down list in the Navigator tab.
- To get more information about a database object, click the arrow next to the object to expand it.

Figure 13-5 How to view the database objects for a schema

How to use SQL Developer with a cloud database

In addition to working with locally hosted databases, SQL Developer can connect to and work with cloud databases such as those hosted by Oracle Autonomous Database. In fact, once you create the connection to the cloud database, working with it is barely different from working with a locally hosted database. The main difference is that running queries may take a little longer because SQL needs to access a database hosted remotely.

How to connect to a cloud database

Figure 13-6 presents a procedure for connecting to the previously created Autonomous Database named PDB1 from SQL Developer. Before you can connect to that database, you'll need to download a credentials wallet from the details page for the Autonomous Database.

A credentials wallet is a secure container for storing credentials. You can choose to download a wallet for either a single instance (one database) or for all databases that you have in your region. Usually, you'll want to choose an instance wallet unless you have a specific use case where you need to access multiple databases in a region. Then, the file that you download will be a zip file with a name like Wallet_PDB1.zip where PDB1 is the name of the database.

Once you download the wallet, setting up the connection is similar to setting up a local connection. The only major difference is that you need to change the connection type to Cloud Wallet and upload the credentials wallet as the Configuration File. Once you've finished entering the required information, you can click the Test button to test the connection. And if it's successful, you can click the Save button to save it in the Connections tab.

A connection to a cloud database

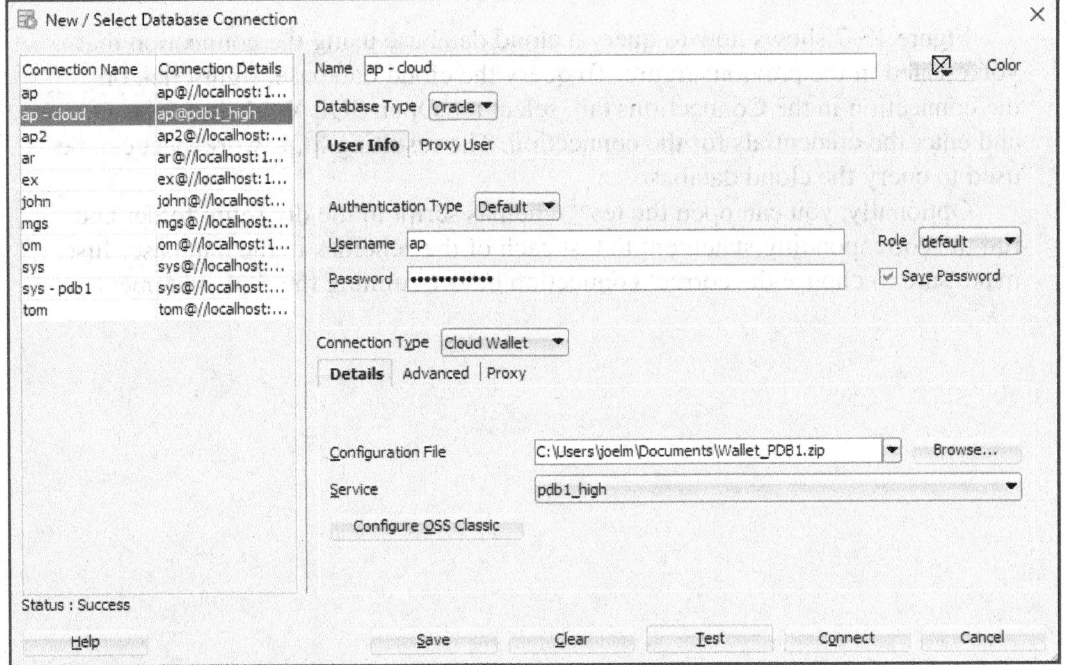

How to connect SQL Developer to a cloud database

1. On the details page for the Autonomous Database to be connected to, click the "Database connection" button.
2. Set the Wallet type to Instance Wallet to create a connection only to the selected database.
3. Click the "Download wallet" button and enter a password you can remember.
4. Click the Download button to download a zip file for the cloud wallet.
5. Start SQL Developer, right-click on the Oracle Connections item, and select New Connection (not New Cloud Connection).
6. Enter a meaningful name for the connection, such as "ap - cloud".
7. Enter the username and password for a user that you've created in the cloud, such as the AP user.
8. Select Cloud Wallet from the Connection Type drop-down.
9. Click the Browse button next to Configuration File and select the zip file you downloaded for the cloud wallet.
10. Click Test to make sure the connection works.
11. Click Save to save the connection.
12. If you want to follow along with the examples, repeat this process for the OM and EX users.

Figure 13-6 How to connect SQL Developer to a cloud database

How to run SQL against a cloud database

Figure 13-7 shows how to query a cloud database using the connection that you created in the previous figure. To query the cloud database, right-click on the connection in the Connections tab, select the Open SQL Worksheet option, and enter the credentials for the connection. The resulting SQL worksheet can be used to query the cloud database.

Optionally, you can open the test_schemas script in the db_setup folder and run the corresponding statement to test each of the schemas in the database. Just make sure to choose the correct connection before running the SQL statement.

SQL Developer after running a statement against the cloud database

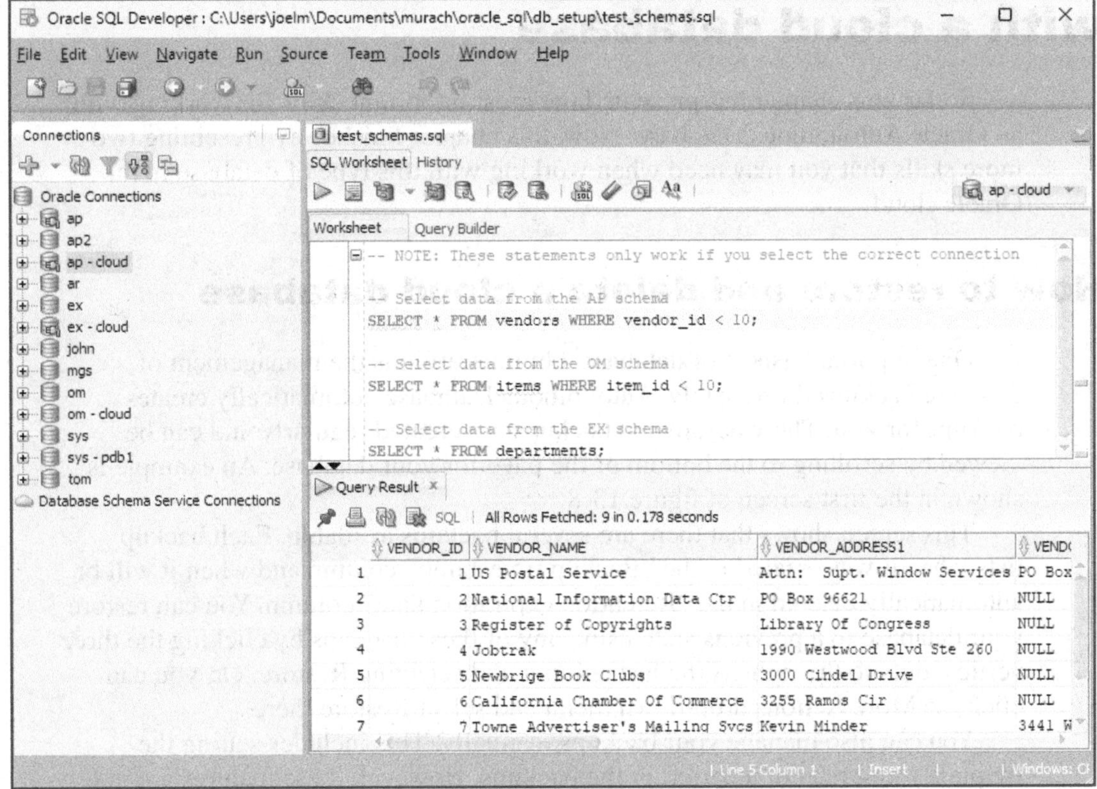

Description

- Once you create a connection to a cloud database, you can use the same skills for running SQL scripts and statements that you use for a local database.

More skills for working with a cloud database

So far, this chapter has presented the most important skills for working with an Oracle Autonomous Database. Now, this chapter finishes by presenting two more skills that you may need when working with this type of database in the Oracle cloud.

How to restore and delete a cloud database

One important aspect of database administration is the management of database backups. Fortunately, Autonomous Database automatically creates backups for you. These automated backups are created regularly and can be viewed by scrolling to the bottom of the page for your database. An example is shown in the first screen of figure 13-8.

This screen shows that there are several backups available. Each backup lists when it was created in the "Backup timestamp" column and when it will be automatically deleted in the "Retention expiration date" column. You can restore your database to a previous state using any of these backups by clicking the three vertical dots to the right of the last column and selecting Restore. Or, you can click the More Actions drop-down menu and select Restore there.

You can also manage your backups manually. This includes setting the frequency and retention period of the backups. However, these features are not currently available in the free tier.

The second example in this figure shows how to delete a database. Deleting a database when you no longer need it is a good practice because it prevents possible misuse of your data. For the sample data used with this book, you don't need to worry much about this. However, keep in mind that free Autonomous Databases will be deleted automatically after a set period of inactivity.

The automated backups for a cloud database

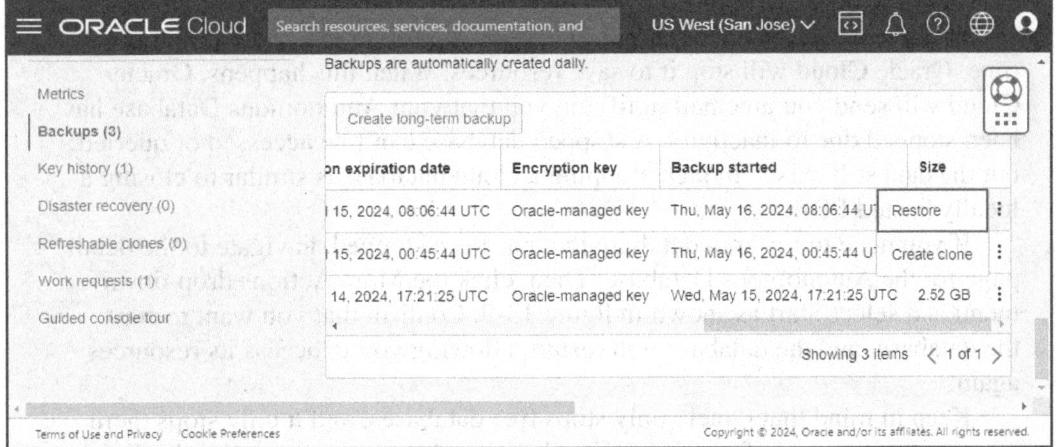

The menu that contains the Terminate item

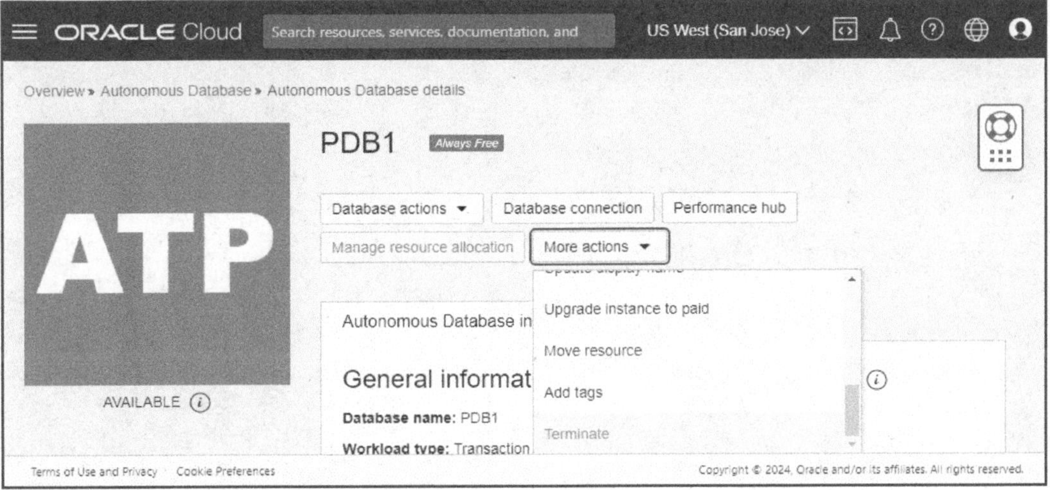

Description

- To view a database's backups, scroll down the details page for the database until you see Backups in the left menu and then click it.
- To restore an automated backup, click the three vertical dots on the right of a backup's name. Alternately, click the More Actions drop-down menu and select Restore.
- Automated backups are retained for 60 days for free-tier databases.
- Free-tier databases do not support manual long-term backups.
- To delete a database, click the More Actions drop-down menu and select Terminate. Then, follow the prompts to delete the database.

Figure 13-8 How restore and delete a cloud database

How to restart a cloud database

If you leave a free Autonomous Database idle for an extended period of time, Oracle Cloud will stop it to save resources. When this happens, Oracle Cloud will send you an email notifying you that your Autonomous Database has been stopped due to inactivity. A stopped database can't be accessed or queried, but the data still exists. In fact, stopping a cloud database is similar to closing a locally hosted PDB.

If you need to restart a database that has been stopped, navigate to the details page for the Autonomous Database. Then, click the More Actions drop-down menu and select Start as shown in figure 13-9. Confirm that you want to start the database, and the database will restart, allowing you to access its resources again.

Keep in mind that Oracle only stops free databases, and it only stops them after a certain period of inactivity. You don't need to worry about Oracle Cloud stopping any paid services without you explicitly telling it to.

A stopped database

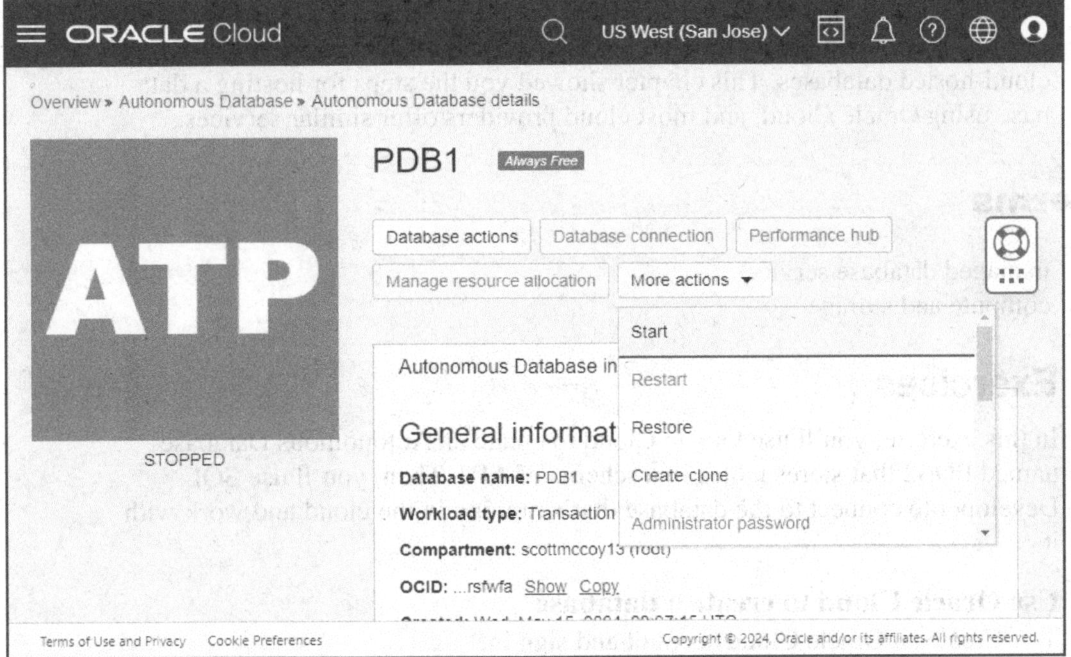

Description

- If you leave a free-tier autonomous database idle for a while, Oracle Cloud will automatically stop it.
- To start the database again, navigate to the database's details page, click the More Actions drop-down menu, and click Start.

Figure 13-9 How to restart a cloud database

Perspective

The world is moving away from locally hosted databases and towards cloud-hosted databases. This chapter showed you the steps for hosting a database using Oracle Cloud, and most cloud providers offer similar services.

Terms

managed database service
compute and storage

Exercises

In this exercise, you'll use Oracle Cloud to create an Autonomous Database named PDB2 that stores tables in a schema of AP2. Then, you'll use SQL Developer to connect to the database that's running in the cloud and work with it.

Use Oracle Cloud to create a database

1. Create an Oracle Cloud account and sign in.
2. Navigate to the Create Autonomous Databases page and create an Autonomous Database named PDB2.
3. Create a user named AP2 for the database with a quota of UNLIMITED and grant that user the role of PDB_DBA.
4. Display an Oracle Cloud worksheet and open the script named create_ap2_schema that's in the scripts/ch13 folder.
5. Run all of the statements in the script. To do that, click the Run Script icon or press F5. This should create the AP2 schema.
6. Display an Oracle Cloud worksheet. In the Navigation tab, select the AP2 user to view the tables in the AP2 schema.

Use SQL Developer to work with the database

7. In SQL Developer, create a connection named AP2 that connects to the Autonomous Database named PDB2 as the user named AP2.
8. Start a new SQL worksheet and enter the following select statement:

```
SELECT * FROM vendors WHERE vendor_id < 11;
```

9. Use the AP2 connection to run the statement and verify it works.
10. Open some of the scripts for the SQL statements presented in chapters 3 through 8 and run some of these statements against the AP2 schema. They should work the same as when you run them against the AP schema.

Use Oracle Cloud to delete the database

11. When you are done testing SQL statements against the AP2 schema, go to the details page for the Autonomous Database and delete the PDB2 database.

Section 4

The essential PL/SQL skills

This section presents the essential skills for using Oracle's procedural language, which is called PL/SQL. These are the skills that will take your SQL capabilities to the next level.

In chapter 14, you'll learn the basics of PL/SQL coding. In chapter 15, you'll learn how to use PL/SQL to manage transactions and locking. In chapter 16, you'll learn how to use PL/SQL to create stored procedures and functions. And in chapter 17, you'll learn how to use PL/SQL to create triggers.

14

How to write PL/SQL code

This chapter shows you how to write scripts that include blocks of PL/SQL (Procedural Language/SQL) code. With the skills you'll learn in this chapter, you'll be able to code scripts that provide functionality similar to procedural programming languages like Java, C#, and Python.

An introduction to PL/SQL **408**
An anonymous PL/SQL block in a script 408
Statements for working with PL/SQL and scripts 410
How to code the basic PL/SQL statements **412**
How to print data to an output window 412
How to declare and use variables 414
How to code IF statements 416
How to code CASE statements 418
How to code loops 420
How to use a cursor 422
How to work with composite variables **424**
How to use collections 424
How to use records 428
How to work with exceptions and errors **430**
How to handle exceptions 430
Predefined exceptions 432
How to drop database objects without displaying errors 434
Perspective **436**

An introduction to PL/SQL

Before you learn the details for writing PL/SQL code, it's helpful to get an overview by looking at a basic block of PL/SQL code in a simple script. In addition, it's helpful to preview the statements that are commonly used within scripts and within blocks of PL/SQL code.

An anonymous PL/SQL block in a script

PL/SQL (*Procedural Language/SQL*) allows you to write procedural SQL code that works with an Oracle database. For example, you can use an IF statement to control the execution of a script, or you can use a loop to execute a statement multiple times.

To include PL/SQL within a script, you can code a *block* of PL/SQL code as shown in figure 14-1. A PL/SQL block has up to three sections. The first is the declarative section, which allows you to declare variables that will hold values you can use in your code. It starts with the DECLARE keyword. The second is the executable section, which contains the body of the PL/SQL block. It is the only section which is required. To start it, you code the BEGIN keyword. Third, if you want to handle errors that might arise when your code runs, you can code an exception-handling section. This section starts with the EXCEPTION keyword.

In this example, the script begins by executing a command that allows SQL Developer to display code output in its Output Script window. Then, it uses a PL/SQL block to print a message that indicates the total balance due for the vendor with an ID of 95.

Since this block doesn't have a name, it is known as an *anonymous PL/SQL block*. In chapter 16, you'll learn how to store a named block of PL/SQL code within the database for reuse.

As you review this script, note that you must code a semicolon at the end of each SQL statement, just as in other SQL scripts. Similarly, you must code a semicolon after the END IF keywords and after the END keyword at the end of the PL/SQL block. Finally, to execute an anonymous PL/SQL block, you must code a front slash (/) after the END keyword for the block.

For now, don't worry if you don't understand the coding details for this script. Instead, focus on the division of the three different sections. Later in this chapter, you'll learn the details that you need to write effective PL/SQL code.

The syntax for an anonymous PL/SQL block

```
[DECLARE
  variable_declaration_statements;]
BEGIN
  body_statements;
[EXCEPTION
  [WHEN OTHERS THEN
    exception_handling_statements;]]
END;
/
```

A script that contains an anonymous PL/SQL block

```
SET SERVEROUTPUT ON;

-- Begin anonymous PL/SQL block
DECLARE
  sum_balance_due_var NUMBER(9, 2);

BEGIN
  SELECT SUM(invoice_total - payment_total - credit_total)
  INTO sum_balance_due_var
  FROM invoices
  WHERE vendor_id = 95;

  IF sum_balance_due_var > 0 THEN
    DBMS_OUTPUT.PUT_LINE('Balance due: $' ||
                         ROUND(sum_balance_due_var, 2));
  ELSE
    DBMS_OUTPUT.PUT_LINE('Balance paid in full');
  END IF;

EXCEPTION
  WHEN OTHERS THEN
    DBMS_OUTPUT.PUT_LINE('An error occured in the script');
END;
/
-- End anonymous PL/SQL block
```

The response from the system

```
Balance due: $171.01
```

Description

- *PL/SQL* (*Procedural Language/SQL*) allows you to write procedural code such as IF statements and loops.
- A *PL/SQL block* can contain a declarative section, an executable section, and an exception-handling section. Only the executable section is required.
- You can use regular SQL inside of a PL/SQL block.
- An *anonymous PL/SQL block* doesn't have a name and isn't stored as a database object.

Figure 14-1 An anonymous PL/SQL block in a script

Statements for working with PL/SQL and scripts

Figure 14-2 begins by summarizing the main PL/SQL statements that are presented in this chapter. These statements can be used within PL/SQL blocks to add functionality that's similar to the functionality provided by procedural languages.

After the PL/SQL statements, this figure presents two Oracle *commands* that are commonly used within scripts that contain blocks of PL/SQL code. First, it presents the CONNECT command that's used to connect to a database. However, you typically don't need to use this command if you set the connection within SQL Developer. As a result, you'll only need this command if a script needs to switch the connection for a script as it executes. Second, this figure presents the SET SERVEROUTPUT command. This command enables printing to the Script Output window of SQL Developer.

After the two Oracle commands, this figure presents two *procedures* that are available from the DBMS_OUTPUT *package* that's included with Oracle Database. You can use these procedures to print data to SQL Developer's Script Output window.

Finally, this figure presents two data types that you can use in your PL/SQL code. PLS_INTEGER stores integers, but takes up less space than the equivalent NUMBER data type and calculates faster because it can use hardware arithmetic. BOOLEAN stores either true, false, or null values. For table columns, this data type only works with Oracle Database 23ai and later, but for PL/SQL, you can use it with earlier versions of Oracle Database such as 21c.

PL/SQL statements for controlling the flow of execution

Statement	Description
IF	Executes statements based on a Boolean condition.
CASE	Executes statements based on the value of a variable or expression.
LOOP	Repeatedly executes statements until an EXIT statement is encountered.
FOR LOOP	Repeatedly executes statements a specific number of times.
WHILE LOOP	Repeatedly executes statements while a Boolean condition is true.
EXECUTE IMMEDIATE	Immediately executes a statement within a script.

Commands for working with scripts

Command	Description
CONNECT user_login	Connects to the database as the specified user.
SET SERVEROUTPUT {ON|OFF}	Enables or disables printing to SQL Developer's Script Output window.

Procedures for printing output to the screen

Procedure	Description
DBMS_OUTPUT.PUT(string)	Prints the specified string without a line break.
DBMS_OUTPUT.PUT_LINE(string)	Prints the specified string followed by a line break.

PL/SQL data types

Data type	Stores
PLS_INTEGER	Integers from -2,147,483,648 to 2,147,483,647
BOOLEAN	True, false, or null.

Description

- A PL/SQL *command* is similar to a SQL statement and is used to perform an action.
- A *procedure* is a stored block of code that performs an action.
- A *package* is a group of related procedures, variables, data types, exceptions, and other code-related objects.
- The DBMS_OUTPUT package that's included by default with Oracle Database includes procedures that can be used to print output to the screen.
- The PLS_INTEGER data type requires less storage than NUMBER and uses hardware arithmetic for faster calculations.
- For table columns, the BOOLEAN data type is only available with Oracle 23ai and later. For PL/SQL, the BOOLEAN data type is available with earlier versions.

Figure 14-2 Statements for working with PL/SQL and scripts

How to code the basic PL/SQL statements

Now that you have a general idea of what a block of PL/SQL code looks like, you're ready to learn the details for working with PL/SQL.

How to print data to an output window

When you develop PL/SQL code, you often need to print text and data to an output window. For example, you may need to print the value of a variable to help debug some code.

Figure 14-3 shows how to print data to the Script Output window of SQL Developer. Here, the script begins by calling the SET SERVEROUTPUT command to enable printing data to the Script Output window. Without it, SQL Developer wouldn't display the output. Then, the script contains a PL/SQL block that uses the PUT_LINE procedure of the DBMS_OUTPUT package to print a message to the Script Output window.

In the Script Output window, the first line shows the message printed by the script, "Test SQL Developer". Then, the Script Output window indicates that the PL/SQL block executed successfully. Note that the message is printed before the PL/SQL block has finished executing.

Once you've set server output to on, you don't need to turn it on again for the rest of the session. However, it doesn't cause any problems to run the SET command again. So, each script presented in this chapter begins by enabling server output to make sure that SQL Developer properly displays any output.

Text displayed in the Script Output window

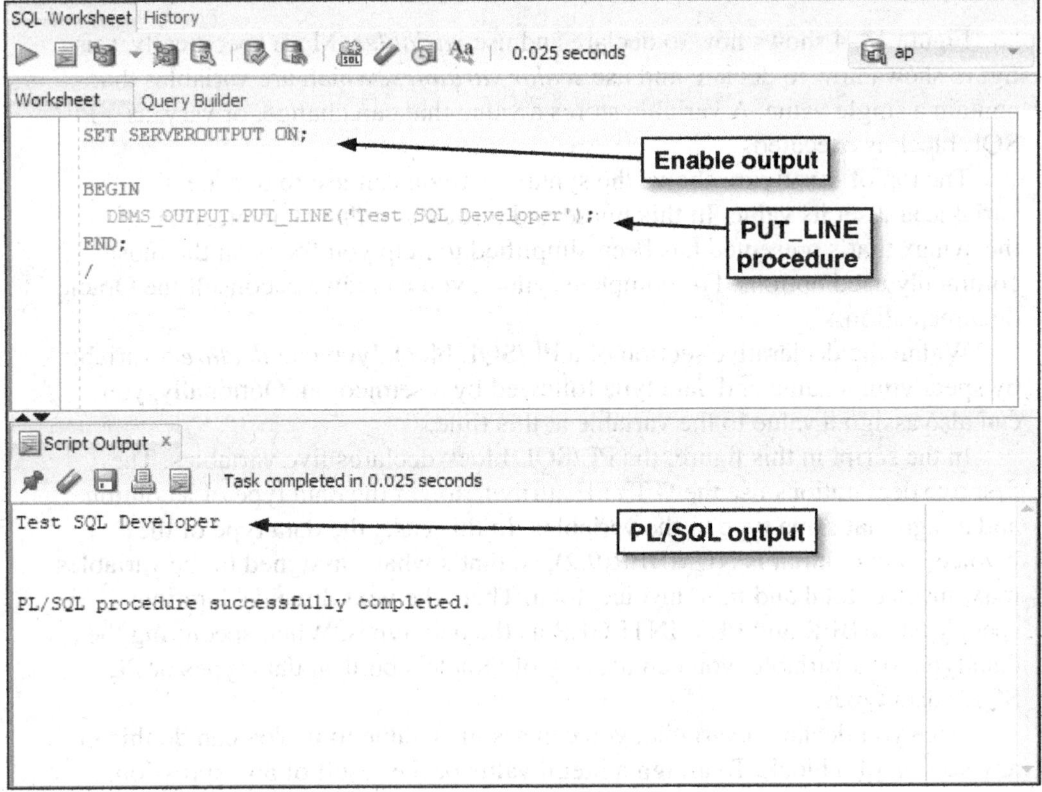

Description

- The PUT_LINE procedure of the DBMS_OUTPUT package allows you to output text from within a PL/SQL block.
- Before you can use the PUT_LINE procedure with SQL Developer's Script Output window, you must enable it by calling the SET SERVEROUTPUT ON command.
- Once you've turned server output on, you don't need to turn it on again for the rest of the session.

Figure 14-3 How to print data to an output window

How to declare and use variables

Figure 14-4 shows how to declare and use *variables*. More specifically, this figure shows how to declare and use *scalar variables*, which are variables that contain a single value. A variable stores a value that can change, or vary, as a PL/SQL block is executed.

The top of this figure shows the syntax that you can use to declare a variable and set its value. In this figure and throughout this chapter, some of the syntax that's presented has been simplified to help you focus on the most commonly used options. For complete syntax, you can always consult the Oracle documentation.

Within the declarative section of a PL/SQL block, you can *declare* a variable by specifying a name and data type followed by a semicolon. Optionally, you can also assign a value to the variable at this time.

In the script in this figure, the PL/SQL block declares five variables. The first two declarations use the %TYPE attribute to get the data type of a column and assign that same type to the variables. In this case, the data type of the invoice_total column is NUMBER(9,2), so that's what's assigned to the variables max_invoice_total and min_invoice_total. Then, the next three declarations specify NUMBER and PLS_INTEGER as the data types. When specifying the data type for a variable, you can use any of Oracle's built-in data types or PL/SQL's data types.

Once you declare a variable, you can assign a value to it. You can do this in any section of a block. To assign a literal value or the result of an expression, code the assignment operator (:=) followed by the literal value or expression. In this script, for example, the last variable declaration uses the assignment operator to assign a value of 95 to the variable named vendor_id_var. In addition, the second statement in the executable section uses the assignment operator to assign the result of a calculation to the variable named percent_difference.

If you want to assign a value that's returned by a SELECT statement to a variable, you can use an INTO clause. In this script, the first statement in the executable section uses an INTO clause to assign three values to three variables. For this to work, the SELECT statement must return exactly one value for each of the variables specified in the INTO clause. In addition, the data types for the columns must be compatible with the corresponding variables.

After all of the variables have been assigned values, a series of statements use the PUT_LINE procedure to display the values of the variables. Finally, the block is concluded with the END statement and then immediately executed with the forward slash.

If you use the %TYPE attribute when you declare the variables, you can be sure that a variable will use the same data type as its column. However, using this attribute requires more typing and results in code that's more difficult to read. As a result, most of the examples in the rest of this chapter don't use the %TYPE attribute to define the data type for their variables.

The syntax for declaring a variable

```
variable_name_1 data_type;
```

The syntax for declaring a variable with the same data type as a column

```
variable_name_1 table_name.column_name%TYPE;
```

The syntax for setting a variable to a selected value

```
SELECT column_1[, column_2]...
INTO variable_name_1[, variable_name_2]...
```

The syntax for setting a variable to a literal value or expression result

```
variable_name := literal_value_or_expression
```

A SQL script that uses variables

```
SET SERVEROUTPUT ON;

DECLARE
  max_invoice_total   invoices.invoice_total%TYPE; -- type is NUMBER(9,2)
  min_invoice_total   invoices.invoice_total%TYPE;
  percent_difference NUMBER;
  count_invoice_id    PLS_INTEGER;
  vendor_id_var       PLS_INTEGER := 95;

BEGIN
  SELECT MAX(invoice_total), MIN(invoice_total), COUNT(invoice_id)
  INTO max_invoice_total, min_invoice_total, count_invoice_id
  FROM invoices WHERE vendor_id = vendor_id_var;

  percent_difference := (max_invoice_total - min_invoice_total) /
                         min_invoice_total * 100;

  DBMS_OUTPUT.PUT_LINE('Maximum invoice: $' || max_invoice_total);
  DBMS_OUTPUT.PUT_LINE('Minimum invoice: $' || min_invoice_total);
  DBMS_OUTPUT.PUT_LINE('Percent difference: %' ||
                        ROUND(percent_difference, 2));
  DBMS_OUTPUT.PUT_LINE('Number of invoices: ' || count_invoice_id);
END;
/
```

The response from the system

```
Maximum invoice: $46.21
Minimum invoice: $16.33
Percent difference: %182.98
Number of invoices: 6
```

Description

- A *variable* is used to store data. Its value can change as the block executes.
- A variable that contains a single value is called a *scalar variable*.
- All variables used in a PL/SQL block must first be *declared* with a name and data type in the declarative section.
- You can assign a value to a variable in any of the three sections.
- The %TYPE attribute returns the data type of a column or variable.

Figure 14-4 How to declare and use variables

How to code IF statements

Figure 14-5 shows how to use an *IF statement* to execute one or more statements based on the value returned by a *Boolean expression*. A Boolean expression is an expression that returns true or false.

The script in this figure uses an IF statement to test the value of a variable. This variable contains the oldest invoice due date in the Invoices table. If this due date is less than the current date, the Boolean expression in the IF clause evaluates to true and the statement in the IF clause prints a message to the output window that indicates that outstanding invoices are overdue.

If the value is equal to the current date, the statement in the ELSIF clause prints a message to the output window that indicates that outstanding invoices are due today. If neither of these conditions are true, the oldest due date must be greater than the current date. As a result, the script prints a message to the output window that indicates that no invoices are overdue.

In this figure, the IF statement only contains one ELSIF clause. However, you can add as many ELSIF clauses as you want. But if you don't need an ELSIF clause at all, you don't have to code one. For example, you could code an IF statement like this:

```
IF first_invoice_due_date < SYSDATE() THEN
  DBMS_OUTPUT.PUT_LINE('Outstanding invoices overdue!');
ELSE
  DBMS_OUTPUT.PUT_LINE('No invoices are overdue.');
END IF;
```

Similarly, the ELSE clause is also optional. As a result, you could code an IF statement like this:

```
IF first_invoice_due_date < SYSDATE() THEN
  DBMS_OUTPUT.PUT_LINE('Outstanding invoices overdue!');
END IF;
```

When you code IF statements, you can nest one IF statement within another, like this:

```
IF first_invoice_due_date <= SYSDATE() THEN
  DBMS_OUTPUT.PUT_LINE('Outstanding invoices are overdue!');
  IF first_invoice_due_date = SYSDATE() THEN
    DBMS_OUTPUT.PUT_LINE('TODAY!');
  END IF;
END IF;
```

In this case, the outer IF statement will be executed when the oldest invoice due date is less than or equal to the current date. However, the nested IF statement will only be executed when the oldest invoice due date is equal to the current date. In other words, if the current date equals the oldest invoice due date, both lines will be printed to the output window.

The syntax of the IF statement

```
IF boolean_expression THEN
  statements;
[ELSIF boolean_expression THEN
  statements;]...
[ELSE
  statements;]
END IF;
```

A script that uses an IF statement

```
SET SERVEROUTPUT ON;

DECLARE
  first_invoice_due_date DATE;

BEGIN
  SELECT MIN(invoice_due_date)
  INTO first_invoice_due_date
  FROM invoices
  WHERE invoice_total - payment_total - credit_total > 0;

  IF first_invoice_due_date < SYSDATE() THEN
    DBMS_OUTPUT.PUT_LINE('Outstanding invoices overdue!');
  ELSIF first_invoice_due_date = SYSDATE() THEN
    DBMS_OUTPUT.PUT_LINE('Outstanding invoices are due today!');
  ELSE
    DBMS_OUTPUT.PUT_LINE('No invoices are overdue.');
  END IF;
END;
/
```

The response from the system

```
Outstanding invoices overdue!
```

Description

- An *IF statement* executes statements depending on whether an expression is true or false.
- If included in an IF statement, the ELSE clause must be coded last. It only executes if all other expressions are false.
- You can only have one IF clause and up to one ELSE clause in an IF statement, but may have many ELSEIF clauses.
- You can nest an IF statement within another IF statement or within other PL/SQL statements.

Figure 14-5 How to code IF statements

How to code CASE statements

In chapter 8, you learned how to code a CASE expression within a SELECT statement. A CASE expression like that usually runs faster than a CASE statement that's coded within a PL/SQL script. As a result, if you can use a CASE expression to solve the task at hand, you should. However, there are times when you will need to use a CASE statement as shown in figure 14-6.

The PL/SQL block in this figure shows how to use a *simple CASE statement* to execute one or more statements depending on the value returned by an expression. To do that, you begin by coding the CASE keyword followed by an expression that returns a value. In this script, the expression is the terms_id_var variable, which returns an integer value that indicates the payment terms for the invoice with an ID of 4.

After the CASE clause, you can code one or more WHEN clauses. Each clause contains statements to be executed if the value of the WHEN clause matches the value of the CASE clause. In this example, there are three WHEN clauses, for the values of 1, 2, and 3. Each of these clauses prints an appropriate message.

With Oracle 23ai and later, you can alternately code a *dangling predicate* instead of an expression for a WHEN clause in a simple CASE statement. A dangling predicate is an expression without a left operand, such as '< 3' (less than 3) or '<> 100' (does not equal 100). Then, the value in the CASE clause is used in place of the missing left operand. If the expression is true, then the statements of the WHEN clause are executed. For example, this WHEN clause could be added to the simple case statement in this figure:

```
WHEN > 3 THEN
  DBMS_OUTPUT.PUT_LINE('Net due more than 30 days');
```

When you use dangling predicates, remember that only the statements in the first matched WHEN clause are executed.

After the WHEN clauses, you can code an optional ELSE clause that's executed if the value that's returned doesn't match any of the values coded in the WHEN clauses. This works much like the ELSE clause of the IF statement.

Alternately, you can use a *searched CASE statement* to execute statements depending on one or more Boolean expressions. This works similarly to an IF statement with a series of ELSIF clauses. For example, you can use a searched CASE statement to replace the IF statement in the previous figure like this:

```
CASE
  WHEN first_invoice_due_date < SYSDATE() THEN
    DBMS_OUTPUT.PUT_LINE('Outstanding invoices overdue!');
  WHEN first_invoice_due_date = SYSDATE() THEN
    DBMS_OUTPUT.PUT_LINE('Outstanding invoices are due today!');
  ELSE
    DBMS_OUTPUT.PUT_LINE('No invoices are overdue.');
END CASE;
```

Conversely, you can easily rewrite the simple CASE statement shown in this figure as an IF statement. Which you use is generally a matter of personal preference and which yields the code that's easiest to understand.

The syntax of the simple CASE statement

```
CASE expression
  WHEN value_1 THEN
    statements;
  [WHEN value_2 THEN
    statements;]...
  [ELSE
    Statements;]
END CASE;
```

A PL/SQL block with a simple CASE statement

```
DECLARE
  terms_id_var NUMBER;
BEGIN
  SELECT terms_id INTO terms_id_var
  FROM invoices WHERE invoice_id = 4;

  CASE terms_id_var
    WHEN 1 THEN
      DBMS_OUTPUT.PUT_LINE('Net due 10 days');
    WHEN 2 THEN
      DBMS_OUTPUT.PUT_LINE('Net due 20 days');
    WHEN 3 THEN
      DBMS_OUTPUT.PUT_LINE('Net due 30 days');
    ELSE
      DBMS_OUTPUT.PUT_LINE('Net due more than 30 days');
  END CASE;
END;
/
```

The response from the system

```
Net due 30 days
```

The syntax of a searched CASE statement

```
CASE
  WHEN boolean_expression THEN
    statements;
  [WHEN boolean_expression THEN
    statements;]...
  [ELSE
    statements;]
END CASE;
```

Description

- A *simple CASE statement* executes statements by matching the value returned by an expression to the values in a series of WHEN clauses.
- A *searched CASE statement* executes statements by searching for an expression that evaluates to true.
- WHEN clauses are evaluated in order. Only the statements in the first clause with a matching variable or true expression will be executed.
- With Oracle 23ai and later, you can use a *dangling predicate*, or expression without a left operand, in place of a value in a WHEN clause of a simple case statement.

Figure 14-6 How to code CASE statements

How to code loops

Figure 14-7 shows how to use three types of *loops* to repeatedly execute a block of statements. This figure starts by showing how to use a *FOR loop* to continue executing while a counter, or *iterand*, is within the specified range. This counter variable is traditionally named i and is not declared in the declarative section of a PL/SQL block. Instead, it is created when the FOR loop executes and is available only to statements inside the loop.

In the example, the code specifies an iterand named i that will start with a value of 1 and end with a value of 3 (1..3). Because the BY clause isn't used, the loop will use the default step size of 1. In other words, the iterand will increase by 1 each time the loop executes. As a result, the statement within the loop is executed three times. This statement within the loop just prints the current value of the iterand to the output window.

The syntax for the FOR loop also shows that you can use the REVERSE keyword to execute a FOR loop in reverse order. For example, you could code the first line of the FOR loop in this figure like this:

```
FOR i IN REVERSE 1..3 LOOP
```

Then, the loop would yield values of 3, 2, and 1.

After the FOR loop, this figure shows how to use a *WHILE loop*. A WHILE loop repeatedly executes statements while a Boolean expression is true. In this example, the WHILE loop executes as long as a counter variable named i is less than 4. However, using a counter in a WHILE loop is more complicated than in a FOR loop. You must first declare the counter variable in the declarative section. Then, you must assign a value to the counter before you begin the loop, and you must increment or decrement the counter within the body of the loop. In this example, the last statement in the loop adds a value of 1 to the counter variable.

Finally, this figure shows how to use a simple loop. A simple loop executes endlessly unless it encounters an EXIT or EXIT WHEN statement. In the example, the last statement in the loop uses an EXIT WHEN statement to exit the loop when i is greater than or equal to 4. Like a WHILE loop, the counter variable must previously be declared and assigned a value. Alternately, this loop could achieve the same result by using an EXIT statement within an IF statement like this:

```
IF i >= 4 THEN
  EXIT;
END IF;
```

If you need to jump to the beginning of a loop, you can use the CONTINUE or CONTINUE WHEN statements. These statements work like the EXIT and EXIT WHEN statements, except that they jump to the beginning of a loop instead of exiting a loop.

The examples in this figure are simple versions of loops. However, Oracle provides options for iterating through much more complex loops, including the ability to set upper and lower bounds on iterands, generating non-linear sequences for counter variables, and looping through pairs of iterands. For more information, consult the Oracle documentation.

The syntax of a FOR loop

```
FOR counter_var IN [REVERSE] counter_start..counter_end [BY step] LOOP
  statements;
END LOOP;
```

A FOR loop

```
FOR i IN 1..3 LOOP                          -- i should not be previously declared
  DBMS_OUTPUT.PUT_LINE('i: ' || i);
END LOOP;
```

The syntax of a WHILE loop

```
WHILE boolean_expression LOOP
  statements;
END LOOP;
```

A WHILE loop

```
i := 1;                                     -- i must already be declared
WHILE i < 4 LOOP
  DBMS_OUTPUT.PUT_LINE('i: ' || i);
  i := i + 1;
END LOOP;
```

The syntax of a simple loop

```
LOOP
  statements;
  EXIT WHEN boolean_expression;
END LOOP;
```

A simple loop

```
i := 1;                                     -- i must already be declared
LOOP
  DBMS_OUTPUT.PUT_LINE('i: ' || i);
  i := i + 1;
  EXIT WHEN i >= 4;
END LOOP;
```

The output for all three loops

```
i: 1
i: 2
i: 3
```

Description

- To execute a SQL statement repeatedly, you can use a *loop*. Oracle provides for three types of loops: *FOR loops*, *WHILE loops*, and *simple loops*.
- The counter variable, or *iterand*, of a FOR loop is created when the loop executes. It is not declared in the declarative section of a PL/SQL block.
- You can use the EXIT and EXIT WHEN keywords to go to the end of a loop.
- You can use the CONTINUE and CONTINUE WHEN keywords to go to the beginning of a loop.
- Loops have many more options for iteration. For more information, consult the Oracle documentation.

Figure 14-7 How to code loops

How to use a cursor

By default, SQL statements work with an entire result set rather than individual rows. However, there are times when you need to be able to work with the data in a result set one row at a time. To do that, you can use a *cursor* as described in figure 14-8.

A cursor allows you to fetch one row at a time from a result set. When used with a FOR loop, a cursor allows you to process rows individually. A cursor may be an *explicit cursor*, which is a SELECT statement that's named and defined in the declarative section of a PL/SQL block. Alternately, a cursor can be an *implicit cursor*, which is when the SELECT statement is coded directly in the header of a FOR loop.

In this figure, the example uses an explicit cursor. As a result, the declarative section of the script begins by declaring a cursor named invoices_cursor that contains two columns from the Invoices table and all of the rows that have a balance due. Then, this script declares a variable named invoice_row. This declaration uses the %ROWTYPE attribute to indicate that the invoice_row variable stores a collection of values with the same column names and data types as a row in the invoices table.

After these two variables have been declared, the executable section loops through each row in the cursor. To do that, the FOR loop uses the invoice_row variable as the iterand, and it uses the IN keyword to specify that the loop should continue for each row in the cursor. Within the FOR loop, an IF statement checks whether the invoice_total value of the current row is greater than 1000. If so, an UPDATE statement adds 10% of the invoice_total column to the credit_total column for the row. Then, it prints a message to show which rows have been updated.

Before you use a cursor to solve a particular problem, you should consider other solutions first. That's because standard database access is faster and uses fewer server resources than cursor-based access. For example, you can accomplish the same update as the script in this figure with this UPDATE statement:

```
UPDATE invoices
SET credit_total = credit_total + (invoice_total * .1)
WHERE invoice_total - payment_total - credit_total > 0
AND invoice_total > 1000
```

The syntax for declaring an explicit cursor

```
CURSOR cursor_name IS select_statement;
```

The syntax for declaring a row variable

```
row_variable_name table_name%ROWTYPE;
```

The syntax for getting a column value from a row variable

```
row_variable_name.column_name
```

A script that uses an explicit cursor

```
SET SERVEROUTPUT ON;

DECLARE
  CURSOR invoices_cursor IS
    SELECT invoice_id, invoice_total  FROM invoices
    WHERE invoice_total - payment_total - credit_total > 0;

  invoice_row invoices%ROWTYPE;

BEGIN
  FOR invoice_row IN invoices_cursor LOOP

    IF (invoice_row.invoice_total > 1000) THEN
      UPDATE invoices
      SET credit_total = credit_total + (invoice_total * .1)
      WHERE invoice_id = invoice_row.invoice_id;

      DBMS_OUTPUT.PUT_LINE('1 row updated where invoice_id = ' ||
                           invoice_row.invoice_id);
    END IF;

  END LOOP;
END;
/
```

The response from the system

```
1 row updated where invoice_id = 3
1 row updated where invoice_id = 6
1 row updated where invoice_id = 8
1 row updated where invoice_id = 19
1 row updated where invoice_id = 34
1 row updated where invoice_id = 81
1 row updated where invoice_id = 88
1 row updated where invoice_id = 113
```

Description

- You can use a *cursor* to fetch individual rows in a result set. When used with a FOR loop, a cursor allows you to individually process each row in the result set.
- The %ROWTYPE attribute creates a variable that can store a group of values corresponding to a row from the specified table.
- An *explicit cursor* is defined and named in the declarative section of a PL/SQL block.
- An *implicit cursor* is defined in the header of a FOR loop and does not have a name.

Figure 14-8 How to use a cursor

How to work with composite variables

Most of the variables you've used so far have been scalar variables, and contain only a single value. However, Oracle also provides for two kinds of *composite variables*: collections and records. These are variables that can contain multiple values.

How to use collections

A *collection* is a group of related values that all have the same data type. There are three types of collections available in PL/SQL: varrays, nested tables, and associative arrays. They work similarly but have some notable differences.

Before you can create a collection, you must first define the data type that your collection will use. You also define any other required aspects of the collection at that time. For example, a varray must specify the number of items that will be in the collection, while an associative array must specify its index type.

This first example in figure 14-9 shows how this works with a *varray* (vector array). In the declarative section, the first statement uses TYPE to declare a new data type named names_array. Each instance of a names_array will contain 3 values, and each of those values will be of the VARCHAR2(25) type. Then, the second statement declares a variable named names of the newly created names_array type. It also uses the assignment operator (:=) to assign the names John, Jane, and Joel to names.

In the executable section, the first line uses the index of the array to access the first value ('John') and change that value to 'Julia' instead. Finally, a FOR loop is used to print each value in the collection. Here, the COUNT method returns the total number of items in the collection. Then, the iterand is used as the index to return each value.

When you declare a varray, it has a fixed size. As a result, if you attempt to use an index to refer to an item outside the range of the varray, Oracle will raise an error. For example, Oracle will raise an error if you refer to a fourth item like this:

```
names(4) := 'Juan';
```

In contrast, when you declare a *nested table* as shown in the second example, you don't declare a size for the list. Instead, you code the TABLE keyword. Other than that, though, nested tables work similarly to varrays. If you want to increase the number of items that are stored by the list, though, you can call the EXTEND method and specify the number of items that you want to add. If, for example, you want to add one item to the names collection presented in the second example, you can code the EXTEND method like this:

```
names.EXTEND(1);
```

Then, you can use the fourth index to set a value for the new item like this:

```
names(4) := 'Juan';
```

The syntax for declaring a varray

```
TYPE type_name IS VARRAY(count) OF data_type;
collection_name type_name [:= type_name(value1 [, value2]...)];
```

The syntax for accessing a value in a collection

```
coll_name(index)
```

A block that uses a varray

```
DECLARE
  TYPE names_array IS VARRAY(3) OF VARCHAR2(25);
  names names_array := names_array('John', 'Jane', 'Joel');

BEGIN
  names(1) := 'Julia';

  FOR i IN 1..names.COUNT LOOP
    DBMS_OUTPUT.PUT_LINE('Name ' || i || ': ' || names(i));
  END LOOP;
END;
/
```

The response from the system

```
Name 1: Julia
Name 2: Jane
Name 3: Joel
```

The syntax for declaring a nested table

```
TYPE type_name        IS TABLE OF data_type;
collection_name       type_name := type_name(value1 [, value2]...);
```

A block that uses a nested table

```
DECLARE
  TYPE names_table         IS TABLE OF VARCHAR2(25);
  names                    names_table := names_table('John', 'Jane', 'Joel');
BEGIN
  names(1) := 'Julia';

  FOR i IN 1..names.COUNT LOOP
    DBMS_OUTPUT.PUT_LINE('Name ' || i || ': ' || names(i));
  END LOOP;
END;
/
```

Description

- A *composite variable* contains a collection of values. PL/SQL has two kinds of composite variables: collections and records.
- You can use a *collection* to store related values of the same data type.
- PL/SQL has three types of collections: *varrays*, *nested tables*, and *associative arrays*.
- Varrays collections always have a specified length while nested table collections do not.
- You can get the length of a collection by using the COUNT method.

Figure 14-9 How to use collections (part 1 of 2)

The third example shows how to use an *associative array*. This type of collection uses either an integer- or character-based index to store *key-value pairs*. The keys must be unique, but do not need to be sequential.

This example begins by creating an associative array type named names_as_array. It stores values with the VARCHAR2(25) type and uses PLS_INTEGER for its index, or key values. Then, a variable named names of this type is declared but does not yet have any values assigned to it, so it's empty.

After the declaration section, the executable section begins by using non-sequential integers as keys for the stored values. Any number of values can be added this way. However, re-using an index will replace the value stored for that key. For example, the first statement assigns 'John' to the value at index 76, but the fourth statement replaces the value with 'Julia'.

After these assignment statements, a FOR loop uses the FIRST and LAST methods of a collection to loop from the first index to the last index. In other words, it loops from 76 to 123. Within the body of the FOR loop, an IF statement uses the EXISTS method to check if a value exists for the specified index. If so, it prints the value for the item to the output window. This prevents an error if you attempt to access a value that doesn't exist in the array.

The fourth example shows how to use a SELECT statement to quickly fill a collection with values. This example begins by declaring a nested table collection of the VARCHAR2(40) type and assigning this collection type to a variable named vendor_names. Then, the executable section uses a SELECT statement with a BULK COLLECT clause to fill this variable with values that are selected from the vendor_name column of the Vendors table. In this example, the SELECT statement only returns the first three rows. However, you can use the same technique to fill a collection with a large number of values.

Note that the syntax for using the BULK COLLECT clause is similar to the syntax for setting a variable to a single selected value. However, with a BULK COLLECT clause the target variable must be defined as one of the collection types.

The syntax for declaring an associative array

```
TYPE type_name       IS TABLE OF data_type [NOT NULL]
                     INDEX BY {PLS_INTEGER|VARCHAR2(size)};
collection_name      type_name := type_name(value1 [, value2]...);
```

A block that uses an associative array

```
DECLARE
  TYPE names_as_array        IS TABLE OF VARCHAR2(25) INDEX BY PLS_INTEGER;
  names                      names_as_array;

BEGIN
  names(76)  := 'John';
  names(100)  := 'Jane';
  names(123)  := 'Joel';
  names(76)  := 'Julia';

  FOR i IN names.FIRST..names.LAST LOOP
    IF names.EXISTS(i) THEN
      DBMS_OUTPUT.PUT_LINE('Name ' || i || ': ' || names(i));
    END IF;
  END LOOP;
END;
/
```

A block that uses a BULK COLLECT clause to fill a collection

```
DECLARE
  TYPE names_table          IS TABLE OF VARCHAR2(40);
  vendor_names              names_table;

BEGIN
  SELECT vendor_name
  BULK COLLECT INTO vendor_names
  FROM vendors
  ORDER BY vendor_id
  FETCH FIRST 3 ROWS ONLY;

  FOR i IN 1..vendor_names.COUNT LOOP
    DBMS_OUTPUT.PUT_LINE('Vendor name ' || i || ': ' || vendor_names(i));
  END LOOP;
END;
/
```

The response from the system

```
Vendor name 1: US Postal Service
Vendor name 2: National Information Data Ctr
Vendor name 3: Register of Copyrights
```

Description

- An associative array collection uses either an integer- or character-based index to create a list of *key-value pairs*. An associative array may have any number of key-value pairs.
- Collections have assorted methods available, such as FIRST, LAST, EXISTS, and EXTEND.
- You can use BULK COLLECT to retrieve a list of values into a collection.

Figure 14-9 How to use collections (part 2 of 2)

How to use records

A collection can only contain items that have the same data type. A *record*, on the other hand, can contain items with many different data types.

To declare a record, you can declare a record type and then declare a variable of that type. When you declare a record type, you list a name and data type for each item, separating each item with a comma. Then, you can access items by coding the name of the variable, the dot operator, and the name of the column.

If this sounds similar to creating a table, that's because it is. Records are typically used to store a row of data that's retrieved from a table or a row that will be inserted into a table.

The first example in figure 14-10 shows how this works. In the declaration section, the first statement creates a type named term_record that specifies three items: a number, a string, and another number. Then, it creates an empty variable of the term_record type named terms_row.

In the declaration section, the names of the record items aren't the same as the names of the columns in the Terms table. However, these items have the same order as the columns in the Terms table. This makes it possible to insert a variable of the term_record type into the Terms table.

In the executable section, the first three statements that set the items of the record to appropriate values for a new row in the Terms table. Then, the fourth statement inserts the values stored in the record into the Terms table. The Terms table shown in this figure shows that the new row was inserted by the first example.

The second example shows how to use the %ROWTYPE attribute to declare a record variable named terms_row. This is a more concise way to declare the terms_row variable than the two declaration statements in the first example. However, when you use this approach, the names of the items are the same as the names of the columns. As a result, you need to use the column names to access the items. For example, you can set the value of the first item like this:

```
terms_row.terms_id := 7;
```

The syntax for declaring a record

```
TYPE type_name      IS RECORD (col_name data_type[, col_name data_type]...)
record_name         type_name := type_name(value1 [, value2]...);
```

A block that uses a record to create and insert a new row

```
DECLARE
  TYPE term_record          IS RECORD (id NUMBER,
                                       descr VARCHAR2(50),
                                       due_days NUMBER);
  terms_row                 term_record;
BEGIN
  terms_row.id := 6;
  terms_row.due_days := 120;
  terms_row.descr := 'Net due 120 days';

  INSERT INTO terms
  VALUES terms_row;
END;
/
```

The Terms table after the new row has been inserted

	TERMS_ID	TERMS_DESCRIPTION	TERMS_DUE_DAYS
1	1	Net due 10 days	10
2	2	Net due 20 days	20
3	3	Net due 30 days	30
4	4	Net due 60 days	60
5	5	Net due 90 days	90
6	6	Net due 120 days	120

Another way to declare a record

```
terms_row           terms%ROWTYPE;
```

Description

- You can use a *record* to store a list of related values of varying data types.
- Records are commonly used to store rows of data for a table.
- You can use the %ROWTYPE attribute to create a record that uses the same column names and data types as the specified table.

Figure 14-10 How to use records

How to work with exceptions and errors

In programming, it's inevitable that your code will eventually cause an exception or error. That's why it's important to anticipate possible exceptions and handle them appropriately, whether that's resubmitting data, rolling back a transaction, or displaying a user-friendly message.

How to handle exceptions

Figure 14-11 begins by showing a script that doesn't handle exceptions. Here, the INSERT statement attempts to insert a duplicate value ('Cash') into a column (account_description) that has been defined with a unique key. This causes Oracle to raise an exception. Since this exception isn't handled, the system displays an error message like the one that's shown.

Although error messages like this can be helpful to a developer, they aren't usually helpful to the end user of an application. As a result, you typically want to handle exceptions before you put your PL/SQL code into production. To do that, you can begin by looking up any errors that you discover during testing.

If, for example, you look up error number 00001, you'll see that it has a name of DUP_VAL_ON_INDEX. Then, you can use the EXCEPTION section of a PL/SQL block to display a more user-friendly message to the user. Or, you can use the EXCEPTION section to perform other error handling tasks, such as writing information about the error to a log table or rolling back a transaction.

The second script in this figure contains two groups of statements for handling exceptions. The first handles the DUP_VAL_ON_INDEX exception that's thrown by the first script. The second handles any other exceptions that might be thrown. Both of these sections print user-friendly messages that describe the error that has occurred. However, the second statement in the WHEN OTHERS clause uses the SQLERRM function to print the error number and a short description.

Oracle checks the WHEN clauses in order, and only executes the first WHEN clause that matches the exception. Since the WHEN OTHERS clause matches all exceptions, it must be coded last. Otherwise, you'll get an error when you attempt to run your script.

To test this script, you can change the values in the INSERT statement. For example, if you enter 'xx' as the first value for the INSERT statement, the script will print the first line shown in the second WHEN clause. Then, it will print the string that's returned by the SQLERRM function, like this:

```
An unexpected exception occurred.
ORA-01722: invalid number
```

However, if you enter valid values, the script will print a message that indicates that 1 row was inserted.

A script that doesn't handle exceptions

```
INSERT INTO general_ledger_accounts VALUES (130, 'Cash');
```

The response from the system

```
SQL Error: ORA-00001: unique constraint (AP.GL_ACCOUNT_DESCRIPTION_UQ) violated
00001. 00000 -  "unique constraint (%s.%s) violated"
*Cause:     An UPDATE or INSERT statement attempted to insert a duplicate key.
            For Trusted Oracle configured in DBMS MAC mode, you may see
            this message if a duplicate entry exists at a different level.
*Action:    Either remove the unique restriction or do not insert the key.
```

The syntax of the error-handling section of a PL/SQL block

```
EXCEPTION
  WHEN most_specific_exception THEN
    statements;
  [WHEN less_specific_exception THEN
    statements;]...
  [WHEN OTHERS THEN
    statements;]
```

A script with error handling

```
SET SERVEROUTPUT ON;

BEGIN
  INSERT INTO general_ledger_accounts VALUES (130, 'Cash');

  DBMS_OUTPUT.PUT_LINE('1 row inserted.');

EXCEPTION
  WHEN DUP_VAL_ON_INDEX THEN
    DBMS_OUTPUT.PUT_LINE('You attempted to insert a duplicate value.');

  WHEN OTHERS THEN
    DBMS_OUTPUT.PUT_LINE('An unexpected exception occurred.');
    DBMS_OUTPUT.PUT_LINE(SQLERRM);
END;
/
```

The response from the system

```
You attempted to insert a duplicate value.
```

Description

- In a PL/SQL block, you can use the error-handling section to handle exceptions.
- Within a WHEN clause in the error-handling section, you can specify the name of a specific exception to handle, or you can use the OTHERS keyword to handle all exceptions that haven't already been handled by a previous WHEN clause.
- The first WHEN clause with a matching error is executed. Other WHEN clauses are then skipped.
- If included, a WHEN OTHERS clause must be coded last in the exception-handling section.
- The SQLERRM function returns a string that contains an error number and short message for the most recent error raised.

Figure 14-11 How to handle exceptions

Predefined exceptions

Figure 14-12 begins by listing five of the *predefined exceptions* that are defined and raised by Oracle. These exceptions should give you an idea of the types of predefined exceptions that are provided by Oracle. For example, the first exception in this list shows the DUP_VAL_ON_INDEX exception that was described in the previous figure. Note that its ORA error number (00001) corresponds with the error number that's shown in the error message in the previous figure.

A list of common predefined exceptions

ORA Error	Exception	Occurs when...
00001	DUP_VAL_ON_INDEX	Your code attempts to store duplicate values in a column that has a unique constraint or index.
01403	NO_DATA_FOUND	A SELECT INTO statement doesn't return any data.
01476	ZERO_DIVIDE	Your code attempts to divide a number by zero.
01722	INVALID_NUMBER	Your code fails to convert a string into a number. This is usually because the characters in the string do not form a valid number.
06502	VALUE_ERROR	Your code encounters an arithmetic, conversion, truncation, or size-constraint error.

Description

- For a complete list of predefined exceptions, consult the Oracle documentation.

Figure 14-12 Predefined exceptions

How to drop database objects without displaying errors

Frequently, you'll need to write scripts that create database objects. For example, the script that you used to create the AP schema creates all of the tables and sequences for the database. However, if you attempt to create an object that already exists, Oracle returns an error and the script doesn't execute correctly. For example, if you try to create the Vendors table and that table already exists, Oracle returns an error that indicates that the table can't be created because it already exists.

To solve this problem prior to Oracle 23ai, you can code a DROP statement that drops the database object before you create it as shown by the first example in figure 14-13. Although that allows the script to successfully create the table, it also causes an error to be displayed if the database object doesn't exist. For example, if the table named Test1 doesn't exist, this script will display an error like the one that's shown.

To solve this additional problem, you can use an anonymous PL/SQL block to handle the error as shown by the second script in this figure. Here, the BEGIN section of the PL/SQL block uses the EXECUTE IMMEDIATE command to execute the DROP TABLE statement that drops the table named Test1. Then, the EXCEPTION section of the PL/SQL block handles this "error" by using the NULL statement to do nothing.

However, if you are using Oracle 23ai or later, you can avoid using PL/SQL at all in this scenario. To do that, you can use the corresponding IF EXISTS and IF NOT EXISTS clauses of most DROP and CREATE statements. This is an easier and better way to suppress error messages. Still, it's good to be aware of the old PL/SQL technique in case you find yourself working with an older version of Oracle.

The syntax of the EXECUTE IMMEDIATE statement

```
EXECUTE IMMEDIATE 'sql_string'
```

A script that will display an error if the object doesn't already exist

```
DROP TABLE test1;
CREATE TABLE test1 (test_id NUMBER);
```

The response from the system

```
Error starting at line : 1 in command -
DROP TABLE test1
Error report -
ORA-00942: table or view does not exist
00942. 00000 -  "table or view does not exist"
*Cause:
*Action:

Table TEST1 created.
```

A script that will execute correctly without displaying an error

```
BEGIN
  EXECUTE IMMEDIATE 'DROP TABLE test1';
EXCEPTION
  WHEN OTHERS THEN
    NULL;
END;
/

CREATE TABLE test1 (test_id NUMBER);
```

The response from the system

```
Table TEST1 created.
```

Description

- To execute DDL statements within a PL/SQL block, you can use the EXECUTE IMMEDIATE statement.
- You can use a NULL statement if a clause requires a statement but you don't want to execute any code.
- With Oracle 23ai and later, you can instead use the IF EXISTS and IF NOT EXISTS clauses of most DROP and CREATE statements to suppress errors.

Figure 14-13 How to drop database objects without displaying errors

Perspective

In this chapter, you learned how to use PL/SQL to write procedural code. You also learned how to declare and use variables, including your own exceptions. You also can control the flow of execution by using IF and CASE statements and FOR, WHILE, and simple loops. You now have a foundation of PL/SQL skills that you can use to build on your skills in the next three chapters of this section.

Terms

PL/SQL (Procedural Language/SQL)
PL/SQL block
anonymous PL/SQL block
command
procedure
package
variable
scalar variable
declare a variable
IF statement
Boolean expression
simple CASE statement
searched CASE statement
dangling predicate
loop
FOR loop
WHILE loop
iterand
cursor
explicit cursor
implicit cursor
composite variable
collection
varray
nested table
associative array
key-value pair
record
predefined exception

Exercises

1. Write a script that declares and sets a variable that's equal to the count of all rows in the Invoices table that have a balance due that's greater than or equal to $5,000. Then, the script should display a message that looks like this:

   ```
   3 invoices exceed $5,000.
   ```

2. Write a script that uses variables to get the count of all of the invoices in the Invoices table that have a balance due and the sum of the balances due for all of those invoices. If that total balance due is greater than or equal to $50,000, the script should display a message like this:

   ```
   Number of unpaid invoices is 40.
   Total balance due is $66,796.24.
   ```

   Otherwise, the script should display this message:

   ```
   Total balance due is less than $50,000.
   ```

3. Write a script that creates a cursor consisting of vendor_name, invoice_number, and balance_due for each invoice with a balance due that's greater than or equal to $5,000. The rows in this cursor should be sorted in descending sequence by balance due. Then, for each invoice, display the balance due, invoice number, and vendor name so it looks something like this:

   ```
   $19,351.18    P-0608    Malloy Lithographing Inc
   ```

4. Enhance your solution to exercise 3 so it shows the invoice data in three groups based on the balance due amounts with these headings:

```
$20,000 or More
$10,000 to $20,000
$5,000 to $10,000
```

Each group should have a heading followed by the data for the invoices that fall into that group. Also, the groups should be separated by one blank line.

15

How to manage transactions and locking

If you've been working with Oracle on your own computer, you've been the only user of your database. In the real world, though, a database may be used by thousands of users at the same time. Then, what happens when two users try to update the same data at the same time? In this chapter, you'll learn how Oracle handles this situation. But first, you'll learn how to combine multiple SQL statements into a single logical unit of work known as a transaction.

How to work with transactions 440
How to commit and roll back transactions 440
How to work with save points 442
How to work with concurrency and locking 444
How concurrency and locking are related 444
How to set the transaction isolation level 446
Best practices for concurrency 448
Perspective 450

How to work with transactions

A *transaction* is a group of SQL statements that you combine into a single logical unit of work. By combining SQL statements like this, you can prevent certain kinds of database errors.

How to commit and roll back transactions

Figure 15-1 starts by presenting a script that includes a PL/SQL block that contains three INSERT statements coded together as a transaction. Within the block, the SET TRANSACTION statement starts a transaction that can read and write data, which is what you want for most transactions. Then, the block includes three INSERT statements. The first adds a new invoice to the Invoices table, and the next two add line items for the invoice to the Invoice_Line_Items table. Next, the COMMIT statement saves, or *commits*, the changes to the database. This ends the current transaction and makes the changes visible to other transactions.

On the other hand, if any of the previous statements raise an error, execution jumps into the exception-handling part of the PL/SQL block. Then, the first statement in the WHEN OTHERS THEN clause uses the ROLLBACK statement to undo, or *roll back*, the current transaction. This also ends the current transaction.

To understand why a transaction is necessary for these three INSERT statements, suppose that you split them into three separate transactions by coding a COMMIT statement after each one. What happens if the third INSERT statement fails? In that case, the Invoices and Invoice_Line_Items tables wouldn't match. Specifically, the sum of the line item amounts in the Invoice_Line_Items table wouldn't be equal to the invoice total in the Invoices table. In other words, data integrity wouldn't be maintained.

In this situation, the solution is to require the statements to all execute successfully together or else all fail together. The key to working with transactions is identifying which statements rely on each other. If you include more statements in a transaction than necessary, you risk rolling back statements you didn't mean to.

If the current session hasn't already started a transaction, Oracle automatically starts one when it runs a DML statement such as an INSERT, UPDATE, or DELETE statement. And once a transaction is started, Oracle adds additional DML statements to the current transaction.

However, many developers consider it a good practice to use a SET TRANSACTION statement like the one shown in this figure to explicitly identify the start of a transaction. This statement raises an error if the current session has already started a transaction. As a result, it provides a way for you to make sure that only the statements you intend are part of the new transaction.

A transaction ends when either a COMMIT or ROLLBACK statement runs. It also ends when a DDL statement such as CREATE TABLE runs by committing all changes in the current transaction.

A script with three INSERT statements coded as a single transaction

```
SET SERVEROUTPUT ON;

BEGIN
  SET TRANSACTION READ WRITE;

  INSERT INTO invoices
  VALUES (115, 34, 'ZXA-080', '30-AUG-24',
          14092.59, 0, 0, 3, '30-SEP-24', NULL);

  INSERT INTO invoice_line_items
  VALUES (115, 1, 160, 4447.23, 'HW upgrade');

  INSERT INTO invoice_line_items
  VALUES (115, 2, 167, 9645.36, 'OS upgrade');

  COMMIT;
  DBMS_OUTPUT.PUT_LINE('The transaction was committed.');
EXCEPTION
  WHEN OTHERS THEN
    ROLLBACK;
    DBMS_OUTPUT.PUT_LINE('The transaction was rolled back.');
END;
/
```

Use a transaction when...

- Two or more INSERT, UPDATE, or DELETE statements affect related data
- Using INSERT and DELETE statements to move rows from one table to another
- The failure of an INSERT, UPDATE, or DELETE statement would violate data integrity

A transaction begins when...

- A SET TRANSACTION statement runs
- A DML statement such as UPDATE, INSERT, or DELETE runs

A transaction ends when...

- A COMMIT statement runs
- A ROLLBACK statement runs
- A DDL statement such as CREATE TABLE, ALTER TABLE, or DROP TABLE runs

Description

- A *transaction* is a group of SQL statements that must all be successfully executed together.
- You can use the SET TRANSACTION statement to start a transaction. This raises an error if another transaction has already been started for the current session.
- By default, Oracle adds all INSERT, UPDATE, and DELETE statements to the current transaction.
- To save, or *commit*, a transaction to the database, you code a COMMIT statement.
- To undo, or *roll back*, a transaction, you code a ROLLBACK statement.

Figure 15-1 How to commit and roll back transactions

How to work with save points

The script in figure 15-2 shows how to use the SAVEPOINT statement to identify one or more *save points* within a transaction. Here, a SAVEPOINT statement identifies a save point before each of the INSERT statements that are included in the script. As a result, the script includes two save points.

This script also shows how to use the ROLLBACK statement to roll back all or part of a transaction. Here, the two ROLLBACK statements roll back the transaction to each of the save points. The first rolls back to the point before the line item was inserted, and the second rolls back to the point before the invoice was inserted.

After the ROLLBACK statements, the script calls the COMMIT statement to commit any changes that have been made. However, the ROLLBACK statements have already rolled back both INSERT statements. As a result, this doesn't commit any changes to the database. To verify this, you can use a SELECT statement to view the rows in the Invoices and Invoice_Line_Items tables that have an invoice_id of 116.

In general, save points are used when a transaction contains so many statements that rolling back the entire transaction would be inefficient. In that case, an application can roll back to the last save point before an error occurred and continue from there. In practice, however, it's better to code transactions with as few statements as possible.

When you roll back to a save point, any save points after the save point you rolled back to are deleted. For example, if you roll back to the before_invoice save point in this figure and then attempt to roll back to the before_line_item save point, Oracle would raise an error. Similarly, when you end a transaction by committing or rolling back a transaction, Oracle deletes all save points for that transaction.

A script that uses save points

```
SAVEPOINT before_invoice;

INSERT INTO invoices
VALUES (116, 48, 'AQ-15x', '15-JUL-24',
        5640.22, 0, 0, 3, '15-AUG-24', NULL);

SAVEPOINT before_line_item;

INSERT INTO invoice_line_items
VALUES (116, 1, 150, 5640.22, 'New chairs');

ROLLBACK TO before_line_item;
ROLLBACK TO before_invoice;

COMMIT;
```

The response from the system

```
SAVEPOINT before_invoice succeeded.
1 row inserted.
SAVEPOINT before_line_item succeeded.
1 row inserted.
ROLLBACK TO succeeded.
ROLLBACK TO succeeded.
COMMIT succeeded.
```

Description

- When you use *save points*, you can roll back a transaction to the beginning of the transaction or to a particular save point.
- You can use the SAVEPOINT statement to create a save point with the specified name.
- You can use the ROLLBACK statement to roll back a transaction to the specified save point. To do that, you can use the TO clause to identify the save point.
- Save points are useful when a single transaction contains so many SQL statements that rolling back the entire transaction would be inefficient.
- When you roll back to a save point, any subsequent save points are deleted.
- When you end a transaction, all save points in that transaction are deleted.

Figure 15-2 How to work with save points

How to work with concurrency and locking

When two or more users have access to the same database, it's possible for them to be working with the same data at the same time. This is called *concurrency*. Although concurrency isn't a problem when two users view the same data at the same time, it can become a problem when one user updates data that other users are also viewing or updating. In the topics that follow, you'll learn how Oracle handles concurrency and some basic skills for working with it.

How concurrency and locking are related

Figure 15-3 presents two transactions that show how Oracle handles concurrency by default. To start, transaction A submits an UPDATE statement that adds 100 to the credit_total column of the invoice that has an invoice_id of 6. Because transaction A hasn't yet committed this change to the database, Oracle places a *lock* on this row. This prevents transaction B from modifying the locked data until transaction A ends and releases the lock.

At this point, transaction B can run the SELECT statement to view the data in the row, even though the row is locked. However, the result set won't include the updated value in the credit_total column. In other words, the SELECT statement can only read changes that have been committed.

On the other hand, the UPDATE statement in transaction B can't update the row due to the lock that transaction A has on the row. As a result, it waits for transaction A to release its lock on the row.

Once transaction A commits the update, it releases its lock on the row, and transaction B runs its UPDATE statement. Since transaction A has committed the updated credit total data to the database, the UPDATE statement in transaction B can read the updated credit total before updating it again. As a result, this UPDATE statement adds another 200 to the previous credit total.

To experiment with concurrency, you can simulate multiple users by starting multiple instances of SQL Developer. For example, you can start two instances of SQL Developer and execute transaction A in the first instance and transaction B in the second instance. Then, you can click the Execute Statement button or press Ctrl+Enter to run one statement at a time. This allows you to slow down the execution of each script. Otherwise, if you click the Run Script button or press F5, the script will run so quickly that you won't be able to get both scripts to access the same row at the same time.

In Oracle, locks are always placed at the most granular level possible. In other words, if an UPDATE statement modifies one row, Oracle attempts to only lock that row. As a result, other rows in the table can still be modified by other transactions. Furthermore, Oracle does not provide for read locks, which prevent another user or transaction from reading data. You should keep this in mind as you code your transactions.

Two transactions that modify the data in the same row

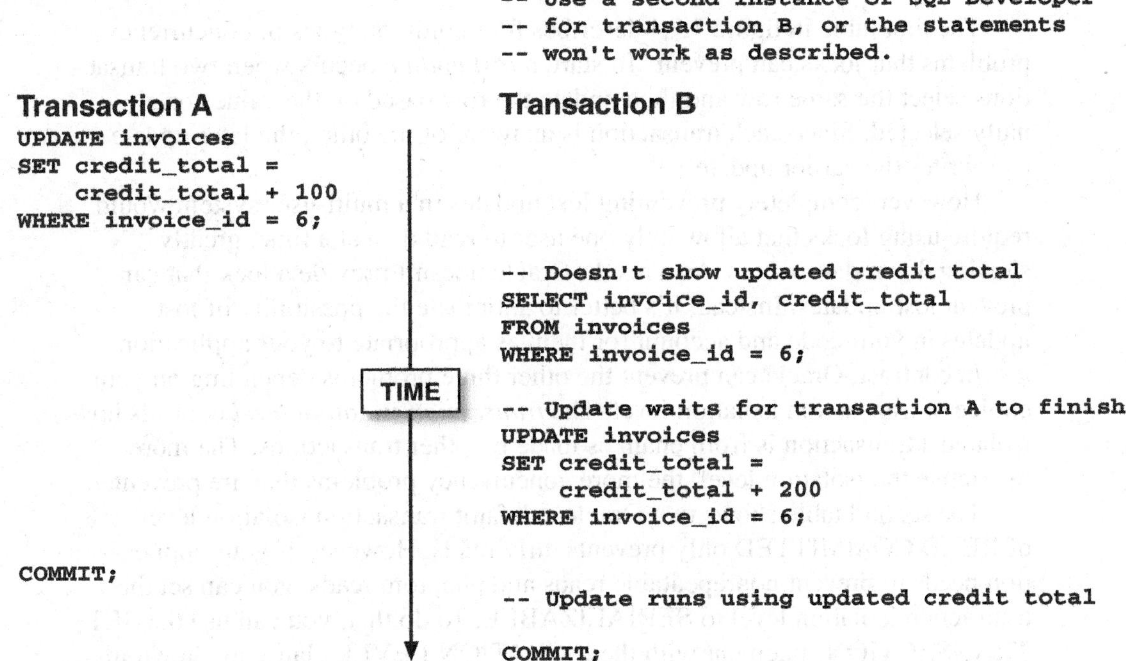

Description

- *Concurrency* is the ability of a system to support two or more transactions working with the same data at the same time.
- A *lock* prevents other users and transactions from modifying the locked data until the lock is released.
- In Oracle, locking is handled automatically.
- Oracle always locks the smallest amount of data possible.
- Oracle does not provide read locks, which prevent two users or transactions from reading the same data at the same time.

Figure 15-3 How concurrency and locking are related

How to set the transaction isolation level

The first table in figure 15-4 describes four common types of concurrency problems that locks can prevent. To start, a *lost update* occurs when two transactions select the same row and then update the row based on the values originally selected. Since each transaction is unaware of the other, the later update overwrites the earlier update.

However, completely preventing lost updates in a multi-user system would require using locks that allow only one user to read data at a time, greatly slowing down data access. As a result, Oracle doesn't provide a lock that can prevent lost updates. Instead, it's better to anticipate the possibility of lost updates in your code and account for them as appropriate to your application.

In contrast, Oracle can prevent the other three problems depending on your choice of transaction isolation level. The *transaction isolation level* controls how isolated a transaction is from changes made by other transactions. The more restrictive the isolation level, the more concurrency problems that are prevented.

The second table shows that Oracle's default transaction isolation level of READ COMMITTED only prevents dirty reads. However, if your application needs to prevent nonrepeatable reads and phantom reads, you can set the transaction isolation level to SERIALIZABLE. To do that, you can use the SET TRANSACTION statement with the ISOLATION LEVEL clause as shown in the example in this figure. This isolates the transaction from any other uncommitted transactions.

The statements in a serializable transaction see the database as it was at the beginning of the transaction plus any changes executed in the transaction. If a serializable transaction attempts to modify data that has been changed and committed after the transaction began, Oracle raises an error.

Since the SERIALIZABLE level eliminates all preventable concurrency problems, you may think that this is always the best option. However, the access time for each transaction is increased because statements must wait as locks are created and released. As a result, you typically only want to use the SERIALIZABLE level for situations in which concurrency problems aren't acceptable.

Four types of concurrency problems

Problem	Description
Lost updates	Occur when two transactions select the same row and then update the row based on the values originally selected. Since each transaction is unaware of the other, the later update overwrites the earlier update.
Dirty reads	Occur when a transaction selects data that hasn't been committed by another transaction. For example, transaction A changes a row. Transaction B then selects the changed row before transaction A commits the change. If transaction A then rolls back the change, transaction B has selected a row that doesn't exist in the database.
Nonrepeatable reads	Occur when two SELECT statements that try to get the same data get different values because another transaction has updated the data in the time between the two statements. For example, transaction A selects a row. Transaction B then updates the row. When transaction A selects the same row again, the data is different.
Phantom reads	Occur when you perform an update or delete on a set of rows at the same time that another transaction is performing an insert or delete that affects one or more rows in that same set of rows. For example, transaction A updates the payment total for each invoice that has a balance due, but transaction B inserts a new, unpaid, invoice while transaction A is still running. After transaction A finishes, there is still an invoice with a balance due.

The transaction isolation levels supported by Oracle

Isolation level	Lost updates	Dirty reads	Nonrepeatable reads	Phantom reads
READ COMMITTED	Allows	Prevents	Allows	Allows
SERIALIZABLE	Allows	Prevents	Prevents	Prevents

The syntax for setting the transaction isolation level

```
SET TRANSACTION ISOLATION LEVEL {SERIALIZABLE|READ COMMITTED}
```

Set the transaction isolation level to SERIALIZABLE

```
SET TRANSACTION ISOLATION LEVEL SERIALIZABLE;
```

Description

- *Lost updates* can't be automatically prevented by a database like Oracle that allows more than one connection to read the same data.
- Oracle always prevents *dirty reads.*
- By default, Oracle allows *nonrepeatable* and *phantom reads* because they speed up data access. However, if you need to prevent these problems, you can set the transaction isolation level to SERIALIZABLE.
- The *transaction isolation level* controls the degree to which transactions are isolated from one another.

Figure 15-4 How to set the transaction isolation level

Best practices for concurrency

Figure 15-5 lists several best practices for coding transactions and managing concurrency. First, keep transactions short. In other words, don't leave transactions open any longer than is necessary. That's because the longer a transaction remains open and uncommitted, the more likely it is that another transaction will need to work with that same resource. Towards that end, don't include SELECT statements or requests for user input within a transaction.

Second, you shouldn't use a higher isolation level than you need. That's because the higher you set the isolation level, the more likely it is that two transactions will be unable to work with the same resource at the same time, slowing down database access.

Third, you should schedule transactions that modify a large number of rows to run when no other transactions, or only a small number of other transactions, will be running. That way, it's less likely that two transactions will try to change the same rows at the same time.

Fourth, you should consider how the SQL statements you write could cause a deadlock. A *deadlock* occurs when neither of two transactions can be committed because each has a lock on a resource needed by the other transaction. This is illustrated by the banking transactions in the example. Here, transaction A updates the savings account first and then the checking account, while transaction B updates the checking account first and then the savings account.

Now, suppose that the first statement in transaction A locks the savings account, and the first statement in transaction B locks the checking account. At that point, a deadlock occurs because transaction A needs the checking account and transaction B needs the savings account, but both are locked.

When this happens, Oracle rolls back the statement that detects the deadlock so the other can proceed. This prevents the database from being endlessly deadlocked, but at the expense of a failed UPDATE statement. So, when you're coding transactions, it's a best practice to update related data in the same sequence.

Keep transactions short

- Keep SELECT statements outside of the transaction except when absolutely necessary.
- Never code requests for user input during a transaction.

Use the lowest possible transaction isolation level

- The READ COMMITTED level, which is the default, is usually sufficient.
- Reserve the use of the SERIALIZABLE level for short transactions that make changes to data where integrity is vital.

Make large changes when you can be assured of nearly exclusive access

- If you need to change millions of rows in an active table, don't do so during hours of peak usage.
- If possible, give yourself exclusive access to the database before making large changes.

Consider locking when coding your transactions

- If you need to code two or more transactions that update the same resources, code the updates in the same order in each transaction.

UPDATE statements that illustrate deadlocking

Transaction A

```
UPDATE savings SET balance = balance - transfer_amount;
UPDATE checking SET balance = balance + transfer_amount;
COMMIT;
```

Transaction B (possible deadlock)

```
UPDATE checking SET balance = balance - transfer_amount;
UPDATE savings SET balance = balance + transfer_amount;
COMMIT;
```

Transaction B (prevents deadlocks)

```
UPDATE savings SET balance = balance + transfer_amount;
UPDATE checking SET balance = balance - transfer_amount;
COMMIT;
```

Description

- A *deadlock* occurs when neither of two transactions can be committed because each transaction has a lock on a resource needed by the other transaction.
- Oracle automatically detects deadlocks and resolves them by rolling back the statement that detected the deadlock.

Figure 15-5 Best practices for concurrency

Perspective

In this chapter, you've learned what an application developer needs to know about transactions, concurrency, and locking. Although Oracle handles the bulk of this work for you, understanding Oracle's default behavior allows you to confidently manage your data even for a database that has many simultaneous connections.

Terms

transaction
commit a transaction
roll back a transaction
save point
concurrency
lock
lost update
dirty read
nonrepeatable read
phantom read
transaction isolation level
deadlock

Exercises

1. Write a set of three SQL statements coded as a transaction to reflect the following change: United Parcel Service has been purchased by Federal Express Corporation and the new company is named FedUP. Rename one of the vendors and delete the other after updating the vendor_id column in the Invoices table.
2. Write a set of two SQL statements coded as a transaction to delete the row with an invoice ID of 114 from the Invoices table. To do this, you must first delete all line items for that invoice from the Invoice_Line_Items table.

16

How to create stored procedures and functions

In chapter 14, you learned how to include anonymous blocks of PL/SQL code in a SQL script. In this chapter, you'll learn how to extend this functionality by storing named blocks of PL/SQL as database objects. These objects can be executed by any database user who has appropriate privileges. This allows you to store procedural logic such as data validation in a central location while controlling how users access the database.

How to code stored procedures 452
How to create a stored procedure 452
How to call a stored procedure 454
How to code input and output parameters 456
How to code optional parameters 458
How to raise a predefined exception 460
How to raise a user-defined exception 462
A stored procedure that inserts a row 464
How to drop a stored procedure 468
How to code functions 470
How to create and call a function 470
A function that uses multiple RETURN statements 472
How to drop a function 474
How to work with packages 476
How to create a package 476
How to drop a package 478
How to use SQL Developer 480
How to view and drop procedures, functions, and packages 480
How to edit and compile procedures and functions 482
How to grant and revoke privileges 482
How to debug procedures and functions 484
Perspective 488

How to code stored procedures

So far, this book has shown how to code anonymous blocks of PL/SQL that aren't stored in a database. Now, you'll learn how to code *stored programs*, which are named blocks of PL/SQL that are stored in the database so you can reuse them. In particular, you'll learn how to code a type of stored program known as a *stored procedure*.

A stored procedure, which can also be called a *sproc* or just a *procedure*, typically performs an action such as modifying the data that's stored within a database. For example, you can use a stored procedure to execute an INSERT, UPDATE, or DELETE statement.

How to create a stored procedure

Figure 16-1 shows how to use the CREATE PROCEDURE statement to create a stored procedure. To start, you code the CREATE keyword, followed by the optional OR REPLACE keywords, followed by the PROCEDURE keyword, followed by the name of the procedure. In this figure, for example, the statement creates a procedure named update_invoices_credit_total. Since this statement includes the OR REPLACE keywords, it creates the procedure if it doesn't already exist, and it replaces the procedure if it does already exist.

After the name of the procedure, you code a set of parentheses. Within the parentheses, you can code one or more *parameters* for the procedure. A parameter is typically used to pass a value to the stored procedure from a calling application. However, a parameter can also be used to pass a value back from the stored procedure to the calling application.

If a procedure accepts more than one parameter, you must use a comma to separate each parameter. The simplest way to declare a parameter is to code the name of the parameter followed by its data type. In this figure, for example, the procedure accepts two parameters. The parameter named invoice_number_param has a data type of VARCHAR2, and the parameter named credit_total_param has a data type of NUMBER.

After declaring any parameters, you code either the IS or AS keyword. It doesn't matter which you use, but it's a good practice to pick one and use it consistently. Then, you code the PL/SQL code to be executed when the procedure is called. In general, this code works like the anonymous blocks of PL/SQL code that you learned how to code in chapter 14. However, if you need to declare variables, you don't use the DECLARE keyword to identify the declarative section. Instead, you just declare the variables immediately after the IS or AS keyword but before the BEGIN keyword.

When you create a stored procedure, Oracle *compiles* the PL/SQL code and stores the compiled code in the database. As part of this process, Oracle's compiler checks the syntax of the code within the procedure. If you've made a coding error, the system responds with an appropriate message and the procedure isn't created.

If using Oracle 23ai or later, you can include an IF NOT EXISTS clause instead of the OR REPLACE clause. In that case, the CREATE PROCEDURE

The simplified syntax of the CREATE PROCEDURE statement

```
CREATE [OR REPLACE] PROCEDURE procedure_name
[(
    parameter_name_1 data_type
    [, parameter_name_2 data_type]...
)]
{IS | AS}
  [declarative_statments;]
BEGIN
  body_statements;
[EXCEPTION
  exception_statements;]
END
```

A script that creates a stored procedure that updates a table

```
CREATE OR REPLACE PROCEDURE update_invoices_credit_total
(
  invoice_number_param  VARCHAR2,
  credit_total_param    NUMBER
)
AS
BEGIN
  UPDATE invoices
  SET credit_total = credit_total_param
  WHERE invoice_number = invoice_number_param;

  COMMIT;
EXCEPTION
  WHEN OTHERS THEN
    ROLLBACK;
END;
/
```

Description

- A *stored procedure* is an executable database object that contains a block of PL/SQL code. A stored procedure can also be called a *sproc* or a *procedure*.
- A stored procedure is a type of *subprogram*, which is a named PL/SQL block that can be reused.
- Stored procedures are *compiled* before they are stored in the database, allowing them to run faster than anonymous scripts that aren't compiled or stored in the database.
- You can use *parameters* to pass one or more values from the calling application to the stored procedure or from the procedure to the calling application.
- To declare a parameter within a stored procedure, you code the name of the parameter followed by its data type. If you declare two or more parameters, you separate the parameters with commas.

Figure 16-1 How to create a stored procedure

statement will only create the procedure if it doesn't already exist. In other words, it won't replace an existing procedure.

How to call a stored procedure

After you create a stored procedure, you can execute, or *call*, a stored procedure as shown in figure 16-2. All of the examples in this figure call the stored procedure created in the previous figure.

The first example uses the CALL statement. This statement passes one value for each of the parameters that are defined by the procedure. Here, the first parameter is a VARCHAR2 literal that specifies the invoice number, and the second parameter is a NUMBER literal that identifies the new amount for the credit total.

When you use the CALL statement, you must pass parameters *by position*. In other words, you must code the parameters in the same order as they are coded in the CREATE PROCEDURE statement. This is the most common way to call stored procedures that have a short list of parameters.

If you are coding a script that calls a procedure, you don't code the CALL keyword. Instead, you just code the name of the procedure and its parameters as shown by the second example in this figure. This example performs the same task as the first example.

When you call a procedure from a script, you can also pass parameters *by name*. To do that, you include the names of the parameters as defined in the CREATE PROCEDURE statement, followed by the association operator (=>), followed by the value that's passed. When you use this technique, you can list parameters in any order. This is illustrated by the third example. If a procedure has many parameters, passing parameters by name is often easier and less error-prone than passing parameters by position.

In chapter 12, you learned how to grant INSERT, UPDATE, and DELETE privileges to specific users. However, if you want to have more fine-grained control over the privileges that you grant to users, you can create stored procedures that perform all of the types of data manipulation that you want to allow within your database. Then, you can grant privileges to execute these stored procedures. For systems where security is critical, this can be an excellent way to prevent both accidental errors and malicious damage to your data.

A statement that calls a stored procedure

```
CALL update_invoices_credit_total('367447', 300);
```

A script that calls a stored procedure

```
BEGIN
  update_invoices_credit_total('367447', 300);
END;
/
```

A script that passes parameters by name

```
BEGIN
  update_invoices_credit_total(
    credit_total_param => 300,
    invoice_number_param => '367447');
END;
/
```

Description

- You can use the CALL statement to *call a procedure*.
- When a procedure accepts parameters, you pass them to the procedure by coding them within the parentheses that follow the procedure name, and by separating each parameter with a comma.
- When you use a CALL statement to call a procedure, you pass parameters *by position*. This means that you list them in the same order that they're listed in the stored procedure.
- When you use a script to call a procedure, you can pass parameters by position or *by name*. When you pass parameters by name, you code the name of each parameter, followed by the association operator (=>), followed by the value. This means that you don't have to list the parameters in the same order as the stored procedure.

Figure 16-2 How to call a stored procedure

How to code input and output parameters

Figure 16-3 shows how to code input and output parameters for a procedure. Procedures provide for three different types of parameters: input parameters, output parameters, and input/output parameters.

An *input parameter* is passed to the stored procedure from the calling code. You can explicitly identify an input parameter by coding the IN keyword after the name of the parameter. In this figure, for example, the first two parameters are identified as input parameters. However, if you omit this keyword, the parameter is assumed to be an input parameter.

Within a procedure, you can use input parameters like variables. However, you can't change the value of the parameter. In this figure, for example, the procedure uses the parameter named invoice_number_param within an UPDATE statement to specify the invoice number for the invoice row to be updated. Then, it uses this parameter within a SELECT statement to get a count of the number of rows that have been updated.

An *output parameter* is returned to the calling code from the stored procedure. To code an output parameter, you must explicitly identify the parameter by coding the OUT keyword after the name of the parameter.

In this figure, for example, the update_count parameter is an output parameter. If the UPDATE statement executes successfully, the SELECT statement stores the count of the rows that were updated in this parameter. Since the invoice number should be unique, this should store a value of 1 in the output parameter. On the other hand, if the UPDATE statement doesn't execute successfully, the SELECT statement in the EXCEPTION section of the procedure stores a value of 0 in the output parameter. Either way, the value of the output parameter is returned to the calling code when the procedure finishes.

To show how a calling application works, this figure includes a script that calls the procedure. Here, script specifies initial values for the two input parameters and a variable named row_count for the output parameter. After the procedure executes, the procedure assigns the value of the output parameter to the variable named row_count. Then, the calling application can use this variable to check how many rows have been updated. In this figure, for example, the script prints the value of this variable to the console. However, it could also use an IF statement to check the value of the variable and perform an appropriate action.

An *input/output parameter* stores an initial value that's passed in from the calling application like an input parameter. However, the procedure can change this value and return it to a calling application like an output parameter. To identify an input/output parameter, you must code the IN OUT keywords after the name of the parameter. Although input/output parameters can be useful in some situations, they can also be confusing, particularly if a procedure is run multiple times. As a result, it generally makes sense to avoid using them when possible.

The syntax for declaring input and output parameters

```
parameter_name_1 [IN|OUT|IN OUT] data_type
```

A stored procedure that uses input and output parameters

```
CREATE OR REPLACE PROCEDURE update_invoices_credit_total
(
  invoice_number_param IN  VARCHAR2,   -- input parameter
  credit_total_param   IN  NUMBER,     -- input parameter
  update_count         OUT INTEGER     -- output parameter
)
AS
BEGIN
  UPDATE invoices
  SET credit_total = credit_total_param
  WHERE invoice_number = invoice_number_param;

  SELECT COUNT(*)
  INTO update_count
  FROM invoices
  WHERE invoice_number = invoice_number_param;

  COMMIT;
EXCEPTION
  WHEN OTHERS THEN
    SELECT 0 INTO update_count FROM dual;
    ROLLBACK;
END;
/
```

A script that calls the stored procedure and uses the output parameter

```
SET SERVEROUTPUT ON;
DECLARE
  row_count INTEGER;
BEGIN
  update_invoices_credit_total('367447', 200, row_count);
  DBMS_OUTPUT.PUT_LINE('row_count: ' || row_count);
END;
/
```

Description

- *Input parameters* accept values that are passed from the calling application. These values cannot be changed by the body of the procedure.
- By default, parameters are defined as input parameters.
- *Output parameters* store values that are passed back to the calling application. These values must be set by the body of the procedure.
- To use an output parameter, the calling application must declare a variable to store its value, and it must include this variable in the parameter list.
- *Input/output parameters* can store an initial value that's passed from the calling application. However, the body of the procedure can change this parameter.
- To identify an input/output parameter, you must code both the IN and OUT keywords.

Figure 16-3 How to code input and output parameters

How to code optional parameters

Figure 16-4 shows how to code an *optional parameter*. When you call a procedure that has optional parameters, you can omit any of its optional parameters. In this figure, for example, the second parameter is an optional parameter. As a result, when you call this procedure, you can omit the second parameter. In that case, the procedure sets the credit total to a default value of 100. Or, you can specify both parameters. In that case, the procedure sets the credit total to the value of the second parameter.

Usually, it makes sense to code all optional parameters at the end of the parameter list. Then, you use the DEFAULT keyword to specify a default value for each of the optional parameters. In this figure, for example, the DEFAULT keyword specifies a default value of 100 for the second parameter.

The first CALL statement in this figure specifies both parameters. As a result, it sets the credit total for the invoice to 200. However, the second CALL statement omits the second parameter. As a result, it sets the credit total for the invoice to the default value of 100.

The syntax for declaring an optional parameter

```
parameter_name_1 [IN|OUT|IN OUT] data_type DEFAULT default_value
```

A CREATE PROCEDURE statement that uses an optional parameter

```
CREATE OR REPLACE PROCEDURE update_invoices_credit_total
(
  invoice_number_param VARCHAR2,
  credit_total_param   NUMBER    DEFAULT 100  -- optional parameter
)
AS
BEGIN
  UPDATE invoices
  SET credit_total = credit_total_param
  WHERE invoice_number = invoice_number_param;

  COMMIT;
EXCEPTION
  WHEN OTHERS THEN
    ROLLBACK;
END;
/
```

A statement that calls the stored procedure

```
CALL update_invoices_credit_total('367447', 200);
```

Another statement that calls the stored procedure

```
CALL update_invoices_credit_total('367447');
```

Description

- *Optional parameters* have a default value. Therefore, they don't require that a value be passed from the calling application.
- It's a good programming practice to code your CREATE PROCEDURE statements so that they list required parameters first, followed by optional parameters. That way, the optional parameters can easily be omitted when calling the procedure.

How to raise a predefined exception

Within a stored procedure, it's generally considered a good practice to prevent errors by checking the parameters before they're used to make sure they're valid. This is referred to as *data validation*. Then, if the data isn't valid, you can execute code that makes it valid, or you can *raise an exception*, which returns an error to the calling application. That way, the application can respond to the error by getting valid data and calling the procedure again.

Figure 16-5 shows how to raise one of the predefined exceptions that are available from Oracle. To do that, you code the RAISE statement followed by the name of the predefined exception. In this figure, for example, the IF statement checks whether the value of the second parameter is less than zero. If so, it raises the predefined VALUE_ERROR exception that's described in chapter 14.

This stored procedure also includes an exception section that can be used to handle all exceptions that might be raised by the code in the procedure. This includes the VALUE_ERROR exception and any other exceptions that might be raised by the UPDATE or COMMIT statements. In this example, the code handles the exception by rolling back any changes that have been made. Then, it raises the exception again by coding the RAISE statement without specifying the name of the exception. This returns the exception to the calling application.

If the calling application doesn't handle this exception, the system displays an error. In this figure, for example, the CALL statement passes a negative value to the second parameter, which causes the VALUE_ERROR exception to be raised. As a result, the system displays an error message that indicates that a value error has occurred.

However, the calling application can also include code that handles the exception. In this figure, for example, the script that calls the procedure includes an EXCEPTION section that handles the VALUE_ERROR exception by printing a message to the console that says, "A VALUE_ERROR occurred." If necessary, this message could be improved to be more user-friendly. In addition, the EXCEPTION section handles all other exceptions by printing a message to the console that says, "An unexpected error occurred." This shows how a calling application can handle each specific exception differently.

The syntax of the RAISE statement

```
RAISE [exception_name]
```

A stored procedure that raises a predefined exception

```
CREATE OR REPLACE PROCEDURE update_invoices_credit_total
(
  invoice_number_param VARCHAR2,
  credit_total_param   NUMBER
)
AS
BEGIN
  IF credit_total_param < 0 THEN
    RAISE VALUE_ERROR;
  END IF;

  UPDATE invoices
  SET credit_total = credit_total_param
  WHERE invoice_number = invoice_number_param;

  COMMIT;
EXCEPTION
  WHEN OTHERS THEN
    ROLLBACK;
    RAISE;
END;
/
```

A statement that calls the procedure

```
CALL update_invoices_credit_total('367447', -100);
```

The response from the system

```
Error report -
ORA-06502: PL/SQL: numeric or value error
ORA-06512: at "AP.UPDATE_INVOICES_CREDIT_TOTAL", line 9
```

A script that calls the procedure

```
SET SERVEROUTPUT ON;
BEGIN
  update_invoices_credit_total('367447', -100);
EXCEPTION
  WHEN VALUE_ERROR THEN
    DBMS_OUTPUT.PUT_LINE('A VALUE_ERROR occurred.');
  WHEN OTHERS THEN
    DBMS_OUTPUT.PUT_LINE('An unexpected error occurred.');
END;
/
```

The response from the system

```
A VALUE_ERROR occurred.
```

Description

- The RAISE statement causes, or *raises*, an exception that's returned to the caller.
- You can use a RAISE statement to raise a predefined exception like the ones described in chapter 14.

Figure 16-5 How to raise a predefined exception

How to raise a user-defined exception

Figure 16-6 shows how to raise a *user-defined error*, which is an error that you create. To do that, you call the RAISE_APPLICATION_ERROR procedure that's available from Oracle. Within the parentheses of this procedure, the first parameter specifies the error number, and the second parameter specifies the error message. Here, the error number must be between -20000 and -20999.

When you use the RAISE_APPLICATION_ERROR procedure, you can specify more helpful and user-friendly messages than the generic messages that are available from the predefined exceptions. In this figure, for example, the error message says, "Credit total may not be negative." At this point, the user or programmer of the calling application should be able to identify and fix the problem.

However, when you raise a user-defined error in this way, a PL/SQL block can't use specific exception handlers until you declare and initialize an exception. To do that, you can declare a variable of the EXCEPTION type. Then, you can use the EXCEPTION_INIT procedure to initialize the exception by specifying its name and error number.

In this figure, for example, the script declares an exception named negative_credit_total and initializes it to the exception with the number of -2001. Then, its exception handler section includes a WHEN clause for this exception that uses the SQLERRM function to display the error number and message for the exception.

Alternately, this script could handle the user-defined exception without declaring and initializing it. To do that, it could use the WHEN OTHERS clause to handle all exceptions. If the user-defined exception has a user-friendly error message, that's often an acceptable approach.

The simplified syntax of the RAISE_APPLICATION_ERROR procedure

```
RAISE_APPLICATION_ERROR(error_number, error_message);
```

A statement that raises a user-defined exception

```
RAISE_APPLICATION_ERROR(-20001, 'Credit total may not be negative.');
```

A statement that calls a procedure that raises a user-defined exception

```
CALL update_invoices_credit_total('367447', -100);
```

The response from the system if the exception is not handled

```
Error report -
ORA-20001: Credit total may not be negative.
ORA-06512: at "AP.UPDATE_INVOICES_CREDIT_TOTAL", line 9
```

The simplified syntax for declaring and initializing an exception

```
exception_name EXCEPTION;
PRAGMA EXCEPTION_INIT(exception_name, error_number);
```

A script that handles a user-defined exception

```
SET SERVEROUTPUT ON;

DECLARE
  negative_credit_total EXCEPTION;                      -- declare exception
  PRAGMA EXCEPTION_INIT(negative_credit_total, -20001);  -- initialize
BEGIN
  update_invoices_credit_total('367447', -100);
EXCEPTION
  WHEN negative_credit_total THEN
    DBMS_OUTPUT.PUT_LINE(SQLERRM);
  WHEN OTHERS THEN
    DBMS_OUTPUT.PUT_LINE('An unexpected error occurred.');
END;
/
```

The response from the system

```
ORA-20001: Credit total may not be negative.
```

Description

- The RAISE_APPLICATION_ERROR procedure raises a *user-defined exception* with the specified error number and message. When you use this procedure, the error number must be between -20000 and -20999.
- In a PL/SQL block, you can handle a user-defined exception by declaring the exception and initializing it to the exception with the specified error number.
- You can use the SQLERRM function to get a string that includes the error number and message for an exception.

Figure 16-6 How to raise a user-defined exception

A stored procedure that inserts a row

Figure 16-7 presents a stored procedure that inserts new rows into the invoices table. This should give you a better idea of how you can use stored procedures.

This procedure uses nine parameters that correspond to nine of the columns in the invoices table. All of these parameters are input parameters, and each parameter is assigned the same data type as the matching column in the invoices table. As a result, if the calling application passes a value that can't be converted to the proper data type, an error will be raised when the procedure is called.

Note that there isn't a parameter for the invoice_id column. That's because the body of the procedure uses a sequence to provide a value for this column. Also, the last five parameters have default values, making them optional. Of these, the last three parameters have been assigned a default value of NULL.

After the AS keyword, the procedure begins by declaring four variables. Of these variables, the first three have data types that correspond with columns in the invoices table. However, the fourth one uses the INTEGER data type to store the number of days before the invoice is due.

Note that all of the variables have a suffix of "_var" while all of the parameters defined earlier have a suffix of "_param". This makes it easy to tell the difference between the parameters that are passed to the procedure from the calling application and the variables that are used within the procedure.

After the BEGIN keyword, the procedure starts by using a simple IF statement to check the value of the parameter for the invoice_total column to see if it is less than zero. If so, the procedure raises a user-defined exception with an appropriate error number and message. This exits the stored procedure and returns the exception to the calling application. Although this figure only codes a single IF statement, it's common to code a series of IF statements like this one to provide more extensive data validation.

After the first IF statement, a SELECT statement stores the next value in the sequence for the invoice_id column in a variable. To do that, this SELECT statement selects the value that's returned by the NEXTVAL pseudocolumn of the sequence into the variable named invoice_id_var.

After the SELECT statement, an IF statement checks the optional parameter for the terms_id column to see if it contains a NULL value. If so, a SELECT statement sets the variable that corresponds to the parameter to an appropriate value. Here, for example, the variable for the terms_id column is set to the value that's stored in the default_terms_id column of the vendors table. Otherwise, this code sets the corresponding variable to the value of the parameter.

Similarly, another IF statement checks the optional parameter for the invoice_due_date column. If the parameter contains a NULL value, it uses the value of the terms_id_var to get the number of days until the invoice is due from the terms table and stores this value in the variable named terms_due_days_var. Then, it calculates a due date for the invoice by adding the number of days until the invoice is due to the invoice date.

After the values have been set for the variables for the invoice_id, terms_id, and invoice_due_date columns, this procedure executes an INSERT statement.

A stored procedure that validates the data in a new invoice

```
CREATE OR REPLACE PROCEDURE insert_invoice
(
  vendor_id_param          invoices.vendor_id%TYPE,
  invoice_number_param     invoices.invoice_number%TYPE,
  invoice_date_param       invoices.invoice_date%TYPE,
  invoice_total_param      invoices.invoice_total%TYPE,
  payment_total_param      invoices.payment_total%TYPE DEFAULT 0,
  credit_total_param       invoices.credit_total%TYPE DEFAULT 0,
  terms_id_param           invoices.terms_id%TYPE DEFAULT NULL,
  invoice_due_date_param invoices.invoice_due_date%TYPE DEFAULT NULL,
  payment_date_param       invoices.payment_date%TYPE DEFAULT NULL
)
AS
  invoice_id_var           invoices.invoice_id%TYPE;
  terms_id_var             invoices.terms_id%TYPE;
  invoice_due_date_var     invoices.invoice_date%TYPE;
  terms_due_days_var       INTEGER;
BEGIN
  IF invoice_total_param < 0 THEN
    RAISE_APPLICATION_ERROR(-20002, 'Invoice total may not be negative.');
  END IF;

  SELECT invoice_id_seq.NEXTVAL INTO invoice_id_var FROM dual;

  IF terms_id_param IS NULL THEN
    SELECT default_terms_id INTO terms_id_var
    FROM vendors WHERE vendor_id = vendor_id_param;
  ELSE
    terms_id_var := terms_id_param;
  END IF;

  IF invoice_due_date_param IS NULL THEN
    SELECT terms_due_days INTO terms_due_days_var
      FROM terms WHERE terms_id = terms_id_var;
    invoice_due_date_var := invoice_date_param + terms_due_days_var;
  ELSE
    invoice_due_date_var := invoice_due_date_param;
  END IF;

  INSERT INTO invoices
  VALUES (invoice_id_var,
          vendor_id_param, invoice_number_param, invoice_date_param,
          invoice_total_param, payment_total_param, credit_total_param,
          terms_id_var, invoice_due_date_var, payment_date_param);

END;
/
```

Figure 16-7 A stored procedure that inserts a row (part 1 of 2)

If this statement executes successfully, the row is inserted into the database. However, since the line items for the invoice haven't yet been inserted, this stored procedure does not use a COMMIT statement to commit changes to the database. As a result, the application that's calling this statement must issue the COMMIT statement. Typically, this would happen after the application uses a similar stored procedure to insert the line items for the invoice. That way, the line items and the invoice are part of the same transaction.

In most cases, a stored procedure like this is called from an application. However, to test a procedure before it's used by an application, you can use CALL statements like the ones in this figure.

The first three CALL statements provide valid values that successfully insert a new row. Of these three statements, the first supplies all of the parameters for the procedure. The second supplies the first eight parameters, but not the ninth. And the third only supplies the first four parameters. This shows that the first four parameters are the only parameters that are required by the procedure. That's because the procedure will either use the default values coded after the parameter, or it will set a default value within the body of the procedure.

The fourth CALL statement provides a negative number for the invoice total parameter. As a result, this CALL statement will cause the stored procedure to raise the user-defined exception with an error number of -2002 and a message of "Invoice total may not be negative". Since the CALL statement doesn't handle this exception, this causes the error message to be displayed. However, if you call the stored procedure from a block of PL/SQL code or from an application, you can include code that handles the exception.

Three statements that call the stored procedure

```
CALL insert_invoice(34, 'ZXA-080', '30-AUG-24', 14092.59,
                    0, 0, 3, '30-SEP-24', NULL);

CALL insert_invoice(34, 'ZXA-080', '30-AUG-08', 14092.59,
                    0, 0, 3, '30-SEP-24');

CALL insert_invoice(34, 'ZXA-080', '30-AUG-24', 14092.59);
```

The response from the system for a successful insert

```
Call completed.
```

A statement that raises an error

```
CALL insert_invoice(34, 'ZXA-080', '30-AUG-24', -14092.59);
```

The response from the system when a validation error occurs

```
Error report -
ORA-20002: Invoice total may not be negative.
ORA-06512: at "AP.INSERT_INVOICE", line 20
```

Description

- If the data for the new row is valid, the procedure inserts the row into the Invoices table. Otherwise, the procedure raises an error.
- If an application calls this procedure, it can handle any errors that are raised by the procedure.
- Since this procedure doesn't include a COMMIT statement, the application that calls this procedure must commit the change to the database. Ideally, the application would commit the data for the new invoice after inserting the line items for the invoice.

Figure 16-7 A stored procedure that inserts a row (part 2 of 2)

How to drop a stored procedure

Figure 16-8 shows how to drop a stored procedure. To do that, you can code the DROP PROCEDURE keywords followed by the name of the procedure. In this figure, the first example uses the CREATE PROCEDURE statement to create a procedure named clear_invoices_credit_total. Then, the second example uses the DROP PROCEDURE statement to drop that procedure.

If you drop a table, sequence, or view used by a procedure, you should be sure to drop the procedure as well. If you don't, the procedure can still be called by any user or application that has been granted the appropriate privileges. Then, an error will occur because the table, sequence, or view that the procedure depends on no longer exists.

The syntax of the DROP PROCEDURE statement

```
DROP PROCEDURE procedure_name
```

A statement that creates a stored procedure

```
CREATE PROCEDURE clear_invoices_credit_total
(
  invoice_number_param  VARCHAR2
)
AS
BEGIN
  UPDATE invoices
  SET credit_total = 0
  WHERE invoice_number = invoice_number_param;

  COMMIT;
END;
/
```

A statement that drops a stored procedure

```
DROP PROCEDURE clear_invoices_credit_total;
```

Description

- To drop a stored procedure from the database, use the DROP PROCEDURE statement.

Figure 16-8 How to drop a stored procedure

How to code functions

In chapter 8, you learned how to use some of the built-in functions that are provided by Oracle. Now, you'll learn how to create your own functions.

In most ways, the code for creating a function works similarly to the code for creating a stored procedure. However, there are two primary differences between stored procedures and functions. First, a function always returns a value or an object. Second, a function can't make changes to the database such as executing an INSERT, UPDATE, or DELETE statement.

How to create and call a function

Figure 16-9 shows how to create a function that returns a single value. That's the type of function that you'll learn to create in this chapter. To start, you code the CREATE FUNCTION statement, followed by the name of the function. If you want to automatically replace an existing function that has the same name, you can include the optional OR REPLACE keywords.

If you're using Oracle 23ai or later, you can code the IF NOT EXISTS keywords instead of the OR REPLACE keywords. Then, the function is created only if a function with that name doesn't already exist.

In this figure, the example shows how to create a function named get_vendor_id. After the name of the function, you code a set of parentheses. Within the parentheses, you code the parameters for the function. In this figure, the function only contains a single parameter of the VARCHAR2 type that's named vendor_name_param. This works similarly to declaring parameters for a stored procedure. The main difference is that it rarely makes sense to use output or input/output parameters for a function. As a result, functions almost always use input parameters as shown by the examples in this chapter.

After the parentheses, you code the RETURN keyword, followed by the data type that's returned by the function. In this figure, the example returns a value of the NUMBER type.

After the declaration of the return type, you code the IS or AS keyword to signal that you are about to begin coding the PL/SQL code for the function. Then, you can declare any variables that will be used internally by the function. In this figure, the PL/SQL code begins by declaring a variable of the NUMBER type named vendor_id_var.

After the BEGIN keyword, this function uses a SELECT statement to get the vendor ID that corresponds with the vendor name. Then, it uses the RETURN statement to return this value to the calling application.

To call a function, you can use it in an expression as if it's one of Oracle's built-in functions. In this figure, the last example shows how to use the function within a SELECT statement to return the vendor ID for the vendor with the name of "Data Reproductions Corp".

The simplified syntax for creating a function

```
CREATE [OR REPLACE] FUNCTION function_name
[(
    parameter_name_1 data_type
    [, parameter_name_2 data_type]...
)]
RETURN data_type
{IS | AS}
  [declarative_statments;]
BEGIN
  body_statements;
[EXCEPTION
  exception_statements;]
END
```

A function that returns the vendor ID that matches a vendor's name

```
CREATE OR REPLACE FUNCTION get_vendor_id
(
   vendor_name_param VARCHAR2
)
RETURN NUMBER
AS
  vendor_id_var NUMBER;
BEGIN
  SELECT vendor_id
  INTO vendor_id_var
  FROM vendors
  WHERE vendor_name = vendor_name_param;

  RETURN vendor_id_var;
END;
/
```

A SELECT statement that calls the function

```
SELECT invoice_number, invoice_total
FROM invoices
WHERE vendor_id = get_vendor_id('Data Reproductions Corp');
```

	INVOICE_NUMBER	INVOICE_TOTAL
1	39104	85.31
2	40318	21842

Description

- A *function* is an executable database object that contains a block of PL/SQL code.
- Like a procedure, a function is a type of stored program.
- A function can accept parameters that work like they do for a procedure.
- A function always returns data.
- The RETURN clause specifies the data type that's returned by a function. This data type can't include a length, precision, or scale.
- A RETURN statement return data to the calling application.
- A function can't make changes to the database.

Figure 16-9 How to create and call a function

A function that uses multiple RETURN statements

The first example in figure 16-10 shows how to create a function that returns a string that depends on the size of the invoice total. This string can be used to group invoices into categories of small, medium, and large. To do that, this function defines a parameter of the NUMBER type named invoice_total_param. In addition, its RETURN clause specifies that the function returns a value of the VARCHAR2 type. Then, the body of the function uses two IF statements to return a null value or a string.

The first IF statement checks if the invoice total is less than zero. If so, the function uses a RETURN statement to return a null value to the calling application. This is one way to end the function and indicate that something went wrong. Alternately, you could raise an exception. This would also end the function and indicate that something went wrong.

The second IF statement begins by checking the size of the invoice total. If the invoice total is greater than 10,000, it returns a string of "Large" and ends the function. If the invoice total is between 1,000 and 10,000, it returns a string of "Medium" and ends the function, Otherwise, the invoice total must be 1,000 or less, so it returns a string of "Small".

As you review this code, note that the function uses multiple RETURN statements. However, Oracle only executes the first RETURN statement since that ends the function.

The second statement shows a SELECT statement that uses this function to return a string that puts each invoice in a group. Note that you could return a similar result set by using a CASE statement like this:

```
SELECT vendor_id, invoice_number, invoice_total,
    CASE
        WHEN invoice_total > 10000
            THEN 'Large'
        WHEN invoice_total > 1000
            THEN 'Medium'
        ELSE 'Small'
    END AS invoice_group
FROM invoices;
```

The main advantage of using the function is that it provides a way to store the logic for assigning a group in a single location. As a result, you can easily call the function to use this logic from multiple SELECT statement. In addition, if the logic for grouping the invoices later changes, you only need to change it in one location.

A function that uses multiple RETURN statements

```
CREATE OR REPLACE FUNCTION get_invoice_group
(
   invoice_total_param NUMBER
)
RETURN VARCHAR2
AS
BEGIN
  IF invoice_total_param < 0 THEN
    RETURN NULL;
  END IF;

  IF invoice_total_param > 10000 THEN
    RETURN 'Large';
  ELSIF invoice_total_param > 1000 THEN
    RETURN 'Medium';
  ELSE
    RETURN 'Small';
  END IF;
END;
/
```

A statement that calls the function

```
SELECT vendor_id, invoice_number, invoice_total,
       get_invoice_group(invoice_total) AS invoice_group
FROM invoices;
```

	VENDOR_ID	INVOICE_NUMBER	INVOICE_TOTAL	INVOICE_GROUP
1	34	QP58872	116.54	Small
2	34	Q545443	1083.58	Medium
3	110	P-0608	20551.18	Large
4	110	P-0259	26881.4	Large

Description

- A function can contain any number of RETURN statements, but only the first RETURN statement encountered is executed.
- A RETURN statement can return a literal value or a variable.

Figure 16-10 A function that uses multiple RETURN statements

How to drop a function

Figure 16-11 shows how to drop a function. To do that, you code the DROP FUNCTION keyword followed by the name of the function. This is illustrated by the third example in this figure, which drops the procedure created by the first example.

The first example creates a function named get_balance_due. This function calculates the balance due for an invoice. To do that, it accepts a parameter that contains an invoice ID value. Then, the body of the function calculates the balance due, stores the result of the calculation in a variable named balance_due_var, and returns that value. Next, the second example shows a SELECT statement that uses this function to return the balance due for the specified invoice ID value.

If you drop a table or view used by a function, you should be sure to drop the related function as well. If you don't, the function can still be called by any user or application that has been granted the appropriate privileges. Then, an error will occur because the table or view that the function depends on no longer exists.

The syntax of the DROP FUNCTION statement

```
DROP FUNCTION function_name
```

A function that calculates balance due

```
CREATE OR REPLACE FUNCTION get_balance_due
(
   invoice_id_param NUMBER
)
RETURN NUMBER
AS
  balance_due_var NUMBER;
BEGIN
  SELECT invoice_total - payment_total - credit_total AS balance_due
  INTO balance_due_var
  FROM invoices
  WHERE invoice_id = invoice_id_param;

  RETURN balance_due_var;
END;
/
```

A statement that calls the function

```
SELECT vendor_id, invoice_number,
       get_balance_due(invoice_id) AS balance_due
FROM invoices
WHERE vendor_id = 37;
```

	VENDOR_ID	INVOICE_NUMBER	BALANCE_DUE
1	37	547479217	116
2	37	547480102	224
3	37	547481328	224

A statement that drops a function

```
DROP FUNCTION get_balance_due;
```

Description

- To delete a function from the database, use the DROP FUNCTION statement.

How to work with packages

A *package* is a container that you can use to organize and group related procedures and functions. In chapter 14, for example, you learned how to use the built-in package named DBMS_OUTPUT to print data to the output window. Now, you'll learn how to store your own procedures and functions within a package.

Although you don't need to store your procedures and functions within a package, there are several advantages. First, packages help you organize your code so all related code is stored in the same package. This has the added benefit of avoiding naming conflicts. Second, packages provide some enhanced functionality. For example, if you declare a variable in the body of a package, that variable is available within the package for the entire session. Third, packages generally provide improved performance since they are loaded into memory when they are first used. Last, package specifications reduce unnecessary recompiling of dependent procedures and functions in other packages.

How to create a package

Figure 16-12 shows how to create a package, which contains two main parts that are created separately. First, you use the CREATE PACKAGE statement to create the *specification* for the package. A specification declares the names, parameters, and return types for all of the subprograms such as procedures and functions that are contained within the package. In addition, it can specify other objects such as cursors or user-defined exceptions.

The example in part 1 of this figure specifies that the package named murach contains one procedure and one function. It specifies two parameters for the procedure. It specifies one parameter for the function. And it specifies the return type for the function. That's all a user needs to know to work with the package.

Second you can use the CREATE PACKAGE BODY statement to create the *body* of a package. The body contains all of the code a user doesn't need to see but that's necessary for the package to work. That means it contains the code that creates any procedures or functions declared in the specification. It may also contain a declarative section for variables that will only be used internally by the package. And it can contain an initialization section for statements that are only to be run once, when the package is accessed for the first time.

Within the body, the names, parameters, and return types of the subprograms and objects must exactly match the names, parameters, and return types declared in the specification. If the specification doesn't match the body for a package, Oracle raises an error when it attempts to compile the package body.

The simplified syntax for defining the specification for a package

```
CREATE [OR REPLACE] PACKAGE package_name {IS|AS}
   subprogram_declarations;
   public_object_declarations;
END [package_name];
/
```

The specification for a package named murach

```
CREATE OR REPLACE PACKAGE murach AS

  PROCEDURE update_invoices_credit_total
  (invoice_number_param VARCHAR2, credit_total_param NUMBER);

  FUNCTION get_vendor_id
  (vendor_name_param VARCHAR2)
  RETURN NUMBER;

END murach;
/
```

The simplified syntax for defining the body for a package

```
CREATE [OR REPLACE] PACKAGE BODY package_name {IS|AS}
  subprogram_definitions;
  object_definitions;
[BEGIN
   statements_to_run_once;]
END [package_name];
/
```

Description

- The *specification* of a package declares all database objects available from the package such as procedures and functions.
- The *body* of a package contains the code for database objects such as procedures and functions.
- Within the body, the names, parameters, and return types must match the names, parameters, and return types in the specification. However, they don't need to be coded in the same order.
- As long as you don't change the specification of a package, you can change the code in the body of a package without needing to change the application that uses the package.
- When you define a procedure or function in a package, you don't need to include the CREATE keyword or the OR REPLACE keywords.

The first example in part 2 implements the specification shown in part 1. To start, it creates a procedure that has a name and parameters that match the specification. Then, it creates a function that has a name, parameters, and a return type that match the specification.

After you've successfully compiled your package, you can use the techniques you learned earlier in this chapter to call a procedure or function that's stored in the package. However, you must preface the name of the procedure or function with the name of the package. For example, the CALL statement shown in this figure calls the procedure that's stored in the murach package, and the SELECT statement uses the function that's stored in the murach package. Note that storing this procedure and function in a package prevents a naming conflict with a procedure and function of the same name that aren't stored in a package.

In this figure, the package contains just one procedure and one function. In practice, though, a package might include dozens of procedures and functions and some of them might be far more complex than the ones in this chapter. A more realistic package might also include global variables that can be shared between procedures and functions.

If you're using Oracle 23ai or later, you can use the IF NOT EXISTS keywords instead of the OR REPLACE keywords in either a CREATE PACKAGE or CREATE PACKAGE BODY statement. Then, these statements will only create the package specification or body if it doesn't exist.

How to drop a package

As you would expect, you can use the DROP PACKAGE statement to drop a package. When you use this statement, the specification and the body for the package are dropped. In this figure, for example, the first DROP statement drops the specification and the body for the package named murach.

In some cases, though, you may only want to drop the body for the package. That way, other procedures or functions that reference the package specification can still be compiled. In that case, you can use the DROP PACKAGE BODY statement to drop only the body of the package.

The body for a package named murach

```
CREATE OR REPLACE PACKAGE BODY murach AS

  PROCEDURE update_invoices_credit_total
  (
    invoice_number_param  VARCHAR2,
    credit_total_param    NUMBER
  )
  AS
  BEGIN
    UPDATE invoices
    SET credit_total = credit_total_param
    WHERE invoice_number = invoice_number_param;

    COMMIT;
  EXCEPTION
    WHEN OTHERS THEN
      ROLLBACK;
  END;

  FUNCTION get_vendor_id
  (vendor_name_param VARCHAR2)
  RETURN NUMBER
  AS
    vendor_id_var NUMBER;
  BEGIN
    SELECT vendor_id
    INTO vendor_id_var
    FROM vendors
    WHERE vendor_name = vendor_name_param;

    RETURN vendor_id_var;
  END;

END murach;
/
```

A statement that calls a procedure that's stored in a package

```
CALL murach.update_invoices_credit_total('367447', 200);
```

A SELECT statement that calls a function that's stored in a package

```
SELECT invoice_number, invoice_total
FROM invoices
WHERE vendor_id = murach.get_vendor_id('IBM');
```

	INVOICE_NUMBER	INVOICE_TOTAL
1	QP58872	116.54
2	Q545443	1083.58

A statement that drops the specification and body for a package

```
DROP PACKAGE murach;
```

A statement that drops only the body for a package

```
DROP PACKAGE BODY murach;
```

Figure 16-12 How to create and drop packages (part 2 of 2)

How to use SQL Developer

To develop effective stored procedures and functions, you must understand the SQL and PL/SQL statements that are presented earlier in this chapter. However, once you understand these statements, SQL Developer provides some tools that can help you compile, edit, and drop the procedures, functions, and packages that you create. In addition, SQL Developer provides an excellent tool for debugging stored procedures and functions.

How to view and drop procedures, functions, and packages

Figure 16-13 shows how to use SQL Developer to view and drop procedures, functions, and packages. To start, you can view the procedures, functions, and packages for a database by connecting to the database and expanding the appropriate folder. Then, you can view the parameters for a procedure or function by clicking on the plus sign (+) to the left of its name.

In this figure, for example, the Packages, Procedures, and Functions folders have been expanded so you can see that some of the procedures, functions, and packages from this chapter are listed. Note that the package named MURACH provides separate objects for the specification and the body.

Once you've displayed one of these database objects, you can drop them by right-clicking on the object and selecting Drop. Then, you can use the resulting dialog box to confirm the drop.

If you want to view the code for a procedure or function, you can do that by clicking on its name. Then, the code for the procedure or function will be displayed in the main window. In this figure, for example, the procedure named update_invoices_credit_total is displayed in the main window. At this point, you can view the code for this procedure.

In addition, you can view other information about the procedure by clicking on one of the other tabs. For example, you can click on the Grants tab to view the roles and users that have been granted privileges on the object. You can click on the Dependencies tab to view the database objects that this object depends on. You can click on the References tab to view the database objects that refer to this object. And you can click on the Details tab to view other details about this object such as the date that the object was created. Before you drop an object, you might want to check the References tab to make sure that other objects don't reference, or depend on, this object.

A procedure after it has been viewed in SQL Developer

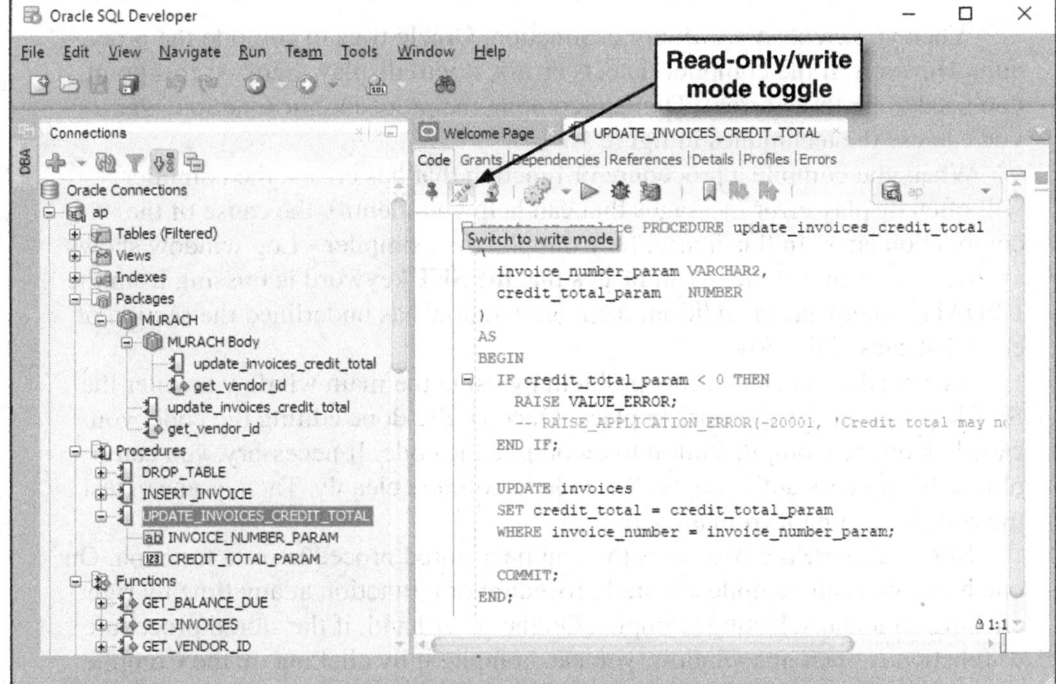

Description

- To view the procedures, functions, and packages for a database, you can use the Connections window in SQL Developer.
- To drop a procedure, function, or package, right-click on the object and select Drop.
- To view the parameters for a procedure or function, expand it by clicking on the plus sign (+) to the left of its name.
- To view the code for a procedure or function, click on its name. Then, the code for the procedure or function will be displayed in the main window.
- To view other information about a procedure or a function, click on its name to display it in the main window. Then, click on one of the other tabs (Grants, Errors, Dependencies, etc.) to get more information.
- To edit a stored procedure or function, click on its name to display it in the main window. Then, if necessary, toggle on write mode.

How to edit and compile procedures and functions

When you create a procedure or function, Oracle tries to compile the procedure. However, if the compiler detects errors, it will display them. At this point, you need to fix these errors. Then, to compile the procedure or function again, you can use the techniques in figure 16-14.

When you compile a procedure or function that has errors, the compiler will often display error messages that can help you identify the cause of the compilation error. In this figure, for example, the Compiler - Log window shows an error message that clearly indicates that the SET keyword is missing from the UPDATE statement. In addition, the main window has underlined the section of code that caused the error.

As a result, you can fix this problem by using the main window to enter the SET keyword in the appropriate place. Once you're done editing the code, you can click on the Compile button to recompile the code. If necessary, you can repeat this process until you get the code to compile cleanly. Then, you can test the code by calling it from a script.

Note that there are two ways to compile a stored procedure or a function. On one hand, you can compile a stored procedure or a function at any time by right-clicking on it and selecting Compile. On the other hand, if the stored procedure or function is open in a window, you can compile it by clicking on the Compile button in the toolbar that's displayed at the top of the screen.

How to grant and revoke privileges

In chapter 12, you learned how to grant and revoke privileges to database objects. Now, this figure shows an easy way to use SQL Developer to grant or revoke privileges to your procedures, functions, and packages. For example, you can grant privileges to a procedure, function, or package by right-clicking on its name and selecting Grant. Then, you can use the resulting dialog box to specify the role or the user and the privileges to be granted.

A procedure with compile-time errors

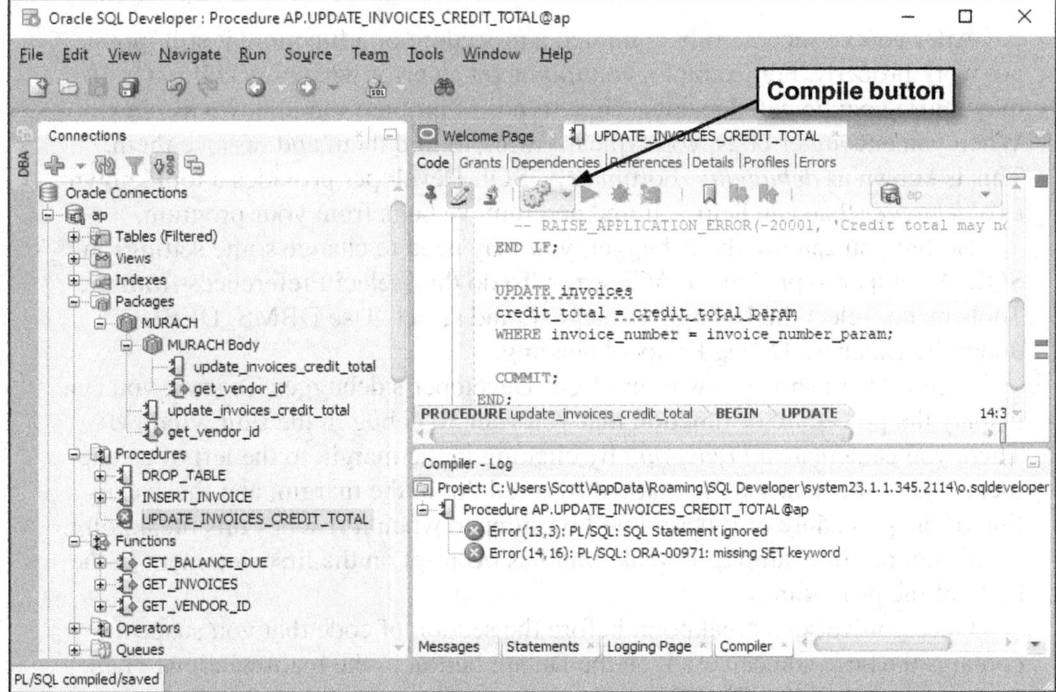

How to compile

- To compile a procedure, function, or package, right-click on its name and select Compile.
- Once you display a procedure or function in the Edit window, you can edit the code that's stored in the database. Then, you can use the Compile button in the toolbar to re-compile the code.
- If the edited code contains errors when you compile it, SQL Developer will display the compile-time errors in a Compiler - Log tab at the bottom of the main window.

How to grant or revoke privileges

- To grant privileges for a procedure, function, or package, right-click on its name and select Grant. Then, use the resulting dialog box to specify the user or role and the privileges to be granted.
- To revoke privileges for a procedure, function, or package, right-click on its name and select Revoke. Then, use the resulting dialog box to specify the user or role and the privileges to be revoked.

Figure 16-14 How to compile procedures and functions

How to debug procedures and functions

After you've successfully compiled a procedure or a function, it still might not work properly. For example, you might get an error when you call it. Or, it may return unanticipated results. These types of problems are known as *bugs*. When you encounter bugs, you typically want to find them and remove them. This is known as *debugging*. Fortunately, SQL Developer provides a tool known as a *debugger* than can help you find and remove bugs from your program.

Before you can use the debugger, you may need to change some settings in SQL Developer to prevent an ACL error. To do this, select Preferences from the Tools menu, select the Debugger category, and select "Use DBMS_DEBUG" under the Database Debug Protocol heading.

Figure 16-15 shows how to use SQL Developer's debugger. To start, you can display the procedure or function that you want to debug in the Edit window. Then, you can create a *breakpoint* by clicking in the margin to the left of a statement. This statement will be marked by a red dot in the margin, and the execution of the procedure or function will be stopped when it reaches this breakpoint. In this figure, for example, a breakpoint has been set on the first statement in the body of the procedure.

Once you've set a breakpoint before the section of code that you suspect contains the bug, you can click on the Debug button in the main window. Then, you can use the Debug PL/SQL dialog box to specify the values for the input parameters that you want to use for debugging. To do that, you code the values in the assignment statements that come after the BEGIN keyword in this dialog box. In this figure, for example, a string value of '367447' has been assigned to the first parameter, and a numeric value of -100 has been assigned to the second parameter. Here, the -100 value provides a way to check the IF statement to make sure that it's working properly.

A procedure with a breakpoint

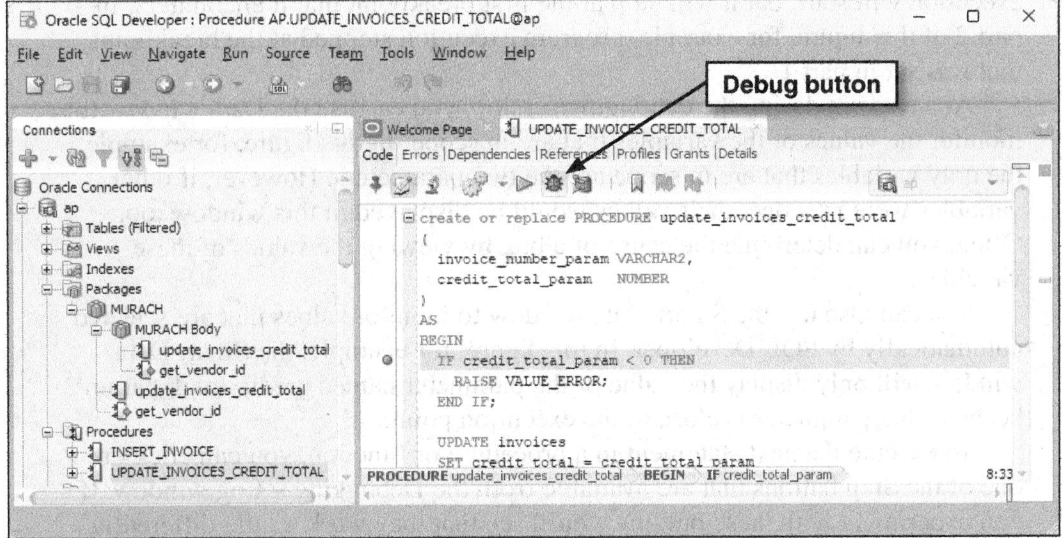

The dialog box for specifying input parameters

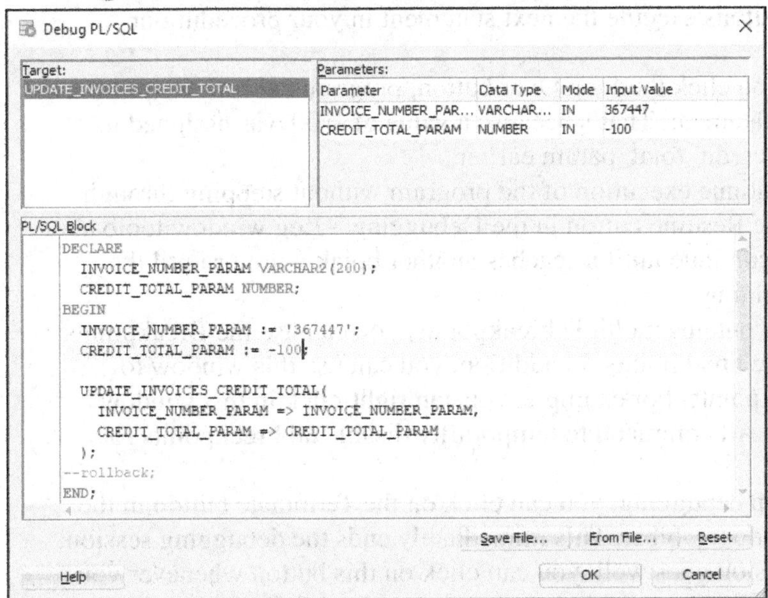

Description

- To set a *breakpoint* where code will stop executing until you start it again, display the procedure or function in the main window. Then, click in the margin to the left of the line of code where you want to set the breakpoint. After that, a red dot will mark the breakpoint.
- To begin debugging, click the Debug button to display the Debug PL/SQL dialog box. Here, you can use the assignment statements that are after the BEGIN keyword to set the input parameters.

Figure 16-15 How to debug procedures and functions (part 1 of 2)

After you specify the values for the input parameters and click OK, program execution will start, but it will stop at the first breakpoint that it encounters. In part 2 of this figure, for example, program execution stopped at the breakpoint that was set in part 1.

At any point during the debugging session, you can use the Data window to monitor the values of the variables that are in scope. In this figure, for example, the only variables that are in scope are the two parameters. However, if other variables were in scope, their values would be displayed in this window too. Often, you can determine the cause of a bug by viewing the values of these variables.

You can also use the Smart Data window to monitor values that are selected automatically by SQL Developer. In this figure, for example, the Smart Data window will only display the value of the parameter named credit_total_param because that parameter is close to the execution point.

To execute the next statement in a procedure or function, you can click on one of the Step buttons that are available from the Debugging – Log window. If you experiment with these buttons, you'll see that they work a little differently if your procedure or function calls another procedure or function. But if your procedure or function doesn't call other procedures or functions, both the Step Over and Step Into buttons execute the next statement in your procedure or function.

In this figure, if you click the Step Over button, program execution would move to the RAISE statement. That's because a value of -100 was assigned to the parameter named credit_total_param earlier.

If you want to continue execution of the program without stepping through it, you can click on the Resume button in the Debugging – Log window toolbar. Then, execution will continue until it reaches another break point or until the program finishes executing.

When a program contains multiple breakpoints, you can use the Breakpoints window view all of the breakpoints. In addition, you can use this window to work with these breakpoints. For example, you can right-click in this window and select the Disable All command to temporarily disable all breakpoints for the application.

If you want to stop debugging, you can click on the Terminate button in the Debugging – Log window toolbar. This immediately ends the debugging session. If your debugging session goes well, you can click on this button whenever you think you've discovered the cause of a bug and are ready to fix it.

A debugging session for a procedure

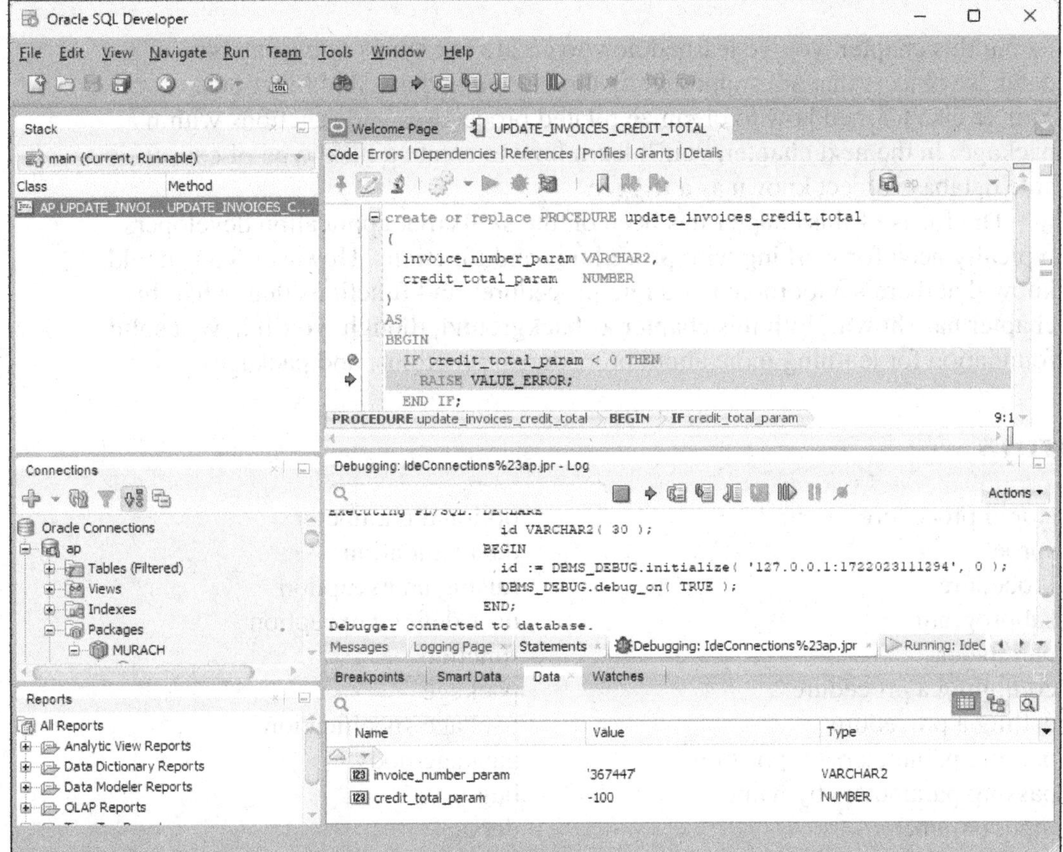

Description

- Before the debugger will work correctly, you may need to select Preferences from the Tools menu, select the Debugger category, and select the "Use DBMS_DEBUG" option.
- To step through the code one statement at a time starting at a breakpoint, click on the Debugging - Log tab. Then, click on the Step Over, Step Into, or Step Out button.
- The Step Over button steps over any called procedures or functions. The Step Into button steps into called procedures and functions. The Step Out button skips the rest of a called procedure or function and returns to the calling procedure or function.
- To view the values of the variables at each step in the execution of a procedure or function, click on the Smart Data or Data tab.
- To resume execution until the next breakpoint or the end of the procedure or function, click on the Resume button. To end debugging, click on the Terminate button.
- To view the breakpoints for a procedure or function, click on the Breakpoints tab. To remove a breakpoint, click on the red marker in the main window or right-click on the breakpoint in the Breakpoints tab and select Delete.

Perspective

In this chapter, you've learned how to create two types of executable database objects that are supported by Oracle: procedures and functions. You've also learned how to organize related procedures and functions within a package. In the next chapter, you'll learn how to create a third type of executable database object known as a trigger.

The focus of this chapter has been on the skills that application developers typically need for working with procedures and functions. However, you should know that there's a lot more to coding procedures and functions than what this chapter has shown. With this chapter as background, though, you'll have a solid foundation for learning more about procedures, functions, and packages.

Terms

stored procedure
sproc
procedure
subprogram
parameter
compiling a procedure
calling a procedure
passing parameters by position
passing parameters by name
input parameter
output parameter
input/output parameter
optional parameter
data validation
raising an exception
user-defined exception
function
package
package specification
package body
bug
debug
debugger
breakpoint

Exercises

1. Create a stored procedure named insert_glaccount that lets a user add a new row to the General_Ledger_Accounts table in the AP schema. This procedure should have two parameters, one for each of the two columns in this table. Then, code a CALL statement that tests the procedure. (Note that this table doesn't allow duplicate account descriptions.)
2. Code a script that calls the procedure that you created in exercise 1 and passes the parameters by name. This procedure should provide exception handling that displays this message if the INSERT statement fails because the account number or account description is already in the table (a DUP_VAL_ON_INDEX error):

   ```
   A DUP_VAL_ON_INDEX error occurred.
   ```

   It should provide this message if any other type of error occurs:

   ```
   An unknown exception occurred.
   ```

3. Code a function named test_glaccounts_description that accepts one parameter that tests whether an account description is already in the General_Ledger_Accounts table. This function should return a value of 1 if the account description is in the table or zero if it isn't. (Note: If a SELECT statement doesn't return any data, it throws a NO_DATA_FOUND exception that your function can handle.)
4. Code a script that calls the function that you created in exercise 1. This script should display this message if the account description is in the table:

   ```
   Account description is already in use.
   ```

   It should provide this message if any other type of error occurs:

   ```
   Account description is available.
   ```

5. Modify the stored procedure that you created in exercise 1 and save it as insert_glaccount_with_test. This procedure should use the function that you created in step 3 to test whether the account description is unique before it issues the INSERT statement. If the account description is a duplicate, this procedure should raise an exception with an error number of -20002 and an error message of

   ```
   Duplicate account description
   ```

6. Modify the script that you created in step 2 so it uses the procedure of exercise 5. This script should use the SQLERRM function to display the error number and message for the exception that's raised if the account description is already in use.

17

How to create triggers

Now that you've learned how to work with procedures and functions, you're ready to learn about another type of executable database object: a trigger. Triggers provide a powerful way to control the SQL statements that modify the data in your database.

Since you can program virtually any logic within the code for a trigger, you can use triggers to enforce data consistency with a flexibility that isn't available from other database features such as constraints. In addition, you can use triggers to log changes to the database, provide for updateable views, and implement business rules.

How to work with triggers...492
How to create a trigger for a table...492
A trigger that enforces data consistency...494
How to use conditional predicates...496
How to create a trigger for a view...498
How to create a system trigger...500
How to enable, disable, rename, or drop a trigger...502
Other skills for working with triggers...504
How to create a compound trigger...504
A trigger that causes the mutating-table error...506
How to solve the mutating-table problem...508
How to use SQL Developer...510
How to view, enable, disable, rename, or drop a trigger...510
How to edit a trigger...512
Perspective...514

How to work with triggers

A *trigger* is a named block of PL/SQL code that is executed, or *fired*, automatically when a certain database action occurs. Most of the time, you'll create triggers that are fired when a DML statement such as an INSERT, UPDATE, or DELETE statement is executed on a table or view. However, it's also possible to create triggers for DDL statements such as CREATE, ALTER, or DROP statements. In addition, it possible to create triggers for database events such as when the database server starts.

How to create a trigger for a table

Figure 17-1 presents the syntax for creating a *simple DML trigger* that fires when INSERT, UPDATE, or DELETE statements are executed on a table. In this figure, the CREATE TRIGGER statement creates a trigger named vendors_before_update_state. Since this statement includes the OR REPLACE keywords, it creates this trigger if it doesn't already exist, and it replaces the trigger if it does already exist.

After the name of the trigger, you code either the BEFORE or AFTER keyword to indicate when the trigger is fired. Then, you code the statement type or types that cause the trigger to fire, separating each statement type with the OR keyword. For an UPDATE statement, you can use the optional OF keyword to specify a list of columns. Then, the trigger is fired only when those columns are modified. If you specify multiple columns, you separate the columns with commas.

In this figure, the trigger is executed before any INSERT statements on the Vendors table or any UPDATE statements that update the vendor_state column of the Vendors table.

After the ON clause, you can optionally specify the FOR EACH ROW clause to indicate that the trigger is a *row-level trigger* that will fire for each row that's modified. If you don't code this clause, the trigger is a *statement-level trigger* that's only executed once for each statement.

If you specify a FOR EACH ROW clause, you can optionally specify a WHEN clause to further control when the trigger is fired. Within the parentheses of the clause, you can code a condition that evaluates to true or false. This condition may use *correlation names* to work with the new and old values of the row that's being inserted or updated. In this figure, for example, the trigger is only fired when the new vendor state isn't uppercase.

When you use the NEW and OLD correlation names, you'll get a null value if you use NEW for a row that's about to be deleted. That's because this row doesn't have any new values. Similarly, you'll get a null value if you use OLD for a row that's about to be inserted. Also, when used outside of a condition in the WHEN clause, you need to preface the correlation names with a colon (:) as shown by the trigger in this figure.

The action for this trigger is a PL/SQL block that contains a single statement. This statement updates the vendor_state column so that state codes are always

The simplified syntax for creating a DML trigger

```
CREATE [OR REPLACE] TRIGGER trigger_name
{BEFORE|AFTER}
{DELETE|INSERT|UPDATE [OF column_list]}
[OR {DELETE|INSERT|UPDATE [OF column_list]}]...
ON table_name
[FOR EACH ROW [WHEN (condition)]]
pl_sql_block
```

Correlation names

Name	Description
OLD	The current row before the statement is executed. Null for INSERT statements.
NEW	The current row after the statement is executed. Null for DELETE statements.

A BEFORE trigger that corrects mixed-case state names

```
CREATE OR REPLACE TRIGGER vendors_before_update_state
BEFORE INSERT OR UPDATE OF vendor_state
ON vendors
FOR EACH ROW WHEN (NEW.vendor_state != UPPER(NEW.vendor_state))
BEGIN
  :NEW.vendor_state := UPPER(:NEW.vendor_state);
END;
/
```

An UPDATE statement that fires the trigger

```
UPDATE vendors
SET vendor_state = 'wi'
WHERE vendor_id = 1;
```

The new row

	VENDOR_NAME	VENDOR_STATE
1	US Postal Service	WI

Description

- A *trigger* is a named block of PL/SQL code that executes, or *fires*, in response to an event.
- You can fire a *simple DML trigger* before or after an INSERT, UPDATE, or DELETE statement is executed on a table or view.
- If you don't specify a FOR EACH ROW clause, the trigger is a *statement-level trigger* that fires once for each statement.
- If you specify a FOR EACH ROW clause, the trigger is a *row-level trigger* that fires once for each row that's modified and meets the optional WHEN condition.
- For a row-level trigger, you can use the OLD and NEW *correlation names* to work with the values for the columns that are stored in the old or new row.
- The NEW and OLD correlation names should be preceded with a colon (:) in the PL/SQL block, but not when used as a row condition.

Figure 17-1 How to create a trigger for a table

stored with uppercase letters. To accomplish this, this statement uses the UPPER function to convert the new value for the vendor_state column to uppercase.

A trigger that enforces data consistency

Triggers are commonly used to enforce data consistency beyond what is possible with constraints. For example, the sum of line item amounts in the Invoice_Line_Items table should always be equal to the corresponding invoice total amount in the Invoices table. However, you can't enforce this constraint individually on the Invoices table or the Invoice_Line_Items table. Instead, you can use a trigger like the one in figure 17-2.

To understand how this trigger works, consider how an invoice is entered into the database. First, you must insert the invoice into the Invoices table. If you tried to insert the line items first, you'd violate referential integrity since the value for the invoice_id column hasn't been inserted into the database yet. Once the invoice has been inserted, you can insert the first line item into the Invoice_Line_Items table.

However, you can't constrain the value for the line-item amount because the constraint would have to be based on the value in the Invoices table. In addition, even if you could implement such a constraint, you would only want to do so once the last line item was inserted. Then, you'd want to be sure that the sum of the line-item amounts equaled the invoice total.

The trigger shown here enforces this rule by firing after an UPDATE statement attempts to update the invoice_total column in the Invoices table. When the trigger fires, it tries to verify that the sum of the line items is equal to the invoice total. If this isn't true, the trigger raises an error. Then, the application that issued the UPDATE statement can handle the exception.

Although this example isn't entirely realistic, you can use triggers like this to enforce business rules or to verify data consistency. Since you can program a trigger to accommodate virtually any situation, triggers are more flexible than constraints. That said, if you can use a constraint to do the same job as a trigger, you should. That's because constraints are simpler to implement, less error-prone, and can be applied to both old and new data.

A trigger that validates line item amounts when updating the invoice total

```
CREATE OR REPLACE TRIGGER invoices_before_update_total
BEFORE UPDATE OF invoice_total
ON invoices
FOR EACH ROW
DECLARE
  sum_line_item_amount NUMBER;
BEGIN
  SELECT SUM(line_item_amt)
  INTO sum_line_item_amount
  FROM invoice_line_items
  WHERE invoice_id = :new.invoice_id;

  IF sum_line_item_amount != :new.invoice_total THEN
    RAISE_APPLICATION_ERROR(-20001,
      'Line item total must match invoice total.');
  END IF;
END;
/
```

An UPDATE statement that fires the trigger

```
UPDATE invoices
SET invoice_total = 600
WHERE invoice_id = 100;
```

The response from the system

```
ORA-20001: Line item total must match invoice total.
```

Description

- Triggers can be used to enforce rules for data consistency that can't be enforced by constraints.

Figure 17-2 A trigger that enforces data consistency

How to use conditional predicates

A simple DML trigger can fire before or after INSERT, UPDATE, or DELETE statements. What if you want the trigger to perform a different action depending on what kind of statement triggered it? To do that, you can use *conditional predicates* to check which kind of statement caused the trigger to fire.

The table at the top of figure 17-3 summarizes the three conditional predicates used by Oracle. As you can see, they correspond to the keywords you use when creating a DML trigger.

To start, the first example shows a CREATE TABLE statement that creates a table named Invoices_Audit. This table contains five columns that store information about the action that occurred on the Invoices table. Of these columns, the first three store values from the Invoices table, and the last two store information about the action that caused the statement to execute.

After the CREATE TABLE statement, this figure shows a CREATE TRIGGER statement that creates a trigger that fires after an INSERT, UPDATE, or DELETE statement is executed on the Invoices table. The body of this trigger uses an IF statement with the INSERTING, UPDATING, and DELETING conditional predicates to check the type of the event that caused the trigger to fire. Then, the trigger inserts data into the new Invoices_Audit table that depends on whether data in the table has been inserted, updated, or deleted.

If, for example, data has been inserted into the table, the first INSERT statement in the trigger is executed. This inserts the new values for the vendor_id, invoice_number, and invoice_total columns into the Invoices_Audit table. In addition, it inserts a string value of 'INSERTED' to indicate that the row has been inserted, and it uses the SYSDATE function to insert the date of the action.

Here, the first INSERT statement inserts the new values for the row that's being inserted, since there aren't any old values for this row. However, the second and third INSERT statements insert the old values for the row that's being updated or deleted since there are old values for these rows.

Although the example that's presented in this figure has been simplified, you can create much more complex audit tables. For example, if you're having a problem updating rows in a database, you can create an audit table and a trigger to store whatever data you want to store about each update. Then, the next time the update problem occurs, you can review the data in the audit table to identify the cause of the problem.

Since the trigger in this figure uses the AFTER keyword, it doesn't insert a row if a statement fails. If, however, you wanted to audit failed statements or debug a suspected problem, you could use the BEFORE keyword instead.

Conditional predicates

Name	Triggered by...
`INSERTING`	An INSERT statement.
`UPDATING [('col_name')]`	An UPDATE statement. You can optionally specify a column name.
`DELETING`	A DELETE statement.

An audit table that records actions on the Invoices table

```
CREATE TABLE invoices_audit
(
  vendor_id           NUMBER           NOT NULL,
  invoice_number      VARCHAR2(50)     NOT NULL,
  invoice_total       NUMBER           NOT NULL,
  action_type         VARCHAR2(50)     NOT NULL,
  action_date         DATE             NOT NULL
);
```

An AFTER trigger that inserts rows into the audit table

```
CREATE OR REPLACE TRIGGER invoices_after_dml
AFTER INSERT OR UPDATE OR DELETE
ON invoices
FOR EACH ROW
BEGIN
  IF INSERTING THEN
    INSERT INTO invoices_audit VALUES
    (:new.vendor_id, :new.invoice_number, :new.invoice_total,
     'INSERTED', SYSDATE);
  ELSIF UPDATING THEN
    INSERT INTO invoices_audit VALUES
    (:old.vendor_id, :old.invoice_number, :old.invoice_total,
     'UPDATED', SYSDATE);
  ELSIF DELETING THEN
    INSERT INTO invoices_audit VALUES
    (:old.vendor_id, :old.invoice_number, :old.invoice_total,
     'DELETED', SYSDATE);
  END IF;
END;
/
```

An INSERT statement that causes the trigger to fire

```
INSERT INTO invoices VALUES
(115, 34, 'ZXA-080', '30-AUG-24', 14092.59, 0, 0, 3, '30-SEP-24', NULL);
```

A DELETE statement that causes the trigger to fire

```
DELETE FROM invoices WHERE invoice_number = 'ZXA-080';
```

The audit table

	VENDOR_ID	INVOICE_NUMBER	INVOICE_TOTAL	ACTION_TYPE	ACTION_DATE
1	34	ZXA-080	14092.59	INSERTED	28-MAY-24
2	34	ZXA-080	14092.59	DELETED	28-MAY-24

Description

- *Conditional predicates* check the type of DML statement that caused the trigger to fire.

Figure 17-3 How to use conditional predicates

How to create a trigger for a view

In chapter 11, you learned how to create updateable views, and you learned that there are some limitations to updating data in updateable views. Now, you'll learn how you can use an INSTEAD OF trigger to work around many of those limitations. To do that, you can code a trigger that executes an INSERT, UPDATE, or DELETE statement on an underlying table instead of executing that statement on the view.

Figure 17-4 shows the syntax for using an INSTEAD OF trigger on a view. If you compare this syntax with the syntax for using a simple DML trigger on a table, you'll see that they're similar.

The INSTEAD OF trigger in this figure allows you to insert a row through a view named IBM_Invoices. This view selects invoices from the Invoices table for the vendor named IBM, which has a vendor ID of 34. Since some of the columns that are required for an INSERT statement aren't included in the view, you can't insert data through this view unless you create a trigger like the one in this figure that inserts the new row into the underlying Invoices table.

This trigger accommodates the missing columns by calculating their values based on some logical assumptions. First, the invoice ID can be retrieved from a sequence. Second, the vendor ID can be assumed to be 34 because that's part of the definition of the view. Third, the terms ID for the invoice can be assumed to be 3 since this is the default terms ID for this vendor. Fourth, the due date for the invoice can be calculated based on the invoice date and the terms that correspond with the terms ID value. In this case, the invoice must be paid within 30 days of the invoice date. In addition, this trigger uses the NEW correlation name to retrieve the values from the new row that the INSERT statement is attempting to insert into the view.

If a stored procedure is available that does what you want it to do, you can call that procedure from a trigger. For instance, you can use a stored procedure named insert_invoice instead of the INSERT statement shown in this figure. To do that, you would call the insert_invoice procedure from the trigger like this:

```
insert_invoice(34, :new.invoice_number, :new.invoice_date,
               :new.invoice_total);
```

This approach has two advantages. First, it makes your code easier to maintain by allowing you to store all logic for inserting an invoice in one location. Second, it makes your code more flexible since the procedure looks up the terms ID and calculates the due date based on that value.

The simplified syntax for an INSTEAD OF trigger for a view

```
CREATE [OR REPLACE] TRIGGER trigger_name
INSTEAD OF
{DELETE|INSERT|UPDATE [OF column_list]}
[OR {DELETE|INSERT|UPDATE [OF column_list]}]...
ON view_name
[FOR EACH ROW [WHEN (condition)]]
pl_sql_block
```

A view that's based on the Invoices table

```
CREATE OR REPLACE VIEW ibm_invoices AS
  SELECT invoice_number, invoice_date, invoice_total
  FROM invoices
  WHERE vendor_id = 34;
```

An INSTEAD OF trigger for the view

```
CREATE OR REPLACE TRIGGER ibm_invoices_instead_of_insert
INSTEAD OF INSERT
ON ibm_invoices
BEGIN
  INSERT INTO invoices VALUES
  (invoice_id_seq.NEXTVAL, 34, :new.invoice_number, :new.invoice_date,
   :new.invoice_total, 0, 0, 3, :new.invoice_date + 30, NULL);
END;
/
```

An INSERT statement that succeeds due to the trigger

```
INSERT INTO ibm_invoices VALUES ('ZXA-080', '30-AUG-14', 14092.59);
```

The view with the new row

	INVOICE_NUMBER	INVOICE_DATE	INVOICE_TOTAL
1	QP58872	25-FEB-24	116.54
2	Q545443	14-MAR-24	1083.58
3	ZXA-080	30-AUG-24	14092.59

Description

- You can use an INSTEAD OF trigger to make views updateable by executing INSERT, UPDATE, or DELETE statements on the underlying tables instead of attempting to insert, update, or delete data through the view.

Figure 17-4 How to create a trigger for a view

How to create a system trigger

So far, you've learned how to use simple DML triggers to work with statements such as INSERT, UPDATE, and DELETE statements. However, you can also use a *system trigger* to work with DDL statements such as CREATE, ALTER, and DROP statements as well as database events such as opening and closing the database. This is illustrated in figure 17-5.

When you code a system trigger, the trigger can respond to DDL events such as the ones shown in the first table. For example, to fire the trigger for any DDL event, you can code the DDL keyword. Or, you can code the keyword for a more specific DDL event such as CREATE or DROP. If you code more than one event, you must separate the different events with the OR keyword.

Within the ON clause, you can specify that you only want the trigger to fire for a specific schema by coding the name of the schema, followed by a dot operator, followed by the SCHEMA keyword. Or, you can code the SCHEMA keyword without explicitly specifying the name of the schema. In that case, the trigger fires for DDL events on the same schema that contains the trigger.

On the other hand, if you want the trigger to fire for all schemas in the current database, you can code the DATABASE keyword. Some events, like AFTER STARTUP, can only be used at the database level.

In this figure, the trigger fires before a CREATE or DROP statement is executed on the AP schema. Then, the body of the trigger raises an error, which prevents the CREATE or DROP statement from executing. In this example, the trigger prevents the CREATE TABLE statement from executing. But it will also prevent any other CREATE statements such as a CREATE VIEW statement from executing.

Fortunately, this trigger doesn't prevent a user with the appropriate privileges from dropping the trigger later. Otherwise, it would be possible to accidentally create a trigger that would prevent you from making any changes to the database objects in your database.

The simplified syntax of a system trigger

```
CREATE [OR REPLACE] TRIGGER trigger_name
{BEFORE|AFTER|INSTEAD OF} ddl_or_db_event [OR ddl_or_db_event]...
ON {[schema_name.]SCHEMA|[PLUGGABLE] DATABASE}
pl_sql_block
```

Common DDL events

Event	Fired when...
`DDL`	Any DDL event occurs.
`CREATE`	An object is created.
`ALTER`	An object is altered.
`DROP`	An object is dropped.
`GRANT`	A GRANT statement is issued.
`REVOKE`	A REVOKE statement is issued.

Common database events

Event	Fired when...
`STARTUP`	An instance of the database is opened. Only valid with AFTER and DATABASE.
`SHUTDOWN`	An instance of the database is closed. Only valid with BEFORE and DATABASE.
`SERVERERROR`	A server error message is logged, after it is safe to fire error triggers. Only valid with AFTER.

A system trigger that prevents creating or dropping database objects

```
CREATE OR REPLACE TRIGGER ap_before_create_drop
BEFORE CREATE OR DROP ON ap.SCHEMA
BEGIN
  RAISE_APPLICATION_ERROR(-20001,
    'You cannot create or drop an object in the AP schema');
END;
/
```

A CREATE TABLE statement that fires the trigger

```
CREATE TABLE test1 (test_id NUMBER);
```

The response from the system

```
ORA-20001: You cannot create or drop an object in the AP schema
```

Description

- A *system trigger* may be either a schema trigger or a database trigger, and is triggered by either a DDL statement or a database event.
- In the ON clause, you can use the SCHEMA keyword to fire the trigger when DDL events occur on the specified schema. If you don't explicitly specify the schema, the statement will use the schema that contains the trigger.
- In the ON clause, you can use the DATABASE keyword to fire the trigger when DDL events occur on any schema in the current database.
- The INSTEAD OF keywords can only be used with CREATE statements.

Figure 17-5 How to create a system trigger

How to enable, disable, rename, or drop a trigger

In some cases, you may want to temporarily disable a trigger before you perform certain operations. For example, you may want to disable the triggers for one or more tables before inserting a large number of rows into the tables. This can help the INSERT statements run faster, and it can let you insert data that isn't allowed by the triggers. Then, when you're done, you can enable the triggers.

Figure 17-6 shows how to code the SQL statements that enable and disable triggers. In addition, it shows how to code the SQL statements that rename or drop a trigger.

To enable or disable a specific trigger, you code the ALTER TRIGGER keywords, followed by the name of the trigger, followed by the ENABLE or DISABLE keyword. In this figure, the first example disables the trigger that was created in the previous figure so objects can be created and dropped within the AP schema. Then, the second example confirms that the trigger has been disabled by creating and dropping a table. Next, the third example enables the trigger so objects can't be created and dropped within the AP schema.

To enable or disable all triggers for a table, you code the ALTER TRIGGER keywords, followed by the name of the table, followed by the ENABLE or DISABLE keyword, followed by the ALL TRIGGERS keywords. In this figure, the fourth example disables all triggers for the Invoices table. Then, it enables all of the triggers for the Invoices table.

To rename a trigger, you code the ALTER TRIGGER keywords, followed by the old name of the trigger, followed by the RENAME TO keywords, followed by the new name for the trigger. In this figure, the fifth example renames a trigger.

To drop a trigger, you code the DROP TRIGGER keywords followed by the name of the trigger. In this figure, the sixth example uses the DROP TRIGGER statement to drop the trigger that was created in the previous figure.

If you drop a table, sequence, or view used by a trigger, you should be sure to drop or disable the trigger as well. If you don't, the trigger can still be fired by a database user. Then, an error will occur because the table, sequence, or view that the trigger depends on no longer exists.

Disable a trigger

```
ALTER TRIGGER ap_before_create_drop DISABLE;
```

Confirm the trigger no longer fires

```
CREATE TABLE test1 (test_id NUMBER);
DROP TABLE test1;
```

Enable the trigger

```
ALTER TRIGGER ap_before_create_drop ENABLE;
```

Disable and enable all triggers for a table

```
ALTER TABLE invoices DISABLE ALL TRIGGERS;
ALTER TABLE invoices ENABLE ALL TRIGGERS;
```

Rename a trigger

```
ALTER TRIGGER invoices_before_update_total
RENAME TO invoices_before_update_inv_tot;
```

Drop a trigger

```
DROP TRIGGER ap_before_create_drop;
```

Description

- To enable or disable a specific trigger, use the ALTER TRIGGER statement with the ENABLE or DISABLE keywords.
- To enable or disable all triggers for a table, use the ALTER TRIGGER statement with the ENABLE or DISABLE keywords and the ALL TRIGGERS keywords.
- To rename a trigger, use the ALTER TRIGGER statement with the RENAME TO keywords.
- To drop a trigger, use the DROP TRIGGER statement.
- With Oracle 23ai and later, you can use the IF EXISTS clause with ALTER TRIGGER and DROP TRIGGER to prevent errors if the named trigger doesn't exist.

Figure 17-6 How to enable, disable, rename, or drop a trigger

Other skills for working with triggers

Sometimes you may need to use a special type of trigger known as a compound trigger. This type of trigger provides an easy way to solve the mutating table error that is difficult to solve with the simple trigger types.

How to create a compound trigger

A *compound trigger* can contain multiple blocks of PL/SQL code called *timing point sections*. Each block executes at a certain point before, during, or after the execution of a statement, and you can include any or all of them in a single compound trigger. Furthermore, you can declare variables that are shared between all of the sections.

When used on a table, the four timing point sections occur before the triggering statement is executed, before each row is modified, after each row is modified, and after the triggering statement has finished executing. Alternately, if the trigger is used on a view, you can code a single INSTEAD OF EACH ROW section.

In figure 17-7, the body of the trigger begins by declaring a variable that stores an integer value. Then, it uses the BEFORE STATEMENT, BEFORE EACH ROW, AFTER EACH ROW, and AFTER STATEMENT sections to display and modify the value of this variable. This shows that all four blocks can access and modify the value that's stored in the variable. In addition, it shows how to code all four blocks for a compound trigger. However, it's common to code a compound trigger that only has two blocks.

When you declare a compound trigger, you use the FOR keyword instead of the BEFORE or AFTER keywords to identify the DML event that causes the trigger to fire. This makes sense because a compound trigger can execute both BEFORE and AFTER a DML statement. In addition, you use the COMPOUND TRIGGER keywords to identify the beginning of the trigger. Then, within each BEFORE or AFTER block, you use the IS keyword to identify the beginning of the PL/SQL block, and you use the corresponding END clause to end each block.

A compound trigger

```
CREATE OR REPLACE TRIGGER invoices_compound_update
FOR UPDATE OF invoice_total, credit_total
ON invoices
COMPOUND TRIGGER
  test_value NUMBER := 1;

  BEFORE STATEMENT IS
  BEGIN
    DBMS_OUTPUT.PUT_LINE('before statement: ' || test_value);
  END BEFORE STATEMENT;

  BEFORE EACH ROW IS
  BEGIN
    test_value := 2;
    DBMS_OUTPUT.PUT_LINE('before row: ' || test_value);
  END BEFORE EACH ROW;

  AFTER EACH ROW IS
  BEGIN
    test_value := 3;
    DBMS_OUTPUT.PUT_LINE('after row: ' || test_value);
  END AFTER EACH ROW;

  AFTER STATEMENT IS
  BEGIN
    test_value := 4;
    DBMS_OUTPUT.PUT_LINE('after statement: ' || test_value);
  END AFTER STATEMENT;
END;
/
```

A statement that fires the trigger

```
UPDATE invoices
SET credit_total = 0
WHERE invoice_id = 100;
```

The response from the system

```
before statement: 1
before row: 2
after row: 3
after statement: 4
```

Description

- A *compound trigger* can fire at multiple *timing points*.
- For a table, the four timing points occur before the triggering statement is executed, before each row is modified, after each row is modified, and after the triggering statement has finished executing.
- You use the COMPOUND TRIGGER keywords to identify the start of the body.
- Within the body, you can declare variables that are shared between the PL/SQL blocks.
- The timing point sections do not need to be coded in any particular order.

Figure 17-7 How to create a compound trigger

A trigger that causes the mutating-table error

If you code a row-level trigger that issues a SELECT statement against the same table that the trigger is defined on, the trigger will compile without returning an error. However, when you execute a statement that fires the trigger, you'll get an error that's known as the *mutating-table error*. This error includes a message that indicates that the table is mutating and that the trigger might not be able to "see" the changes. As a result, the trigger fails to execute properly.

To understand this error more thoroughly, let's assume that you have a business rule that says that the credit total for an invoice should never be greater than the maximum invoice total for a vendor. Unfortunately, this business rule is too complex to implement with a constraint. At first glance, though, it seems like you should be able to implement this business rule by coding a trigger like the one shown in figure 17-8.

Here, the declaration for the trigger indicates that the trigger executes before an UPDATE statement is executed against each row of the Invoices table. Then, the body of the trigger declares a variable and uses a SELECT statement to store the maximum invoice total for the vendor in the variable. Finally, this trigger uses an IF statement to check if the new credit total is greater than the maximum invoice total that's stored in the variable. If so, the trigger raises an appropriate error, which prevents the UPDATE statement from executing.

When you run this CREATE TRIGGER statement, it compiles cleanly. However, when you issue an UPDATE statement like the one in this figure, Oracle returns the mutating-table error because the trigger is attempting to read data from the table at the same time that the UPDATE statement is attempting to update the table.

At this point, you may wonder why this trigger causes the mutating-table error while the trigger presented in figure 17-2 does not. If you compare these triggers, you'll see that they're both row-level triggers based on the Invoices table and that they both use a SELECT statement within the body of the trigger. However, the trigger in figure 17-2 selects data from a different table than the table that the trigger is defined. As a result, it doesn't cause the mutating-table error.

If your trigger doesn't need to use the NEW or OLD correlation names, you can solve the mutating-table error by removing the FOR EACH ROW clause from the trigger. In that case, the row-level trigger will be converted to a statement-level trigger, and the mutating-table error won't occur. Unfortunately, the trigger shown in this figure needs to use NEW to check the new credit total for the row.

A trigger that causes the mutating-table error

```
CREATE OR REPLACE TRIGGER invoices_before_update_total_2
BEFORE UPDATE OF invoice_total, credit_total
ON invoices
FOR EACH ROW
DECLARE
  max_invoice_total NUMBER;

BEGIN
  SELECT MAX(invoice_total)
  INTO max_invoice_total
  FROM invoices
  WHERE vendor_id = :new.vendor_id;

  IF (:new.credit_total > max_invoice_total) THEN
    RAISE_APPLICATION_ERROR(-20001,
   'invoice_credit value may not exceed the maximum invoice_total value.');
  END IF;
END;
/
```

A statement that fires the trigger

```
UPDATE invoices
SET credit_total = 0
WHERE invoice_id = 100;
```

The response from the system

```
ORA-04091: table AP.INVOICES is mutating, trigger/function may not see it
```

Description

- The *mutating-table error* is raised when a trigger attempts to execute a SELECT statement on the same table that the trigger is defined on.

Figure 17-8 A trigger that causes the mutating-table error

How to solve the mutating-table problem

Figure 17-9 shows how to use compound triggers to solve the mutating-table problem. This trigger works because it uses a statement-level trigger to execute the SELECT statement that retrieves the data before the triggering statement is executed, and it stores this data in variables. Then, it uses a row-level trigger to process each row using the data that has been stored the variables. As a result, the row-level trigger never attempts to retrieve data from the table while it is mutating.

The trigger shown in this figure is a compound trigger that's fired when the invoice_total or credit_total column of the Invoice table is updated. This trigger begins by declaring some variables that store data about the maximum invoice total for each vendor. In particular, the first two statements declare nested tables that store the vendor IDs and maximum invoice totals for each vendor. Then, the third statement declares an associative array. This array is later used to store the maximum invoice totals for each vendor and to index these values by vendor ID. Finally, the next three statements assign these three collections to variables named vendor_ids, max_invoice_totals, and ids_totals.

After this trigger declares these variables, it executes the code in the BEFORE STATEMENT block. This block begins by declaring two NUMBER variables, one to store the vendor ID and one to store the maximum invoice total. Then, this code uses a SELECT statement to select the vendor IDs and the maximum invoice totals into their corresponding variables. To accomplish this task, this SELECT statement uses the BULK COLLECT keywords. In addition, it uses a GROUP BY clause to create a summary query. As a result, this SELECT statement will only return one row for each vendor.

After the SELECT statement, a FOR loop stores the vendor IDs and maximum invoice totals in the variable named ids_totals. Within the loop, the first two statements use the counter variable for the loop to return the values for the vendor ID and the maximum invoice total. Finally, the last statement in the loop stores the maximum invoice total in the id_totals variable, and it uses the vendor ID value as the index.

After the BEFORE STATEMENT block, the code in the BEFORE EACH ROW block begins by declaring a NUMBER variable to store the maximum invoice total that's retrieved from the indexed table. Then, it uses the new vendor ID value to retrieve the maximum invoice total for the vendor. Finally, it uses an IF statement to check if the new credit total value is greater than the maximum invoice total value. If so, this trigger raises an appropriate error.

A compound trigger that solves the mutating-table problem

```
CREATE OR REPLACE TRIGGER invoices_before_update_total_2
FOR UPDATE OF invoice_total, credit_total
ON invoices
COMPOUND TRIGGER
  TYPE vendor_ids_table             IS TABLE OF NUMBER;
  TYPE max_invoice_totals_table     IS TABLE OF NUMBER;
  TYPE ids_totals_table             IS TABLE OF NUMBER
                                    INDEX BY PLS_INTEGER;

  vendor_ids                        vendor_ids_table;
  max_invoice_totals                max_invoice_totals_table;
  ids_totals                        ids_totals_table;

  BEFORE STATEMENT IS
    vendor_id NUMBER;
    max_invoice_total NUMBER;
  BEGIN
    SELECT vendor_id, MAX(invoice_total)
    BULK COLLECT INTO vendor_ids, max_invoice_totals
    FROM invoices
    GROUP BY vendor_id;

    FOR i IN 1..vendor_ids.COUNT() LOOP
      vendor_id := vendor_ids(i);
      max_invoice_total := max_invoice_totals(i);
      ids_totals(vendor_id) := max_invoice_total;
    END LOOP;
  END BEFORE STATEMENT;

  BEFORE EACH ROW IS
    max_invoice_total NUMBER;
  BEGIN
    max_invoice_total := ids_totals(:new.vendor_id);
    IF :new.credit_total > max_invoice_total THEN
      RAISE_APPLICATION_ERROR(-20001,
        'credit_total may not exceed the maximum
        invoice_total for a vendor.');
    END IF;
  END BEFORE EACH ROW;

END;
/
```

A statement that fires the trigger

```
UPDATE invoices
SET credit_total = 1000
WHERE invoice_id = 100;
```

The response from the system

```
ORA-20001: credit_total may not exceed the maximum invoice_total for a
vendor.
```

Figure 17-9 How to solve the mutating-table problem

How to use SQL Developer

To be able to develop effective triggers, you must understand the SQL and PL/SQL statements that they are created with. However, once you understand these statements, SQL Developer provides some tools that can help you work with the triggers that you create.

How to view, enable, disable, rename, or drop a trigger

Figure 17-10 shows how to use SQL Developer to view, rename, drop, enable, and disable triggers. To start, you can view the triggers for a database by connecting to the database and expanding the appropriate folder in the Connections window. In this figure, the Triggers folder of the AP schema is expanded so you can see all of the triggers that were presented in this chapter.

Triggers can be stored in packages just as procedures and functions can be. As a result, if you want to view a trigger that's stored in a package, you must begin by expanding the package that contains the trigger.

If you want to view the code for a trigger, you can do that by clicking on its name. Then, the code for the trigger will be displayed in the main window. In this figure, the trigger named invoices_after_dml is displayed in the main window.

To view other information about the trigger, click on one of the other tabs. For example, you can click on the Dependencies tab to view the database objects that this trigger depends on. Or, you can click on the Details tab to view other details about the trigger, such as the date that the trigger was created.

Once you've displayed a trigger, you can enable or disable it by right-clicking on it in the Connections window and selecting Enable or Disable. Then, you can use the resulting dialog box to confirm the action. Similarly, you can rename or drop a trigger by right-clicking on it and selecting Rename or Drop Trigger. Then, you can use the resulting dialog box to provide a new name for the trigger or to confirm the drop.

Finally, if you want to edit the trigger as shown in the next figure, you may need to switch to write mode. To do that, you can click on the Switch button in the toolbar that's displayed above the main window. In this figure, the mouse cursor is positioned over this button and about to click on it.

A trigger that's being viewed in the main window

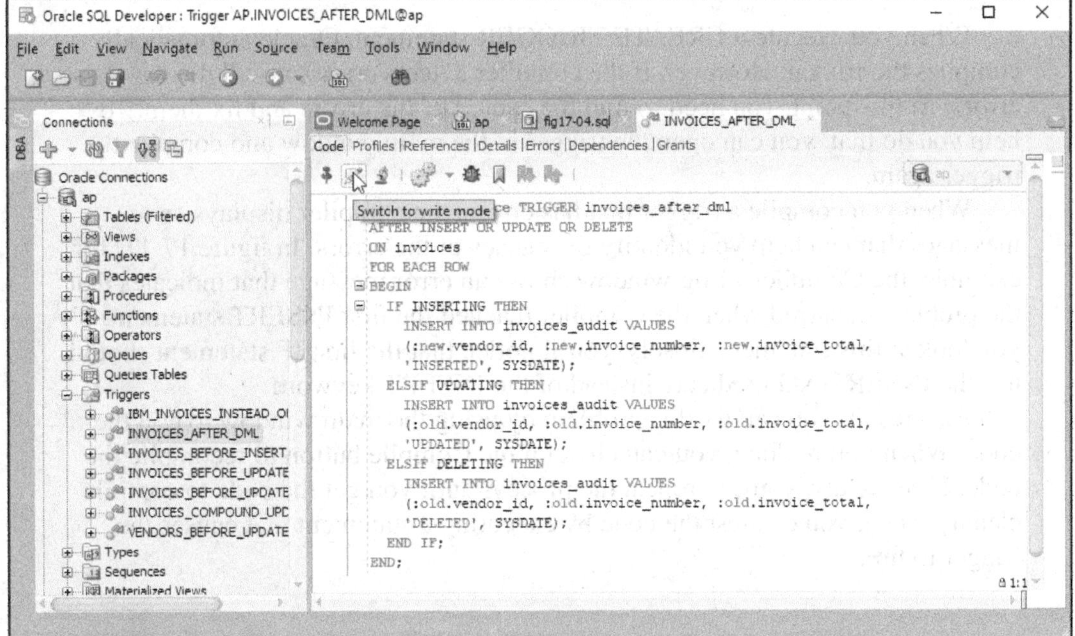

Description

- To view the triggers for a database, expand the corresponding folder in the Connections window.
- To view the code for a trigger, click on its name to display its code in the main window.
- To view other information about a trigger, click on its name to display its code in the main window. Then, click on one of the other tabs to get more information.
- To enable or disable a trigger, right-click on it and select Enable or Disable.
- To rename a trigger, right-click on it and select Rename.
- To drop a trigger, right-click on it and select Drop Trigger.

Figure 17-10 How to view, enable, disable, rename, or drop a trigger

How to edit a trigger

When you execute a CREATE TRIGGER statement, Oracle automatically compiles the trigger. However, if the compiler detects errors, it will display these errors. At this point, you need to find the causes of the errors and fix them. To help you do that, you can open the trigger in the main window and compile the trigger again.

When you compile a trigger that has errors, the compiler displays error messages that can help you identify the causes of the errors. In figure 17-11, for example, the Compiler - Log window shows an error message that indicates that the problem occurred when the compiler reached the first INSERT statement. If you look at this statement closely, you'll notice that the first IF statement should use the INSERTING predicate instead of the INSERT keyword.

As a result, you can fix this problem by using the main window to edit the code. When you're done, you can click on the Compile button to recompile the code. If necessary, you can repeat the process until you get the code to compile cleanly. Then, you can test the code by executing a statement that causes the trigger to fire.

A trigger that's being edited in the main window

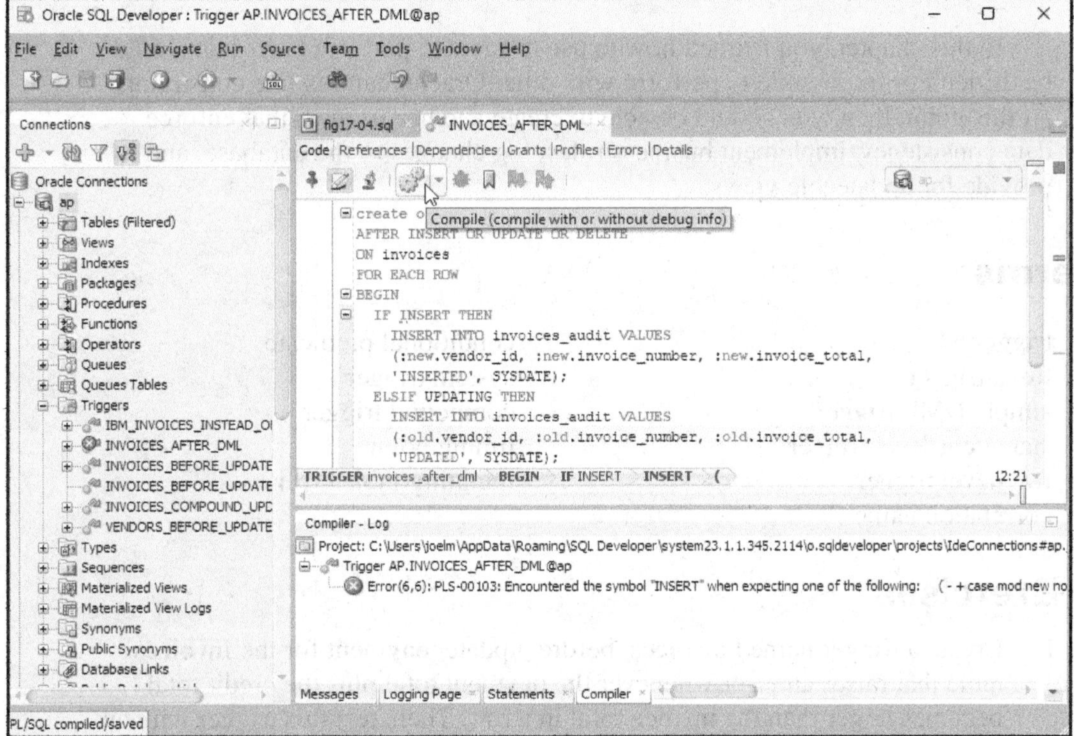

Description

- To edit a trigger, click on its name in the Connections window to display it in the main window. Then, if the trigger is in read-only mode, click on the Switch button in the toolbar to switch to write mode.
- To compile the edited trigger, click the Compile button in the toolbar.
- If the edited code contains errors when you compile it, SQL Developer displays the compile-time errors in a Compiler - Log tab at the bottom of the main window.

Figure 17-11 How to edit a trigger

Perspective

In this chapter, you learned how to use triggers to perform tasks that would be difficult or impossible to perform with other Oracle features like constraints. At this point, then, you should be able to create and use triggers that enforce data consistency, implement business rules, log changes to the database, and provide for updateable views.

Terms

trigger
fire a trigger
simple DML trigger
statement-level trigger
row-level trigger
correlation name
conditional predicate
system trigger
compound trigger
timing point
mutating-table error

Exercises

1. Create a trigger named invoices_before_update_payment for the Invoices table that raises an error whenever the payment total plus the credit total becomes larger than the invoice total in a row. Then, test this trigger with an appropriate UPDATE statement. (Note that you could code a check constraint to accomplish the same task.)
2. Create a trigger named invoices_after_update_payment for the Invoices table that displays the vendor name, invoice number, and payment total in the output window whenever the payment total is increased. Then, test this trigger with an appropriate UPDATE statement. (A trigger like this could be modified so it stores the data in an audit table.)
3. Use figure 11-4 of chapter 11 to create an updateable view named balance_due_view. Then, create and test an INSTEAD OF trigger named balance_due_view_insert that lets the user insert a new row through the view by inserting a row into the underlying Invoices table. Within this trigger, you can call the stored procedure named insert_invoice from chapter 16.

Appendix A

How to set up Windows for this book

Before you begin reading this book, we recommend that you install Oracle Database XE and Oracle SQL Developer. Both of these software products are available for free from the Oracle website, and you can install them on your computer as described in this appendix.

After you install these products, we recommend that you download the files for this book that are available from the Murach website (www.murach.com) and set up your system as described in this appendix. Once you've done that, you're ready to gain valuable hands-on experience by running the SQL scripts for the examples presented in this book. In addition, you can get more practice by doing the exercises that are at the end of each chapter.

How to install Oracle Database XE 516
How to install Oracle SQL Developer 518
How to download the files for this book 520
How to connect as the sysdba user 522
How to create the schemas for this book 524
How to import connections for the schemas 526
How to make sure your system is set up correctly 528
How to stop and start the database service 530

How to install Oracle Database XE

Oracle Database Express Edition (XE) is a lightweight version of Oracle Database that's free to use. You can use it to learn how to work with an Oracle database before committing to a paid option. That's why this book assumes that you have installed the Express Edition of Oracle on your computer.

Once you've downloaded the files for Oracle Database XE, you can install it as described in figure A-1. As part of this procedure, you will need to specify a password for the system user. When you do, *make sure to remember the password that you enter.* You will need to use this password again later.

It's best to install Oracle Database when you're at home or wherever you'll be doing the exercises for this book. That's because the installation process takes some time to execute, and you may need to tell your firewall to allow Java to access your home network.

If you encounter an error during installation and are presented with the option to either retry or abort, click the Retry button. Oracle's installer is robust and retrying may solve the problem.

Oracle Database 23ai was released in May 2024, but an Express Edition of 23ai had not been released at the time of this writing. That's why this figure shows the dialog for installing the previous version of Oracle Database, 21c. If an Express Edition of Oracle Database 23ai or later is available to you now, you should be able to use it with this book. That's because Oracle Database is backwards compatible. As a result, any code that works for an older version should also work with newer versions.

The Oracle Database XE Downloads page

`https://www.oracle.com/database/technologies/xe-downloads.html`

The dialog for installing Oracle Database Express Edition

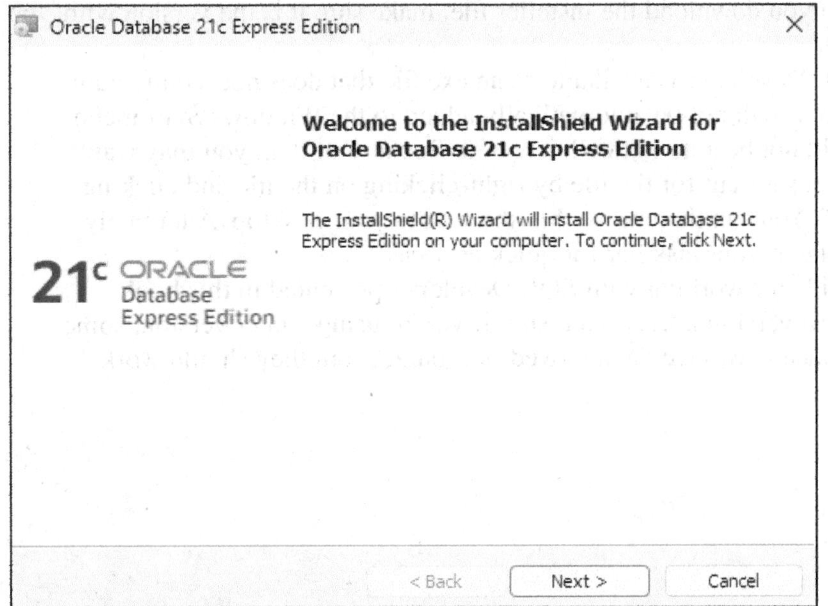

How to install Oracle Database XE (Express Edition)

1. Find the download page for Oracle Database XE on the Oracle website. You can visit the URL above or search the internet for "oracle database xe download".
2. Click the button to download the corresponding zip file.
3. Find the zip file on your hard drive and extract all the files in it into a temporary folder.
4. In the temporary folder, find the file named setup.exe and run it.
5. Respond to the resulting dialog boxes. You can accept the default options.
6. When these dialog boxes ask you to specify a password for the system and sys accounts, *make sure to remember the password that you enter.*

Notes

- If your firewall displays a dialog, allow access on private networks.
- If you encounter an error and are presented with the option to retry or abort, click the retry option first. It may fix the issue and allow the installation process to continue.
- Oracle Database 23ai was released in May 2024, but an Express Edition of 23ai had not been released at the time of this writing. All of the code presented in this book should work as described for Oracle 21c as well as newer versions of Oracle such as 23ai.

Figure A-1   How to install Oracle Database XE

How to install Oracle SQL Developer

Oracle SQL Developer is a free program that makes it easy to work with Oracle databases. To install SQL Developer, you can use the procedure shown in figure A-2. When you download the installer file, make sure it is the version with JDK 11 included.

Because SQL Developer is available as an exe file that does not require additional installation, it will not be automatically added to the Windows Start menu and a shortcut will not be automatically created. Because of that, you may want to create your own shortcut for the file by right-clicking on the file and clicking "Create shortcut". You can then move this shortcut to your desktop. Alternately, you can pin the file to your task bar for quick access.

All of the skills for working with SQL Developer presented in this book were tested against version 23.1. As a result, if you're using a later version, some parts of the interface may have been moved or renamed, but they should work similarly.

The URL for Oracle SQL Developer

```
https://www.oracle.com/database/sqldeveloper/technologies/download/
```

The SQL Developer files after they have been downloaded and unzipped

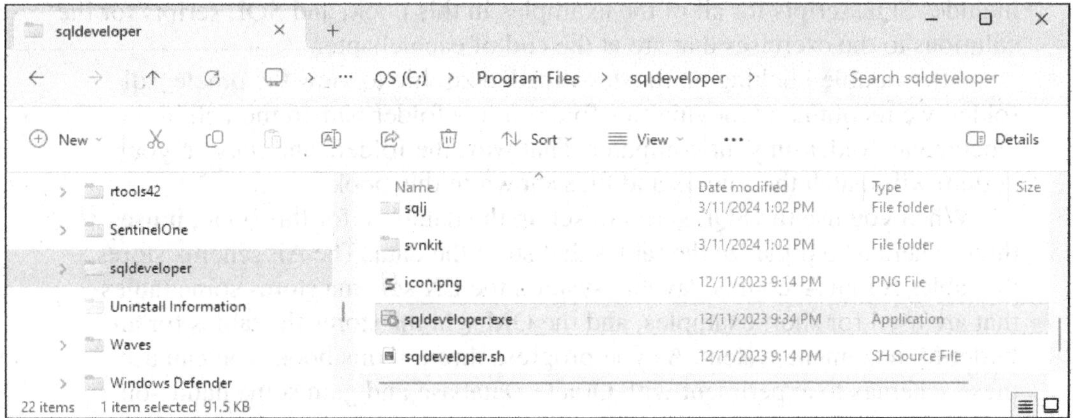

How to install Oracle SQL Developer

1. Find the download page for SQL Developer on the Oracle website. You can visit the URL shown above or search the internet for "oracle sql developer download."
2. Scroll to find the download link for the Windows 64-bit version with JDK 11 included and click it. You will be prompted to log in or create an Oracle account if you don't already have one. Creating an account is free.
3. Find the downloaded zip file and double-click it to view its files. You should see a folder named sqldeveloper.
4. Copy the sqldeveloper folder into the C:\Program Files folder.
5. You should now be able to start SQL Developer by double-clicking on the sqldeveloper.exe file in the C:\Program Files\sqldeveloper.

Notes

- To make it easy to start SQL Developer, you may want to right-click the sqldeveloper.exe file and click "Pin to taskbar". Once you do that, you can start SQL Developer just like you start other programs available from the task bar.
- To add an icon for SQL Developer to the desktop, you can right-click the sqldeveloper.exe file and click "Create shortcut". Then, you can move the shortcut that's created to the desktop.

Figure A-2 How to install Oracle SQL Developer

How to download the files for this book

Figure A-3 shows how to download the files for this book. This downloadable zip file includes a SQL script that sets up the database for this book. It also includes SQL scripts for all of the examples in this book, and SQL scripts for the solutions to the exercises that are at the end of each chapter.

After double-clicking on the downloaded zip file to view the oracle_sql folder, we recommend moving this folder into a folder named murach in the Documents folder on your computer. That way, the folders and files on your system will match the folders and files shown in this book.

When you use the SQL script to set up the database for this book, it uses three schemas to organize the tables that store the data. The AP schema stores the tables for an Accounts Payable system, the EX schema stores some tables that are used for short examples, and the OM schema stores the tables for an Order Management system. As you progress through this book, you can use these schemas to experiment with Oracle Database and gain some hands-on experience.

The recommended folder for the files

`Documents\murach\oracle_sql`

The files for this book

Folder	Contains
`db_setup`	The SQL script that sets up the database for this book.
`book_scripts`	The SQL scripts for all of the examples presented in this book.
`ex_solutions`	The SQL scripts for the solutions to the exercises at the end of each chapter.

The schemas for this book

Schema	Description
`ap`	The AP (Accounts Payable) schema. This schema is used by most examples in this book.
`ex`	The EX (Examples) schema. This schema contains several tables that are used for short examples.
`om`	The OM (Order Management) schema. This schema is used by a few examples in this book.

How to download the files

1. Go to www.murach.com.
2. Find the page for *Murach's Oracle SQL and PL/SQL (3rd Edition).*
3. Scroll down to the "FREE downloads" tab and click it.
4. Click the Download Now button for the zip file.
5. On your computer, find the downloaded zip file and double-click it. This should display a folder named oracle_sql.
6. Use File Explorer to create a folder named murach in your Documents folder.
7. Copy the oracle_sql folder into the newly created murach folder.

Description

- All of the files described in this book are contained in a zip file that can be downloaded from www.murach.com.
- An Oracle database can contain one or more schemas to organize the tables that store the data. This book uses three schemas in its examples: AP, EX, and OM.

Figure A-3 How to download the files for this book

How to connect as the sysdba user

To use SQL Developer to run SQL statements, you'll need to create a connection to your installation of Oracle Database. To do that, you can follow the procedure presented in figure A-4. This will create a connection named sys with the sysdba (System Database Administrator) role.

This is your first chance to find out if your installation of Oracle Database was successful. If you get an error when you test the connection even after following the steps exactly, it is likely Oracle Database is not running for some reason. In that case, you may need to manually start the database service as described in figure A-8. However, it could also be because your firewall may be blocking your connection to the database. In that case, you may need to change the settings in your firewall software.

The dialog for creating the sys connection

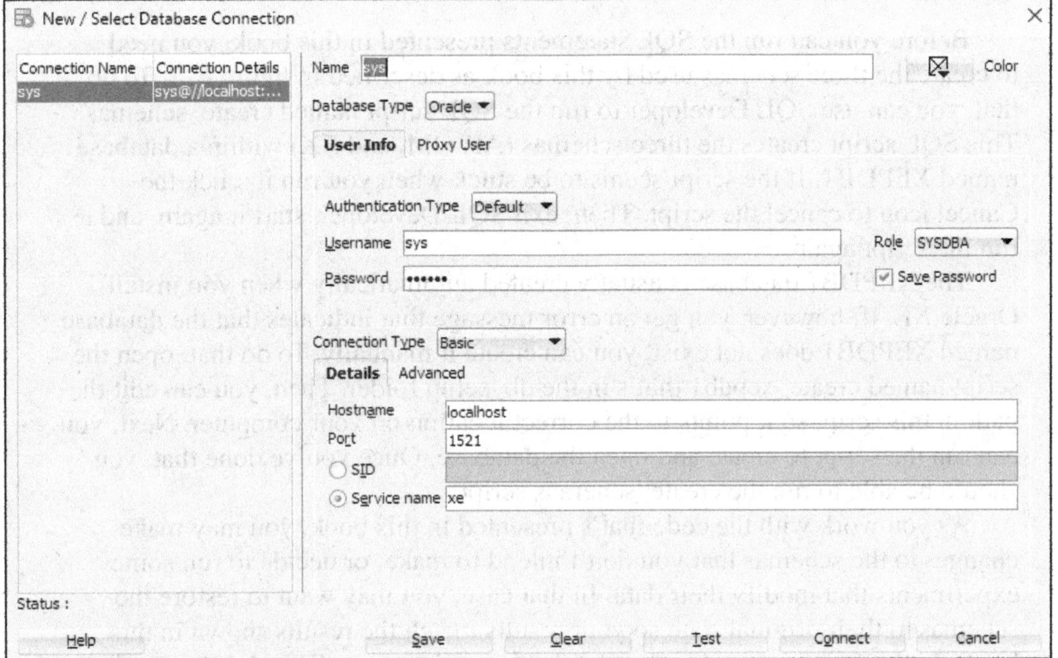

How to connect as the sysdba user

1. Start SQL Developer.
2. Right-click on Oracle Connections in the Connections window and select New Connection to display the dialog box for creating database connections.

   In this window:

   - Enter "sys" for the connection name.
   - Enter "sys" as the username and select SYSDBA (System Database Administrator) from the Role drop-down.
   - Enter the password you entered when installing Oracle Database. If you don't want to be prompted for the password every time you connect, check the Save Password box.
   - Select "Service name" and enter "xe" for the name of the service.
3. Click the Test button to test the connection. If the connection works, SQL Developer displays a success message below the Connections window.
4. Click the Save button to save the connection. When you do, SQL Developer adds the connection to dialog box and the Connections window.
5. You can now connect as the sysdba user using the sys connection.

Figure A-4 How to connect as the sysdba user

How to create the schemas for this book

Before you can run the SQL statements presented in this book, you need to create the three schemas used by this book as described in figure A-5. To do that, you can use SQL Developer to run the SQL script named create_schemas. This SQL script creates the three schemas (AP, OM, and EX) within a database named XEPDB1. If the script seems to be stuck when you run it, click the Cancel icon to cancel the script. Then, exit SQL Developer, start it again, and run the script again.

The XEPDB1 database is usually created automatically when you install Oracle XE. If, however, you get an error message that indicates that the database named XEPDB1 does not exist, you can create it manually. To do that, open the script named create_xepdb1 that's in the db_setup folder. Then, you can edit the path in this script so it points to the correct location on your computer. Next, you can run the script to create and open the database. Once you've done that, you should be able to run the create_schemas script.

As you work with the code that's presented in this book, you may make changes to the schemas that you don't intend to make, or decide to run some experiments that modify their data. In that case, you may want to restore the schemas to their original state so your results match the results shown in this book. To do that, just run the create_schemas script again. This deletes the three schemas described in this appendix and then recreates them. However, you may need to first exit SQL Developer and start it again before running the script.

The folder that contains the create_schemas.sql file

`Documents\murach\oracle_sql\db_setup`

SQL Developer after executing the create_schemas.sql file

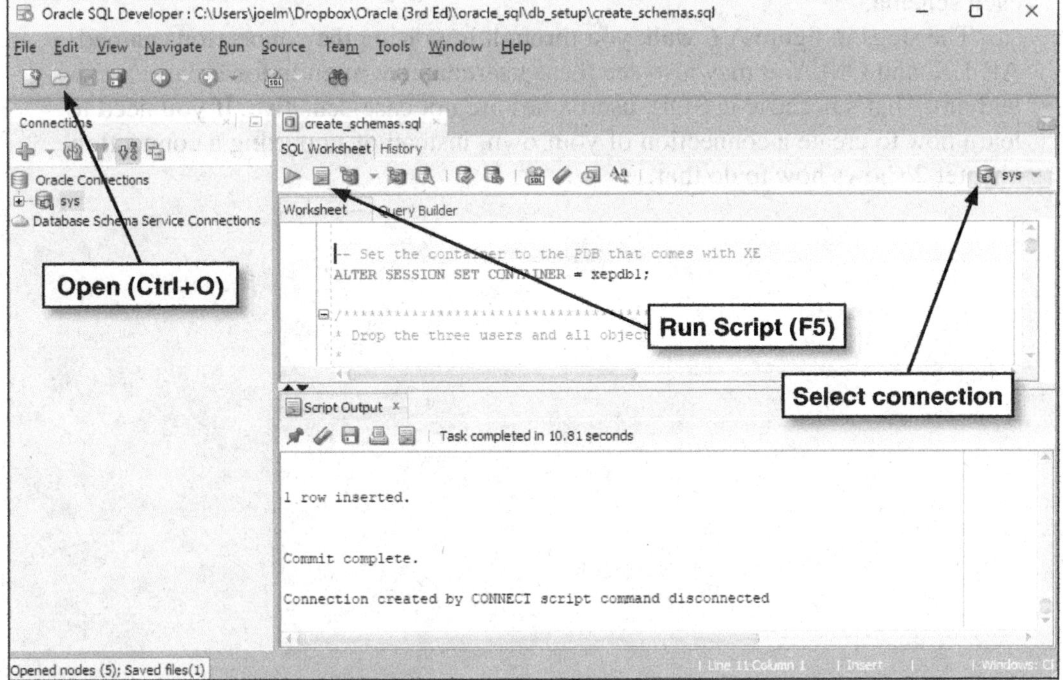

How to create the schemas

1. Open the create_schemas script located in the db_setup folder. To do that, click the Open button or press Ctrl+O and use the resulting dialog box.
2. Select the sys user you created in the previous figure from the drop-down list on the right side of the text editor toolbar.
3. Run the script. To do that, you can click the Run Script button or press F5. When you do, the Output window displays messages that indicate whether the script executed successfully. It may take a little while to finish due to the script's length.

   If the script seems to be stuck, click the Cancel icon to cancel the script. Then, exit SQL Developer, start it again, and run the script again.

If you get an error that says XEPDB1 does not exist...

1. Open the script named create_xepdb1 in the db_setup folder.
2. Edit the path in the script so it matches the path to oradata/XE on your computer.
3. Run the script. This should create the XEPDB1 database.
4. Run the create_schemas script again. It should work correctly.

Figure A-5 How to create the schemas for this book

How to import connections for the schemas

To work with a schema, you must first be connected to it. Since there are three schemas for this book, we recommend creating three connections, one for each schema.

The steps in figure A-6 walk you through importing the connections named AP, EX, and OM. You may also see these usernames written in lowercase: ap, ex, and om. That's because in SQL, usernames are not case-sensitive. If you need to learn how to create a connection of your own, instead of importing a connection, chapter 2 shows how to do that.

The dialog for importing connections

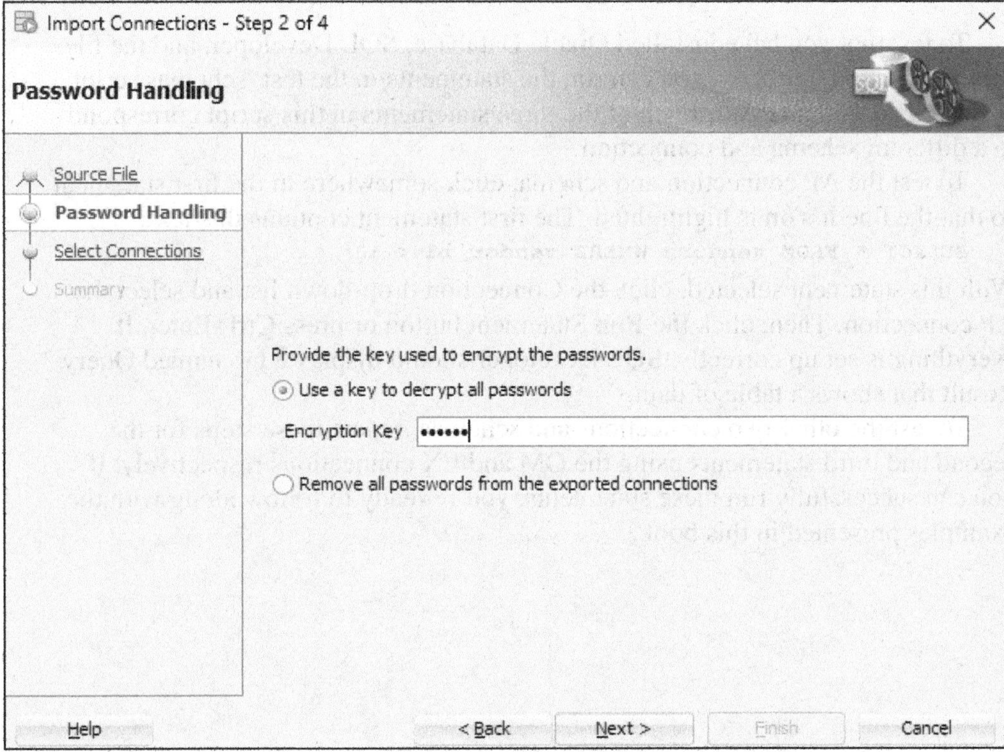

How to import connections for the schemas

1. Right-click on Oracle Connections in the Connections window and select "Import Connections".
2. When the dialog asks for a filename, select the connections.json file located in the oracle_sql\db_setup folder. Then, click Next.
3. Select "Use a key", enter "murach" as the encryption key, and click Next.
4. Check the boxes for all three connections (ap, om, and ex), select "Replace" for duplicate connections, and click Next.
5. Review that the information you've entered is correct and click Finish.
6. Check the Oracle Connections window to make sure all three connections are displayed.

Figure A-6 How to import connections for the schemas

How to make sure your system is set up correctly

To test that you have installed Oracle Database, SQL Developer, and the files for this book all correctly, you can run the statements in the test_schemas script as described in figure A-7. Each of the three statements in this script correspond to a different schema and connection.

To test the AP connection and schema, click somewhere in the first statement so that the line it's on is highlighted. The first statement contains this text:

```
SELECT * FROM vendors WHERE vendor_id < 10;
```

With this statement selected, click the Connection drop-down list and select the AP connection. Then, click the Run Statement button or press Ctrl+Enter. If everything is set up correctly, SQL Developer should display a tab named Query Result that shows a table of data.

To test the other two connections and schemas, repeat these steps for the second and third statements using the OM and EX connections respectively. If you can successfully run these statements, you're ready to follow along with the examples presented in this book.

SQL Developer after running a statement against the AP schema

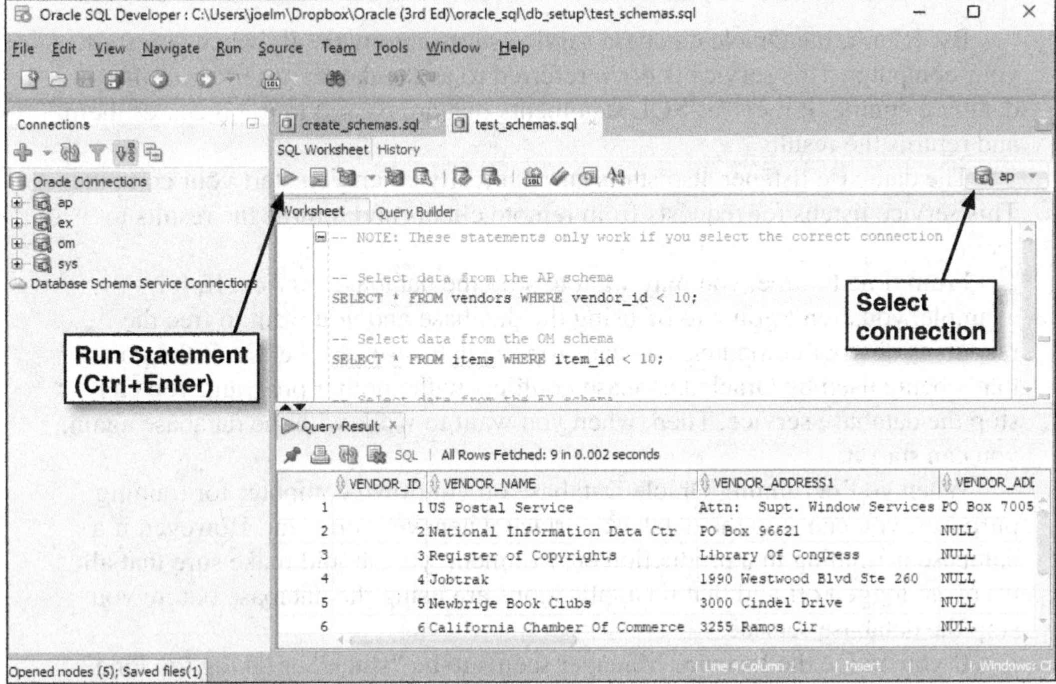

How to run a statement against the AP schema

1. Open the test_schemas script located in the db_setup folder.
2. Use the drop-down list on the upper-right side of the main window to select the connection named AP that you imported in the previous figure.
3. Put your cursor on the first statement in this script and press Ctrl+Enter or click the Run Statement button. This should run the statement at the cursor and display a list of nine vendors.

How to run a statement against the EX and OM schemas

- Use the same procedure as for the AP schema, but use the EX and OM connections to run the second and third statements against the OM and EX schemas respectively.

Figure A-7 How to make sure your system is set up correctly

How to stop and start the database service

By default, the Oracle database service starts automatically when you start your computer. This service is often referred to as the database server, or the database engine. It receives SQL statements that are passed to it, processes them, and returns the results.

The database listener also starts automatically when you start your computer. This service listens for requests from remote clients and returns the results to them.

From time to time, you may want to stop the database service. If, for example, you aren't going to be using the database and you want to free the resources on your computer, you can stop the database service. Or, if the port that's being used by Oracle Database conflicts with another program, you can stop the database service. Then, when you want to work with the database again, you can start it.

When you're running Oracle Database on your own computer for training purposes, you can stop the database service whenever you want. However, if a database is running in a production environment, you should make sure that all users are logged off and that no applications are using the database before you stop the database service.

If you notice the Instance Manager seems to be "stuck" or taking too long to shut down the service, that might be because you have connections to Oracle Database open on your system. For example, SQL Developer may have an open connection to the database service. To fix this, first close all connections by exiting SQL Developer. Then, try shutting down the service again.

The dialog for starting and stopping Oracle services

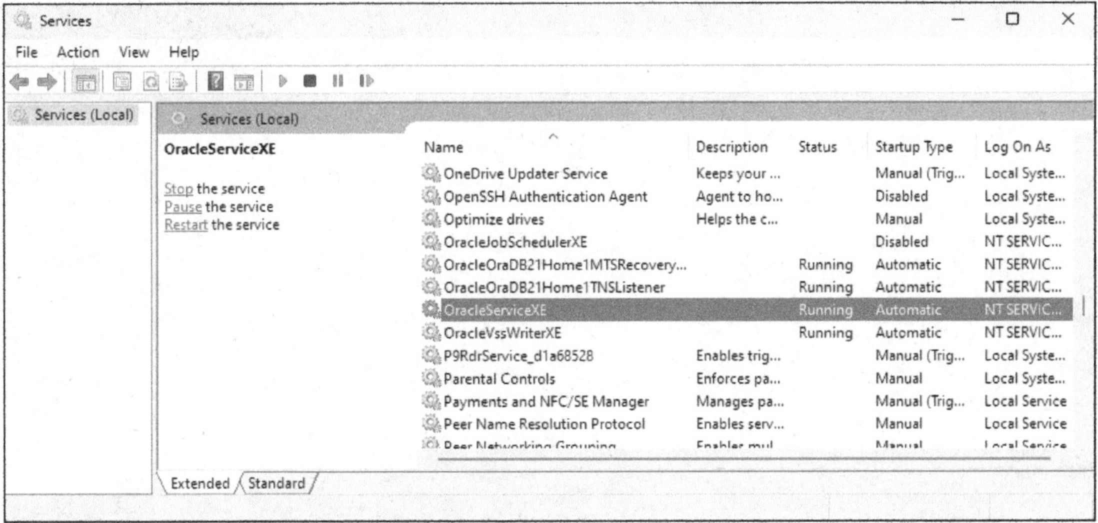

How to start and stop an Oracle service

1. Click the Windows Start button, type "services", and press Enter.
2. In the Services dialog, right-click on the service and select Start or Stop.

How to control when a service starts

- To control when a service starts, right-click the service and select Properties. Then, use the drop-down menu to change the Startup type.

Description

- By default, the Oracle services start when you start your computer.
- To make sure the Oracle XE server works as described in this book, make sure the service for the database server (OracleServiceXE) and the listener are both running.

Figure A-8 How to start and stop the database service

Appendix B

How to set up macOS for this book

Before you begin reading this book, we recommend that you download the files for this book that are available from the Murach website (www.murach.com). In addition, we recommend that you install Oracle SQL Developer. This product is available for free from the Oracle website, and you can install it on your computer as described in this appendix.

After you've installed SQL Developer, we recommend that you create an Oracle database in the cloud and connect to it as described in this appendix. Once you've done that, you'll be ready to gain valuable hands-on experience by following along with the examples presented in this book. In addition, you'll be able to gain more practice by doing the exercises at the end of each chapter.

How to download the files for this book 534
How to install Oracle SQL Developer 536
How to set up the database for this book 538
How to create the connections for this book 540
How to make sure your system is set up correctly 542

How to download the files for this book

Figure B-1 shows how to download the files for this book. This downloadable zip file includes a SQL script that sets up the database for this book. It also includes SQL scripts for all of the examples in this book and SQL scripts for the solutions to the exercises that are at the end of each chapter.

After double-clicking on the zip file to view the oracle_sql folder, we recommend copying this folder into a folder named murach in your Documents folder. That way, the folders and files on your system will match the folders and files shown in this book.

When you use the SQL script to set up the database for this book, it uses three schemas to organize the tables that store the data. The AP schema stores the tables for an Accounts Payable system, the EX schema stores some tables that are used for short examples, and the OM schema stores the tables for an Order Management system.

The recommended folder for the files

`Documents/murach/oracle_sql`

The files for this book

Folder	Contains
`db_setup`	The SQL script that sets up the database for this book.
`book_scripts`	The SQL scripts for all of the examples presented in this book.
`ex_solutions`	The SQL scripts for the solutions to the exercises at the end of each chapter.

The schemas for this book

Schema	Description
`ap`	The AP (Accounts Payable) schema. This schema is used by most examples in this book.
`ex`	The EX (Examples) schema. This schema contains several tables that are used for short examples.
`om`	The OM (Order Management) schema. This schema is used by a few examples in this book.

How to download the files

1. Go to www.murach.com.
2. Find the page for *Murach's Oracle SQL and PL/SQL (3rd Edition).*
3. Scroll down to the "FREE downloads" tab and click it.
4. Click the Download Now button for the zip file.
5. Find the downloaded zip file and double-click it. This should display a folder named oracle_sql.
6. Use Finder to create a folder named murach in your Documents folder.
7. Copy the oracle_sql folder into the murach folder.

Description

- All of the files described in this book are contained in a zip file that can be downloaded from www.murach.com.
- An Oracle database can contain one or more schemas to organize the tables that store the data. This book uses three schemas in its examples: AP, EX, and OM.

Figure B-1 How to download the files for this book

How to install Oracle SQL Developer

Oracle SQL Developer is a free program that makes it easy to work with Oracle databases. To install SQL Developer, you can use the first procedure shown in figure B-2. When you download the installer file, make sure it is the version with the JDK included.

If your operating system denies SQL Developer access to the Documents folder, you can use the second procedure shown in this figure to allow SQL Developer to access this folder. Although this procedure can be a little tricky to follow, it should solve this issue.

The URL for Oracle SQL Developer

```
https://www.oracle.com/database/sqldeveloper/technologies/download/
```

How to install Oracle SQL Developer

1. **Find the download page.** To do that, you can visit the URL shown above or search the internet for "oracle sql developer download" to find the page on the Oracle website for downloading SQL Developer.
2. **Download the zip file for SQL Developer.** To do that, scroll down and find the link to a Mac version that includes JDK 11, and click on the one that's right for your processor. For Intel processors, you can use the Mac OSX version. You will be prompted to log in or create an Oracle account if you don't already have one. Creating an account is free.
3. **Double-click the zip file.** To do that, use Finder to navigate to the Downloads folder and double-click on the zip file. This should create an application file named SQLDeveloper.
4. **Move SQLDeveloper into the Applications folder.** To do that, you can drag the SQLDeveloper file from the Downloads folder to the Applications folder.
5. **Start SQL Developer.** To do that, double-click on SQLDeveloper in the Applications folder.
6. **Open a script for this book.** To do that, click the Open button in the toolbar and use the resulting dialog to select this script:

   ```
   Documents/murach/oracle_sql/db_setup/test_schemas.sql
   ```

   If the operating system denies access to the Documents folder, you can follow the steps below to grant that access.

How to allow SQL Developer to access the Documents folder

1. **Display the System Preferences dialog.** To do that, you can select System Preferences from the Apple menu.
2. **Display the Full Disk Access options.** To do that, double-click the Security & Privacy icon, click the Privacy tab, and select the Full Disk Access on the left side of the dialog.
3. **Find the bash command.** To do that, switch to Finder, select Go To Folder from the Go menu, and enter "/bin/bash". This should display the bash command in a folder.
4. **Add the bash command to the list of apps.** To do that, switch back to the System Preferences window. Then, click on the Lock icon in the bottom left of the dialog and enter your password to unlock the options. Next, drag the bash command from Finder into the list of apps.

Note

- To make it easy to start SQL Developer, you may want to add it to your dock. To do that, start SQL Developer, Ctrl-click the SQLDeveloper icon in the dock, select Options, and select "Keep in Dock". Once you do that, you can start SQL Developer just like you start other programs available from the dock.

Figure B-2 How to install Oracle SQL Developer

How to set up the database for this book

At the time of this writing, there was no way to install Oracle Database natively on macOS. Although it's possible to install Oracle Database on macOS by using a virtual platform such as Docker, that process is complicated.

Because of that, we recommend that macOS users set up the database for this book by creating a free Oracle database in the cloud as described in figure B-3. To do that, you can follow along with the beginning of chapter 13 to create a cloud database named PDB1 that has schemas named AP, OM, and EX. Fortunately, that's relatively easy to do. In addition, working with a database in the cloud is a valuable skill.

As you work with the code that's presented in this book, you may make changes to the schemas that you don't intend to make, or decide to run some experiments that modify their data. In that case, you may want to restore the schemas to their original state so your results match the results shown in this book. To do that, just run the create_schemas script again. This deletes the three schemas described in this appendix and then recreates them.

Oracle Database 23ai was released in May 2024, and all of the code presented in this book should work with that version. If you're creating a new Oracle database in the cloud, you can use Oracle Database 23ai. However, you can also use a newer version if one is available to you. That's because Oracle Database is backwards compatible. As a result, any code that works for an older version should also work with newer versions.

The Oracle Cloud portal

cloud.oracle.com

The Autonomous Database Details page

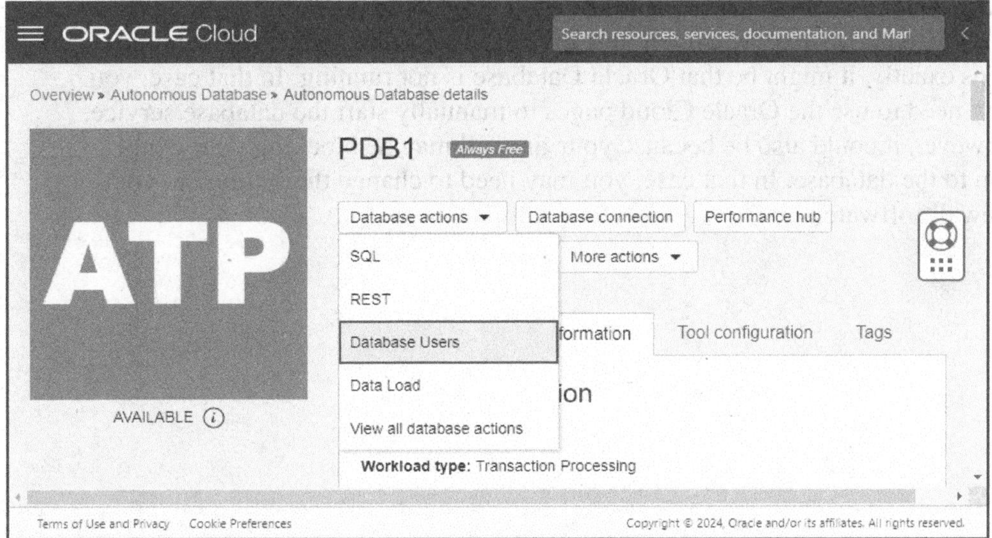

How to create the database and the schemas

1. Create a database named PDB1 by following the instructions in figures 13-1 and 13-2 of chapter 13.
2. Create users named AP, OM, and EX by following the directions in figure 13-3.
3. Create the tables for the AP, OM, and EX schemas by following the directions in figure 13-4 to run the script named create_schemas_in_cloud that's in the db_setup folder.
4. View the tables for the AP, OM, and EX schemas by following the instructions in figure 13-5.

How to restore the schemas

- Follow the directions in figure 13-4 to run the create_schemas_in_cloud script again. This should delete the schemas and recreate them.

Note

- This book often instructs you to right-click, because that's common in Windows. On macOS, right-clicking is not enabled by default. Instead, you can use Ctrl-click instead of right-click.

Figure B-3 How to set up the database for this book

How to create the connections for this book

To work with a schema, you must first be connected to it. Since there are three schemas for this book, we recommend creating three connections, one for each schema. To do that, you can follow the procedure presented in figure B-4.

If you get an error when you test the connection even after following the steps exactly, it might be that Oracle Database is not running. In that case, you may need to use the Oracle Cloud pages to manually start the database service. However, it could also be because your firewall may be blocking your connection to the database. In that case, you may need to change the settings in your firewall software.

The dialog for creating the connections

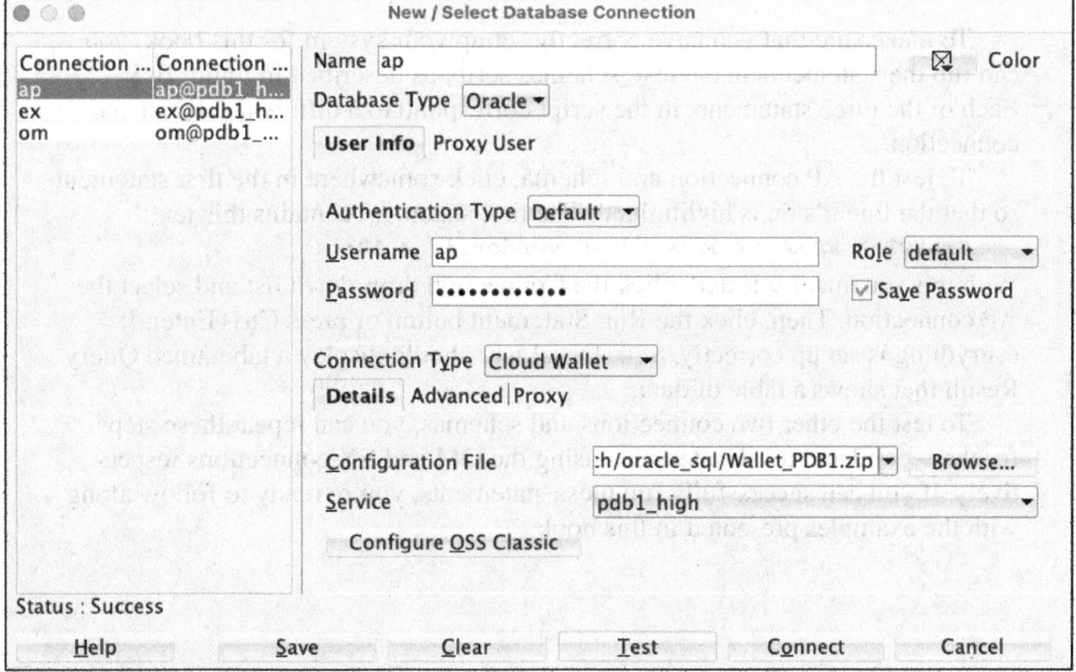

How to download the zip file for the cloud wallet

1. On the Autonomous Database Details page, click the "Database connection" button that's to the right of the "Database actions" button.
2. Set the Wallet type to Instance Wallet.
3. Click the "Download wallet" button and enter a password you can remember.
4. Click the Download button to download a zip file for the cloud wallet.

How to create the connections for the AP, OM, and EX users

1. Start SQL Developer.
2. In the Connections window, Ctrl-click on Oracle Connections and select New Connection.
3. Select Cloud Wallet from the Connection Type dropdown.
4. Click the Browse button to the right of the Configuration File item and select the zip file for the cloud wallet.
5. Enter AP as the name for the connection.
6. Enter AP as the username and enter the password that you created for that user.
7. Click the Test button. It should display a status of "Success".
8. Click the Save button to save the connection.
9. Repeat steps 5-8 for the OM and EX users.

Figure B-4 How to create the connections for this book

How to make sure your system is set up correctly

To make sure that you have correctly set up your system for this book, you can run the statements in the test_schemas script as described in figure B-5. Each of the three statements in the script correspond to a different schema and connection.

To test the AP connection and schema, click somewhere in the first statement so that the line it's on is highlighted. The first statement contains this text:

```
SELECT * FROM vendors WHERE vendor_id < 10;
```

With this statement selected, click the Connection drop-down list and select the AP connection. Then, click the Run Statement button or press Ctrl+Enter. If everything is set up correctly, SQL Developer should display a tab named Query Result that shows a table of data.

To test the other two connections and schemas, you can repeat these steps for the second and third statements using the OM and EX connections respectively. If you can successfully run these statements, you're ready to follow along with the examples presented in this book.

SQL Developer after running a statement against the AP schema

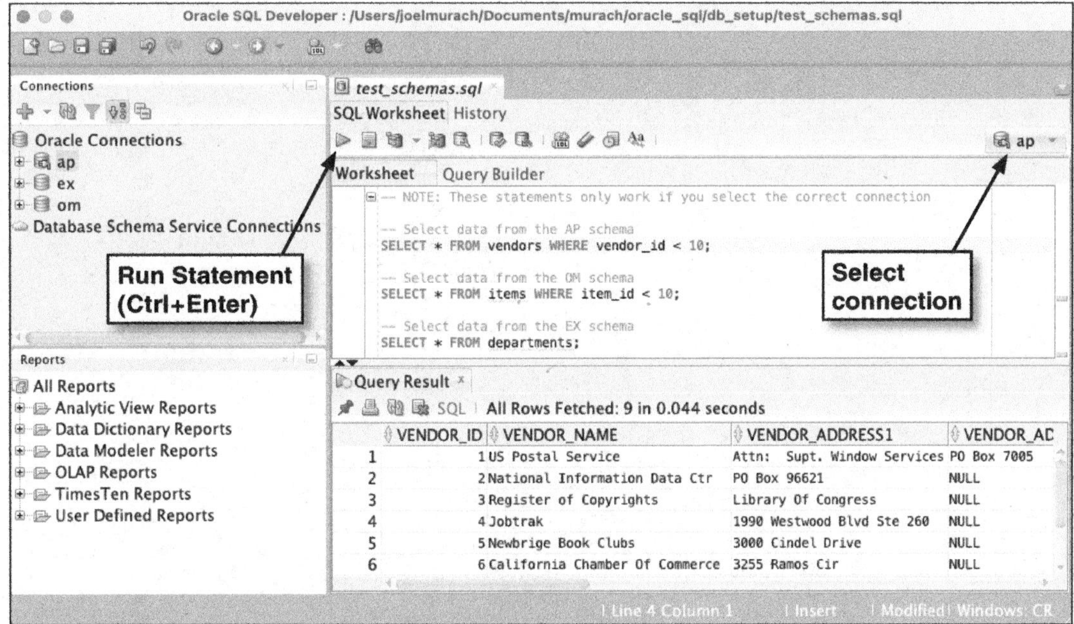

How to run a statement against the AP schema

1. Start SQL Developer.
2. Open the test_schemas script located in the oracle_sql/db_setup folder.
3. Use the drop-down list on the upper-right side of the main window to select the connection named AP that you imported in the previous figure.
4. Put your cursor on the first statement in this script and press Ctrl+Enter or click the Run Statement button. This should run the statement at the cursor and display a list of nine vendors.

How to run a statement against the OM and EX schemas

- Use the same procedure for the AP schema, but use the OM and EX connections to run the second and third statements against the OM and EX schemas respectively.

Figure B-5 How to make sure your system is set up correctly

Index

%ROWTYPE attribute, 422-423, 428-429
%TYPE attribute, 414-415

A

ABS function, 226-227
ADD_MONTHS function, 234-235
addition operator (+), 60-61
admin connection, 376-377
AFTER EACH ROW section, 504-505
AFTER keyword, 496-497
AFTER STATEMENT section, 504-505
aggregate function, 128-131
ALL keyword, 54-55, 64-65, 168-169
ALL TRIGGERS keywords, 502-503
ALTER INDEX statement, 21
ALTER SEQUENCE statement, 21, 316-317
ALTER SESSION statement, 236-237
ALTER TABLE statement, 21, 306-309
ALTER TRIGGER statement, 502-503
ALTER USER statement, 21, 358-359
American National Standards Institute, 16-17
analytic clause, 144-145
analytic function, 144-145
anchor member, 186-187
AND operator, 68-69
ANSI (American National Standards Institute), 16-17
antijoin, 122-123
ANY keyword, 170-171
API (application programming interface), 6-7
application programming interface, 6-7
application server, 8-9
arithmetic expression, 60-61
arithmetic operators, 60-61
AS keyword, 56-57
ASC keyword, 78-79
ASCII
  character code, 252-253
  function, 252-253
associative array, 426-427
asterisk (*) operator, 52-53
attribute, 266-267
Autonomous Database service, 388-389
AVG function, 128-131

B

back-end processing, 6-7
base table, 334-335
BEFORE EACH ROW section, 504-505
BEFORE keyword, 496-497
BEFORE STATEMENT section, 504-505
BETWEEN operator, 72-73
BINARY_DOUBLE type, 222-223
BINARY_FLOAT type, 222-223
book files
  install on macOS, 534-535
  install on Windows, 520-521
Boolean
  data type, 212-213
  expression, 416-417
BOOLEAN type, 222-223, 410-411
breakpoint, 484-486
built-in data type, 212-213
BULK COLLECT, 426-427, 508-509

C

calculated value, 54-55
CALL statement, 454-455
carriage return character code, 252-253
Cartesian product, 116-117
cascaded delete, 302-303
CASE expression, 254-255, 418-419
CASE keyword, 410-411
CAST function, 246-247
CDB (container database), 294-295
CDB$ROOT, 294-295
CEIL function, 226-227
cell, 10-11
CHAR type, 214-215
character data type, 212-213
character set, 214-215
check constraint, 304-305
check system setup
  macOS, 542-543
  Windows, 528-529
CHR function, 252-253
client, 4-5
client/server system, 4-5
cloud, 4-5
cloud computing platform, 4-5
COALESCE function, 256-257

column, 10-11
alias, 56-57, 80-81
attributes, 298-299
constraint, 298-299
function, 128-129
column-level constraint, 300-301
comment
block, 24-25
single-line, 24-25
COMMIT statement, 208-209, 440-441
common character functions, 216-219
common table expression (CTE), 184-185
common user, 294-295, 354-355
comparison operators, 66-67
compile, 452-453
complex search conditions, 138-139
composite
index, 280-281, 312-313
primary key, 10-11
variable, 424-425
compound condition, 68-69
compound join condition, 94-95
compound trigger, 504-505
COMPOUND TRIGGER statement, 504-505
compute, 388-389
concatenate, 58-59
concatenation operator (||), 58-59
concurrency, 444-445
concurrency problems, 446-447
conditional predicate, 496-497
DELETING, 496-497
INSERTING, 496-497
UPDATING, 496-497
connect as sysdba on Windows, 522-523
CONNECT BY clause, 188-189
CONNECT statement, 356-357, 410-411
constraint, 308-309
container database, 294-295
CONTINUE keyword, 420-421
CONTINUE WHEN keywords, 420-421
convert
character to character code, 252-253
data types, 246-247
correlated subquery, 172-173
correlation name, 492-493
COUNT function, 128-131
create admin user, 356-357
CREATE ANY TABLE privilege, 362-363
create book schemas (Windows), 524-525
create database (macOS), 538-539
create database connections (macOS), 540-541
CREATE FUNCTION statement, 470-471
CREATE INDEX statement, 21, 312-313
CREATE PACKAGE BODY statement, 476-479
CREATE PACKAGE statement, 476-479
CREATE PROCEDURE
privilege, 362-363
statement, 452-453
CREATE ROLE statement, 360-361
CREATE SEQUENCE, 21
privilege, 362-363
statement, 314-317
CREATE SESSION privilege, 362-363
CREATE TABLE
privilege, 362-363
statement, 21-23, 298-299, 356-357
CREATE TABLE AS statement, 194-195
CREATE TRIGGER statement, 492-493
CREATE USER statement, 21, 354-355
CREATE VIEW
privilege, 362-363
statement, 338-341
cross join, 116-117
CUBE clause, 140-143
CURRENT ROW keyword, 146-147
CURRENT_DATE function, 234-235
CURRENT_TIMESTAMP function, 240-241
CURRVAL pseudocolumn, 314-315
cursor, 422-423

D

dangling predicate, 418-419
data access API, 6-7
data definition language (DDL), 20-21
data elements, 268-275
foreign key, 274-275
identify, 268-269
identify columns, 272-273
identify tables, 272-273
index, 280-281
primary key, 274-275
subdivide, 270-271
data manipulation language (DML), 20-21
data structure, 266-2677
data type, 14-15, 212-213
data validation, 460-461
database
object, 30-31
schema, 30-31
server, 8-9
system, 266-267

database administrator, 20-21
database management system, 6-7
date literal, 66-67
DATE type, 230-231
datetime
 extract components, 244-245
 format elements, 232-233
 search, 236-237
 time search, 250-251
datetime functions, 234-235
datetime operator
 -, 234-235
 +, 234-235
DBA (database administrator), 20-21
DBA_USERS view, 374-375
DBMS (database management system), 6-7
DBMS_DEBUG option, 484-486
DBMS_OUTPUT package, 410-411, 476-477
DBMS_OUTPUT.PUT, 410-411
DBMS_OUTPUT.PUT_LINE, 410-413
DBTIMEZONE column, 230-231
DDL, 20-21
deadlock, 448-449
debug, 484-486
debugger, 484-486
declarative referential integrity, 276-277
declare variable, 414-415
DEFAULT
 column attribute, 298-299
 keyword, 458-459
 profile, 358-359
default value, 14-15
default views for user information, 374-375
DELETE
 privilege, 362-363
 statement, 21, 206-207
delete
 multiple rows, 206-207
 single row, 206-207
 with subquery, 206-207
DELETING conditional predicate, 496-497
denormalization, 290-291
denormalize a data structure, 290-291
DENSE_RANK function, 152-155
derived data, 288-289
DESC keyword, 78-79
dirty read, 446-447
DISTINCT keyword, 54-55, 64-65, 130-131
division operator (/), 60-61
DML, 20-21
DML trigger, 492-493
double-precision, 222-223
DROP ANY TABLE privilege, 362-363
DROP FUNCTION statement, 474-475
DROP INDEX statement, 21, 312-313
DROP PACKAGE statement, 478-479
DROP PROCEDURE statement, 468-469
DROP ROLE statement, 360-361
DROP SEQUENCE statement, 21
DROP TABLE
 privilege, 362-363
 statement, 21, 310-311
DROP TRIGGER statement, 502-503
DROP USER statement, 21, 358-359
Dual table, 84-85
DUP_VAL_ON_INDEX exception, 430-433

E

EER diagram, 324-325
ELSIF clause, 416-417
empty analytic clause, 144-145
ENABLE TRIGGER statement, 502-503
enhanced entity-relationship (EER) diagram, 324-325
enterprise system, 4-5
entity, 266-267
entity-relationship modeling, 266-267
equal operator (=), 66-67
equijoin, 112-113
ER modeling, 266-267
error, 430-431
exception, 430-431
EXCEPTION section, 460-461
EXCEPTION_INIT procedure, 462-463
EXECUTE IMMEDIATE, 410-411
 command, 434-435
EXECUTE privilege, 362-363
EXISTS method, 174-175, 426-427
explicit cursor, 422-423
explicit inner join, 90-91
expression, 54-55
EXTEND method, 424-427
EXTRACT function, 244-245

F

FETCH keyword, 82-83
field, 10-11
filter, 50-51
FIRST method, 426-427

first normal form, 282-285
fixed-length string, 214-215
FLOAT type, 222-223
floating-point numbers, 222-223
  search, 228-229
FLOOR function, 226-227
FOLLOWING keyword, 146-147
FOR LOOP, 410-411, 420-421
FORCE keyword, 338-341
foreign key, 10-13, 274-277, 302-303
format
  datetime, 232-233
  number, 224-225
frame, 146-149
freeform language, 24-25
FROM_TZ function, 240-241
front end, 6-7
front-end processing, 6-7
full outer join, 102-107
function
  argument, 62-63
  call, 470-471
  create, 470-471
  drop, 474-475
  multiple return statements, 472-473
function-based index, 312-313

G

generate ID values, 318-319
GENERATED keyword, 318-319
GRANT ALL PRIVILEGES statement, 356-357
grant and revoke
  roles, 380-381
  system privileges, 382-383
grant privileges, 364-365
GRANT statement, 21, 364-365
greater than operator (>), 66-67
greater than or equal to operator (>=), 66-67
GROUP BY clause, 132-135
GROUP keyword, 146-149
GROUPING function, 258-259

H

HAVING clause, 132-137
hierarchical query clause, 188-189

I

IDENTIFIED clause, 354-355
IDENTITY clause, 318-319
IF EXISTS clause, 434-435
IF NOT EXISTS clause, 434-435
IF statement, 410-411, 416-417
implicit cursor, 422-423
implicit inner join, 100-101
implicit outer join, 108-109
import connections, 526-527
IN keyword, 70-71, 456-457
index, 10-11, 280-281, 312-313
INITCAP function, 216-219
inline constraint, 300-301
inline view, 176-177
inner join, 90-101, 110-111
INNER keyword, 90-91
input parameter, 456-457
input/output parameter, 456-457
insert
  default values, 198-199
  multiple values, 200-201
  null values, 198-199
  single row, 196-197
  with subquery, 200-201
INSERT
  privilege, 362-363
  statement, 21-23, 196-201
INSERT INTO statement, 196-201
INSERTING conditional predicate, 496-497
install Oracle Database XE, 516-517
install Oracle SQL Developer
  macOS, 536-537
  Windows, 518-519
instance, 266-267
INSTEAD OF phrase, 498-499
INSTR function, 216-219
INTERSECT keyword, 122-123
interval
  formats, 238-239
  functions, 242-243
INTERVAL DAY TO SECOND type, 230-231
INTERVAL YEAR TO MONTH type, 230-231
INVALID_NUMBER exception, 432-433
IS NULL clause, 76-77
iterand, 420-421

J

join condition, 90-91
join operator ((+)), 108-109
JSON data type, 212-213

K

key-value pair, 426-427
keyword, 50-51

L

LAN (local area network), 4-5
large object data type, 212-213
LAST method, 426-427
LAST_DAY function, 234-235
left outer join, 102-105
LENGTH function, 216-219
less than operator (<), 66-67
less than or equal to operator (<=), 66-67
LEVEL pseudocolumn, 188-189
LIKE operator, 74-75
line feed character code, 252-253
literal value, 58-59
local area network, 4-5
local user, 294-295, 354-355
LOCALTIMESTAMP function, 240-241
lock, 444-445
logical operators, 68-69
LOOP, 410-411
lost update, 446-447
LOWER function, 216-219
LPAD function, 216-219
LTRIM function, 216-219

M

managed database service, 386-389
many-to-many relationship, 12-13
mask, 74-75
MAX function, 128-131
MAXVALUE keyword, 316-317
MIN function, 128-131
MINUS keyword, 122-123
MINVALUE keyword, 316-317
MOD function, 62-63, 226-227
MONTHS_BETWEEN function, 234-235
multiplication operator (*), 60-61
multi-table
  join, 98-99
  outer join, 106-107
multi-tenant database, 294-295
multivalued dependencies, 282-283
mutating-table error, 506-507
  solution, 508-509

N

named window, 150-151
national character set, 214-215
natural join, 114-115
NATURAL keyword, 114-115
NCHAR
  function, 252-253
  type, 214-215
nested subquery, 160-161
nested table, 424-425
network, 4-5
NEW correlation name, 492-493
NEXT_DAY function, 234-235
NEXTVAL pseudocolumn, 314-315
NO_DATA_FOUND exception, 432-433
non-primary key, 10-11
noncorrelated subquery, 172-173
nonrepeatable read, 446-447
normal forms, 278-279
  Boyce-Codd, 282-283
  domain-key, 282-283
  fifth, 282-283
  first, 282-285
  fourth, 282-283
  second, 282-283, 286-287
  sixth, 282-283
  third, 282-283, 288-289
normalize a data structure, 282-283
normalization, 278-279
normalized data, 278-279
not equal operator (<>), 66-67
NOT EXISTS operator, 174-175
NOT NULL column constraint, 298-299
NOT operator, 68-69
NTILE function, 152-155
NULL statement, 434-435
null value, 14-15, 76-77
number format elements, 224-225
NUMBER type, 222-223
numeric data type, 212-213, 222-229

numeric functions, 226-227
NUMTODSINTERVAL function, 242-243
NUMTOYMINTERVAL function, 242-243
NVARCHAR2 type, 214-215
NVL function, 216-219, 256-257
NVL2 function, 216-219, 256-257

O

object privileges, 362-363
OFFSET keyword, 82-83
OLD correlation name, 492-493
one-to-many relationship, 12-13
one-to-one relationship, 12-13
ONLY keyword, 82-83
open-source DBMS, MySQL, 18-19
optional parameter, 458-459
OR operator, 68-69
OR REPLACE
  clause, 338-341
  keywords, 452-453, 470-471
ORA_CIS_PROFILE, 358-359
ORA_STIG_PROFILE, 358-359
Oracle Cloud, 386-402
  account, 386-387
  Autonomous Database service, 388-389
  back up database, 400-401
  create database, 388-389
  create tables, 392-393
  create user, 390-391
  delete database, 400-401
  portal, 386-387
  restart database, 402-403
  restore database, 400-401
  run script, 392-393
  SQL Developer, 396-397
  view database objects, 394-395
Oracle Database documentation, 42-43
Oracle Database XE (Windows install), 516-517
Oracle Mobile Authenticator, 386-387
Oracle SQL Developer
  install for macOS, 536-537
  install for Windows, 518-519
ORDER BY clause, 78-81
order of precedence, 60-61
ORDER SIBLINGS BY clause, 188-189
orphaned record, 276-277
OUT keyword, 456-457
outer join, 102-109, 110-111
out-of-line constraint, 300-301
output parameter, 456-457
OVER
  clause, 150-151
  keyword, 144-145

P

package, 476-479
package body, 476-477
package specification, 476-479
parameter, 452-453
parse string, 220-221
partition, 144-145
PARTITION BY
  clause, 150-153
  keyword, 144-145
PDB$SEED, 294-295
PDB (pluggable database), 294-297
  close, 296-297
  create, 296-297
  drop, 296-297
  list open mode, 296-297
  open, 296-297
  set container, 296-297
peer, 146-147
peer group, 146-147
phantom read, 446-447
PL/SQL, 408-423
  control statements, 410-411
  script commands, 410-411
  anonymous block, 408-409
  block, 408-409
  collection, 424-425
  data types, 410-411
  loops, 420-421
  predefined exceptions, 432-433
  printing procedures, 410-411
  record, 428-429
  scalar variables, 414-415
  searched CASE statement, 418-419
  simple CASE statement, 418-419
  variables, 414-415
PLS_INTEGER type, 410-411
pluggable database, 294-295
POWER function, 226-227
PRECEDING keyword, 146-149
precision, 222-223
predefined exception, 432-433, 460-461
primary key, 10-11, 274-275
primary key constraint, 300-301

PRIOR operator, hierarchical query clause, 188-189
private synonym, 368-369
profile, 358-359
proprietary DBMS
  DB2, 18-19
  Microsoft SQL Server, 18-19
  Oracle Database, 18-19
pseudocode, 182-183
pseudocolumn, 188-189
public synonym, 368-369

Q

qualified column name, 90-91
qualified table name, 90-91
query, 6-7, 22-23
quota, 354-355
QUOTA keyword, 354-355

R

raise an exception, 460-461
RAISE statement, 460-461
RAISE_APPLICATION_ERROR procedure, 462-463
RANGE keyword, 146-149
RANK function, 152-155
ranking functions, 152-155
RDBMS (relational database management system), 16-17
READ COMMITTED isolation level, 446-447
read-only view, 342-343
record, 10-11, 428-429
recursion, 186-187
recursive member, 186-187
recursive query, 186-187
REFERENCES keyword, 302-303
referential integrity, 276-277
  constraint, 302-303
  violations, 277
REGEXP_LIKE operator, 74-75
regular expression, 74-75
relational database, 10-11
relational database management system, 16-17
rename table, 310-311
REPLACE function, 216-219
RESTART keyword, 316-317
result set, 22-23
RETURN clause, 470-473
revoke privileges, 366-367
REVOKE statement, 21, 366-367
right outer join, 102-105
role, 360-361
ROLE_SYS_PRIVS view, 374-375
ROLE_TAB_PRIVS view, 374-375
roll back, 208-209, 440-441
ROLLBACK statement, 208-209, 440-443
ROLLUP clause, 140-141,
ROUND function, 62-63, 130-131, 226-227, 234-235
row, 10-11
row limiting clause, 82-83
ROW_NUMBER function, 152-155
row-level trigger, 492-493
ROWS keyword, 146-149
RPAD function, 216-219
RTRIM function, 216-219

S

save point, 442-443
SAVEPOINT statement, 442-443
scalar function, 62-63
scalar variables, 414-415
scale, 222-223
schema, 30-31, 294-295
scientific notation, 222-223
scripts, 320-321
search condition, 50-51
searched CASE expression, 254-255, 418-419
second normal form, 282-283, 286-287
SELECT clause, 52-55
  column specification, 54-55
  privilege, 362-363
self-join, 96-97
semijoin, 122-123
sequence, 314-315
SERIALIZABLE isolation level, 446-447
server, 4-5
SESSIONTIMEZONE column, 230-231
set operators, 122-123
SET SERVEROUTPUT command, 410-413
SET TRANSACTION ISOLATION LEVEL statement, 446-447
SET TRANSACTION statement, 440-441
SIGN function, 226-227
significant digits, 222-223
simple CASE statement, 418-419
simple loop, 420-421

single-precision, 222-223
SOME keyword, 170-171
sort
  by alias, 80-81
  by column number, 80-81
  by expression, 80-81
sort strings in numeric sequence, 248-249
spatial data type, 212-213
SQL Developer, 28-29
  connect to cloud database, 396-397
  column definitions, 32-33
  compile function, 482-483
  compile procedure, 482-483
  compile trigger, 512-513
  connect as sysdba, 522-523
  connection tab, 28-29
  create admin connection, 376-377
  create database connection, 28-29
  database objects, 30-31
  debug, 484-486
  disable trigger, 510-511
  drop trigger, 510-511
  edit function, 482-483
  edit procedure, 482-483
  edit trigger, 512-513
  enable trigger, 510-511
  grant and revoke roles, 380-381
  grant and revoke system privileges, 382-383
  grant privileges, 482-483
  handle syntax error, 36-37
  open script, 38-39
  open SQL worksheet, 34-35
  Oracle Cloud, 396-397
  output tab, 40-41
  query cloud database, 398-399
  rename trigger, 510-511
  revoke privileges, 482-483
  run script, 40-41
  run statements, 34-35
  save script, 38-39
  table data, 32-33
  view trigger, 510-511
  work with functions, 480-481
  work with indexes, 322-323
  work with packages, 480-481
  work with procedures, 480-481
  work with sequences, 322-323
  work with tables, 322-323
  work with users, 378-379
  work with views, 350-351
SQL dialect, 16-17
SQL extension, 16-17
SQL query, 6-7
SQL statement, 20-21
SQLERRM function, 430-431, 462-463
SQRT function, 226-227
start database service on Windows, 530-531
START WITH clause, 188-189, 316-317
statement level trigger, 492-493
step over, 484-486
stop database service on Windows, 530-531
storage, 388-389
stored procedure, 452-453
  call, 454-455
  drop, 468-469
  input parameter, 456-457
  input/output parameter, 456-457
  named parameter, 454-455
  optional parameter, 458-459
  output parameter, 456-457
  parameter, 452-453
  positional parameter, 454-455
stored program, 452-453
string
  constant, 58-59
  data types, 214-221
  expression, 58-59
  literal, 58-59
  pattern, 74-75
structured query language (SQL), 6-7
subprogram, 452-453
subquery, 70-71, 160-189
  ALL keyword, 166-167
  ANY keyword, 166-167
  compare to joins, 162-163
  compare with expression, 166-167
  complex queries, 180-183
  in FROM clause, 176-177
  in search conditions, 164-165
  in SELECT clause, 178-179
  SOME keyword, 166-167
  ways to introduce, 160-161
  with IN operator, 164-165
subquery factoring clause, 184-185
SUBSTR function, 62-63, 216-221
subtraction operator (-), 60-61
SUM function, 128-131
summary query, 128-129

summary row, 140-141
SYNONYM keyword, 368-369
synonym, 368-369
syntax, 50-51
syntax error, 36-37
syntax reference, 50-51
SYS_EXTRACT_UTC function, 240-241
SYSDATE function, 62-63, 234-235
system privileges, 362-363
SYSTEM tablespace, 354-355
system trigger, 500-501
SYSTIMESTAMP function, 240-241

T

tab character code, 252-253
table, 10-11
  alter, 306-309
  create, 298-299
  drop, 310-311
table alias, 92-93, 96-97, 176-177
table relationships, 12-13
table scan, 280-281
table-level constraint, 300-301
tablespace, 296-297
  SYSTEM, 354-355
  TEMP, 354-355
  USERS, 354-355
TEMP tablespace, 354-355
temporal data type, 212-213, 230-245
third normal form, 282-283, 288-289
timestamp formats, 238-239
timestamp functions, 240-241
TIMESTAMP type, 230-231
TIMESTAMP WITH LOCAL TIMEZONE type, 230-231
TIMESTAMP WITH TIMEZONE type, 230-231
timezone_names view, 238-239
timing point section, 504-505
TO_CHAR function, 62-63, 246-247
TO_DATE function, 246-247
TO_DSINTERVAL function, 242-243
TO_NUMBER function, 246-247
TO_TIMESTAMP function, 240-241
TO_TIMESTAMP_TZ function, 240-241
TO_YMINTERVAL function, 242-243
transaction, 208-209, 440-441
  best practices, 448-449
  isolation level, 446-447
transitive dependencies, 282-283
trigger, 492-495
  AFTER, 496-497, 500-501
  BEFORE, 496-497
  compound, 504-505
  CREATE, 500-501
  DDL, 500-501
  disable, 502-503
  DROP, 500-503
  enable, 502-503
  for a view, 498-499
  GRANT, 500-501
  INSTEAD OF, 498-499
  rename, 502-503
  REVOKE, 500-501
  SERVERERROR, 500-501
  SHUTDOWN, 500-501
  STARTUP, 500-501
TRIM function, 216-219
TRUNC function, 226-227, 234-235
truncate table, 310-311

U

UNBOUNDED FOLLOWING keyword, 146-147
UNBOUNDED PRECEDING keyword, 146-147
uncorrelated subquery, 172-173
UNION ALL keyword, 118-121
  recursive query, 186-187
UNION keyword, 118-121
UNIQUE column constraint, 298-299
unique key, 10-11
UNLIMITED TABLESPACE privilege, 362-363
unnormalized data, 278-279
updatable view, 346-347
UPDATE, 21
  multiple columns, 202-203
  multiple rows, 202-203
  privilege, 362-363
  single column, 202-203
  single row, 202-203
  statement, 202-203
  with subquery, 204-205
updateable view, 342-343
UPDATING conditional predicate, 496-497
UPPER function, 216-219
user profile, 358-359
USER_RESOURCE_LIMITS view, 374-375
USER_ROLE_PRIVS view, 374-375
USER_SYS_PRIVS view, 374-375

USER_TAB_PRIVS view, 374-375
USER_USERS view, 374-375
user-defined exception, 462-463
USERS tablespace, 354-355
USING keyword, 112-113

V

VALUE_ERROR exception, 432-433, 460-461
VALUES clause, 196-197
VARCHAR2 type, 214-215
variable-length string, 214-215
varray, 424-425
view, 334-351
  benefits of using, 336-337
  create, 338-341
  delete through, 346-347
  insert through, 346-347
  read-only, 342-343
  SQL Developer, 350-351
  updatable, 342-343, 346-347
  WITH CHECK OPTION clause, 344-345
view privileges and roles, 374-375
viewed table, 334-335
virtual server, 4-5
virtual table, 334-335

W

WAN (wide area network), 4-5
web server, 8-9
WHEN OTHERS statement, 430-431
WHEN OTHERS THEN statement, 440-441
WHERE clause, 52-53, 64-77
WHILE LOOP, 410-411, 420-421
wide area network, 4-5
wildcard, 74-75
window, 144-147
WINDOW clause, 150-151
WITH ADMIN OPTION clause, 364-365
WITH CHECK OPTION statement, 344-345
WITH GRANT OPTION clause, 364-365
WITH clause (subquery factoring), 184-185
WITH READ ONLY clause, 342-343
WITH TIES keyword, 82-83

X, Y, Z

XML data type, 212-213
ZERO_DIVIDE exception, 432-433

What software you need

- Oracle Database XE or the Oracle Cloud Autonomous Database service
- Oracle SQL Developer

This software is available for free, and appendixes A (Windows) and B (macOS) show how to install it.

How to download the code for this book

1. Go to www.murach.com.
2. Find the page for *Murach's Oracle SQL and PL/SQL (3rd Edition)*.
3. Scroll down to the FREE Downloads tab and click it.
4. Click the Download Now button to download the zip file.
5. Find the downloaded zip file and double-click it. This should display a folder named oracle_sql.
6. Create a folder named murach in your Documents folder.
7. Move the oracle_sql folder into the murach folder.

For more detailed instructions, please see appendix A (Windows) or B (macOS).

What the download contains

- A script that sets up the database for this book by creating three schemas
- Scripts for the SQL and PL/SQL examples that are presented throughout this book
- Solutions to the exercises that are at the end of each chapter

To view detailed instructions for setting up the database, please see appendix A (Windows) or B (macOS).